Progress in Mathematics
Volume 64

Hyo Chul Myung

Malcev-Admissible Algebras

1986

Birkhäuser
Boston · Basel · Stuttgart

Hyo Chul Myung
Department of Mathematics
University of Northern Iowa
Cedar Falls, IA 50614
U.S.A.

Library of Congress Cataloging in Publication Data
Myung, Hyo Chul, 1937–
Malcev-admissible algebras.
(Progress in mathematics ; vol. 64)
Bibliography: p.
Includes indexes.
1. Lie-admissible algebras. I. Title.
II. Series: Progress in mathematics (Boston, Mass.) ;
vol. 64.
QA252.3.M98 1986 512′.55 86-12923

CIP-Kurztitelaufnahme der Deutschen Bibliothek
Myung, Hyo Chul:
Malcev admissible algebras / Hyo Chul Myung.—
Boston ; Basel ; Stuttgart : Birkhäuser, 1986.
(Progress in mathematics ; Vol. 64)
ISBN 3-7643-3345-6
NE: GT

ISBN 0-8176-3345-6
ISBN 3-7643-3345-6

Printed in the U.S.A.

TO MY CHILDREN

KAREN
PEGGY
JANE
AND
MICHAEL

PREFACE

Since the brief introduction of Lie-admissible algebras by A. A. Albert in 1948, very little has been known about the structure of these algebras until the subject recently came to the attention of physicists. During the last few years the theory of Lie-admissible algebras has seen a considerable growth both in theoretical mathematics and its applications. Many of the mathematical papers on the subject were published in physics journals. The present book is concerned with the mathematical side of the theory of Lie-admissible algebras and is based on those results published since 1978. Included in this book are Malcev-admissible algebras which are a natural generalization of Lie-admissible algebras but which are largely represented by octonions and their variants. It turns out that virtually all results about Lie-admissible algebras can be extended to Malcev-admissible algebras. This fact provided the title of the book "Malcev-Admissible Algebras".

The main objective is to present a self-contained and detailed account of the theory of Malcev-admissible algebras which has been developed in the past few years. Thus this book does not presuppose any acquaintance with Malcev-admissible algebras, however, we assume that the reader is familiar with the standard theory of Lie and Malcev algebras and with a very few results about other nonassociative algebras.

Given an algebra A over a field of characteristic $\neq 2$ with multiplication denoted by xy, we associate an anticommutative algebra A^- with multiplication $[x,y] = xy - yx$ defined on the vector space A. Then, A is termed Lie-admissible or Malcev-admissible if A^- is a Lie or Malcev algebra. Beginning with Albert's problem of classifying all flexible Lie-admissible algebras A with A^- semisimple, a common theme of the study of Lie-admissible algebras in both mathematical and physical settings has been to maintain a given Lie algebra structure on A^-. Thus, the study of Lie-admissible algebras has originated from a Lie algebra point of view. If A is any Lie algebra with product $[x,y]$ and if $x \circ y$ denotes any commutative product defined on the vector space A, then the algebra $(A,*)$ with a new product $x * y = \frac{1}{2}[x,y] + x \circ y$ defined on A has the property that $(A,*)^- = A^-$. Moreover, every Lie-admissible algebra arises in this manner. This simple point illustrates the fundamental fact that every Lie algebra can occur as the A^- of a host of Lie-admissible algebras by simply varying the commutative product $x \circ y$. This suggests that the assumption of Lie-admissibility alone is too broad a hypothesis to yield a fruitful structure theory. These observations also apply to Malcev and Malcev-admissible algebras.

In addition to Malcev-admissibility, the basic assumption imposed on A in this book is the flexible identity $(xy)x = x(yx)$, the third power identity $(xx)x = x(xx)$, or power-associativity (each element generates an associative subalgebra) which first appeared in the work of Albert [1,2] and which has also been useful in the study of other nonassociative algebras. The flexible identity is a strengthened form of the third power identity, and, under appropriate restrictions on the characteristic, the third and fourth power identities are equivalent to power-associativity. This last fact concerns the main topic in Chapter 2 where we determine all power-associative Malcev-admissible algebras A by determining all third and fourth power-associative products on A, when A^- is isomorphic to the attached minus algebra of an $n \times n$ matrix algebra, of an octonion algebra, to a simple Malcev algebra, or more generally to a semisimple Malcev algebra. It turns out in all cases that the product on such an algebra A has the form of the Lie product in A^- plus a commutative product defined by linear forms and symmetric bilinear forms on A. This result is extended to finite-dimensional central simple alternative algebras of characteristic $\neq 2,3$. The situation when A^- is semisimple with more than two simple summands is more complicated and requires the use of a graph theoretical approach to describe power-associativity of A in terms of the relations between the linear forms on summands of A^-.

The problem in Chapter 2 was motivated by some recent works in particle physics. The first physical application stems from an attempt to introduce power-associative products on the real associative envelope A of spin $\frac{1}{2}$, 1, or $\frac{3}{2}$ matrices, which leave the original Lie product in

A^- unchanged. Power-associativity in these broader underlying algebras is necessary to have a well defined notion of the exponential of a spin matrix. For an arbitrary spin number s for which the real associative envelope A is the $(2s + 1) \times (2s + 1)$ matrix algebra over the reals for an integer s and is the $(s + \frac{1}{2}) \times (s + \frac{1}{2})$ matrix algebra over the real quaternions for a half integer s, such power-associative products on A are determined as special cases by our general result on central simple associative algebras. The second originates from an attempt to use a flexible Lie-admissible algebra A to generalize the Heisenberg equation whose solutions require the underlying algebra A to be power-associative. Results in Chapter 2 also generalize several earlier works on Malcev-admissible algebras, including the so-called para-octonion and pseudo-octonion algebras.

The remainder of this book deals exclusively with flexible Malcev-admissible algebras. Some basic definitions and results are given in Chapter 1, and this is the only chapter where we have made an attempt to investigate flexible Malcev-admissible algebras A of arbitrary dimension and of arbitrary characteristic $\neq 2$. When A^- has a Cartan decomposition relative to an abelian Cartan subalgebra H such that the adjoint mappings $ad_h : x \to ad_h(x) = [h,x]$ for $h \in H$ act diagonally on each root space, the multiplication in A is determined by that in H, being a commutative algebra. As such, Kac-Moody algebras, classical Lie algebras, and generalized Witt algebras are well known algebras having this type of Cartan decompositions. A principal result in Chapter 1 is that if H is a nil subalgebra of A of finite nil-index and the center of A^- is zero, then A is a Malcev algebra. When H is a nil subalgebra of nil-index ≤ 4 and A^- has a one-dimensional center, the multiplication in A has the form of the Lie product in A^- plus a commutative product given by a symmetric invariant form on A^-. We make a brief application of these results to Kac-Moody algebras and generalized Witt algebras. Included in this chapter is the result that if A is a flexible Malcev-admissible nilalgebra of dimension ≤ 4 and if A^- is nilpotent, then A is nilpotent and $A^4 = 0$. The proof of this result essentially classifies all such algebras.

Chapters 3-5 are devoted to the study of finite-dimensional flexible Malcev-admissible algebras A over a field F of characteristic 0

(except Section 4.5). Given a semisimple subalgebra S of A^- (such a subalgebra exists by Levi's theorem when A^- is not solvable), the derivation algebra Der S of S is inner and a semisimple Lie algebra. Thus, the flexible identity makes Der S act as derivations on A which is therefore regarded as a Der S-module and which consequently decomposes into the direct sum of irreducible Der S-submodules V_i. It is this fact which enables us to determine the products between the summand V_i's by describing all Der S-module homomorphisms of $V_i \otimes V_j$ into V_k for all i, j, k. The latter problem is intimately related to the decomposition of $V_i \otimes V_j$ into irreducible Der S-submodules, which is not known for the general case. As is well known, when S is split over F, the dimension of the F-space $\mathrm{Hom}_{\mathrm{Der}\, S}(V_i \otimes V_j, V_k)$ of all Der S-module homomorphisms of $V_i \otimes V_j$ into V_k equals the number of irreducible summands of $V_i \otimes V_j$ isomorphic to V_k. For a split simple Lie algebra L, in Chapter 3 we give an explicit decomposition of the special case $L \otimes L$ and a principle to count $\dim_F \mathrm{Hom}_L(L \otimes V, V)$ for an arbitrary irreducible L-module V, without decomposing $L \otimes V$. These results along with the decompositions of $U \otimes V$ for some low dimensional irreducible modules U, V play a fundamental role throughout our investigation in Chapters 3-5.

In Chapter 3 we give some basic facts about Malcev modules and show that a Malcev module V for a split semisimple Malcev algebra S can be regarded as a Lie module for Der S and that the irreducible summands of V as an S-module coincide with those as a Der S-module. All flexible Malcev-admissible algebras A with A^- semisimple (not necessarily split over F) are classified in Chapter 3. As an application of this we determine all real flexible Malcev-admissible algebras with A^- simple.

Chapter 4 is concerned with the structure of flexible Malcev-admissible algebras A for the general case where the solvable radical R of A^- is nonzero. Given a Levi factor S of A^- with decomposition $S = S_1 + \cdots + S_n$ into simple ideals, we first refine the Levi decomposition $A^- = S + R$ by decomposing R into the direct sum of certain S-submodules of R. In fact, it suffices for our investigation to assume that S is a split semisimple subalgebra of A^- and R is an S-submodule of A complementary to S such that $[R,R] \subseteq R$. Given a subset Γ of $\{1,\cdots,n\}$, we first consider the sum R_Γ of all irreducible S-submodules W of R such that $\Gamma = \{j \mid 1 \le j \le n$ and $[S_j,W] \ne 0\}$, called the support of W. Then, A decomposes as $A = \Sigma_{i=1}^n S_i + \Sigma_{\Gamma\ne\phi} R_\Gamma + R_0$ where R_0 is the sum of trivial submodules, and the refinement remains unchanged when A is regarded as a Lie module for Der S = Der $S_1 + \cdots +$ Der S_n. We establish containment relations between the submodules S_i and R_Γ under the product in A which are dictated by the module structure. These relations play main roles for the principal results in Chapter 4 which determine all flexible Malcev-admissible algebras A such that either $[S,R] = 0$, or A is simple and R is abelian.

In case $[S,R] = 0$, R is a subalgebra of A, the multiplication in each S_i is the one determined for simple A^- plus a multiple of an element a of R with $[a,R] = 0$ by the Killing form on S_i, $S_iS_j = 0$ for $i \ne j$, and the multiplication between S_i and R is given by $xa = ax = \tau_i(a)x$ for $x \in S_i$, $a \in R$, where τ_i is a linear form on R. This result serves as a reduction theorem in the sense that it reduces the study of the structure of flexible Malcev-admissible algebras of characteristic 0 to the subclass of those algebras A such that no simple subalgebra of A^- centralizes the radical of A^-. However, a general structure theory for that subclass is not known. When A is simple and R is abelian, every nontrivial irreducible S-submodule of R must be isomorphic to some S_i. This allows A to decompose as $A = S_1 \otimes P_1 + \cdots + S_n \otimes P_n + R_0$ where each P_i is a trivial S-module and S acts on the first component of $S_i \otimes P_i$. Then, R_0 is a commutative subalgebra of A and $(S_i \otimes P_i)(S_j \otimes P_j) = 0$ for $i \ne j$. The remaining products are associated with two commutative products "$*$", "$\bullet$" defined on P_i, a bilinear mapping $\phi : P_i \times P_i \to R_0$, and a linear mapping $T : R_0 \to P_i$ for each i which are subject to a set of conditions forced by flexible Malcev-admissibility. If $x \otimes a$, $y \otimes b \in S_i \otimes P_i$ and $z \in R_0$, then $(x \otimes a)(y \otimes b) = [x,y] \otimes a * b + x \# y \otimes a \bullet b + K(x,y)\phi(a,b)$ and $z(x \otimes a) = (x \otimes a)z = x \otimes (a * T(z))$, where $x \# y = xy + yx - \frac{2}{n+1}(\mathrm{tr}\ xy)\ I$ if $S_i = \mathfrak{sl}(n+1)$ is a Lie algebra of type $A_n (n \ge 2)$ and $x \# y = 0$ otherwise, and $K(\ ,\ )$ is the Killing form on S_i. The use of tensor products for the product in A is similar to the Tit's construction of exceptional Lie algebras. The multiplications described in both cases provide important sources for the construction of simple flexible Malcev-admissible algebras.

Section 4.5 is devoted to the construction of some simple flexible Malcev-admissible algebras which are designed to illustrate the diversity of Malcev algebras that can occur as the algebra A^- of a simple flexible Malcev-admissible algebra A. The construction is based on quadratic algebras and on quasi-classical Malcev algebras that are by definition Malcev algebras with a nondegenerate symmetric invariant bilinear form. In the final section of Chapter 4, an example is given to indicate that it is not feasible for flexible Malcev-admissible algebras to develop a structure theory of the type that works so well for such classes of algebras as associative, more generally, alternative, Jordan, Lie, or Malcev algebras.

In Chapter 5 we classify under some restrictions all flexible Malcev-admissible algebras of dimension ≤ 8 over an algebraically closed field F of characteristic 0, when A^- is not solvable. Since the case for dimension ≤ 4 has been determined in Chapter 4, the classification begins with dimension 5. Certain nonassociative algebras which have appeared in physics exhibit many symmetries, or equivalently many automorphisms. Since the automorphism group of a real or complex algebra A is a Lie group whose Lie algebra is $\mathrm{Der}\, A$, the largeness of $\mathrm{Der}\, A$ reflects the symmetries of A. Here we treat some special cases in low dimensions. Included in Sections 5.3 and 5.4 are a class of algebras of dimension 7 which are acted on by $\mathfrak{sl}(3)$ as derivations, and a class of algebras of dimension 8 which are acted on by G_2 as derivations. The former case is motivated by the color algebra used by Domokos and Kövesi-Domokos [1], and in both cases we give a condition for the algebra to be Malcev-admissible in terms of constant relations. In the final two sections, two classes of algebras A of dimension 15 which are respectively acted on by $\mathfrak{sl}(3)$ and G_2 as derivations are constructed. These two constructions are also motivated from physics, and flexible Lie-admissibility for case $\mathfrak{sl}(3) \subseteq \mathrm{Der}\, A$ and flexible Malcev-admissibility for case $G_2 \subseteq \mathrm{Der}\, A$ are described in terms of constant conditions. Two specializations of A^- in the first case include G_2 as a subalgebra and the Lie algebra $\mathfrak{sl}(4)$. A special case of the second gives a flexible Malcev-admissible algebra A such that A^- has an abelian radical of dimension 8, which also arises from the construction in Chapter 4.

It was not our intention to make this book exhaustive, but, as the first book in the subject to the best of our knowledge, we have made an effort to treat the most important lines of current developments of the subject. The completeness and coherence of the structure theory of Malcev-admissible algebras even with additional identities are, at present, far from those of the three striking classes of nonassociative algebras, alternative, Jordan and Lie algebras. There seems to be, however, a trend to study nonassociative algebras (of characteristic 0) based on representations of Lie algebras. It is this trend that has inspired us to write a book on the subject.

We are greatly indebted to many people who offered numerous suggestions and discussions at workshops and conferences on the subject and through personal communications. Especially, various critical comments by G. M. Benkart, S. Okubo, J. M. Osborn, A. A. Sagle and R. M. Santilli and their work on the subject have been most influential for undertaking the writing. Many thanks are also due to J. S. Cross who assisted with this project by carefully reading the entire handwritten manuscript and providing numerous suggestions. Mrs. Ginny Diercks typed the entire manuscript, and we have no words to describe our indebtedness for her painstaking efforts tempered with great patience throughout the tedious typing process.

The writing of this book was initiated and completed under the support of two grants, the Distinguished Scholar Award in the spring of 1984 and the Professional Development Leave in the fall of 1985 at the University of Northern Iowa. A partial release of time from the Department of Mathematics in the fall of 1984 was immensely helpful in keeping the project on schedule. We gratefully acknowledge all this support without which the project could not be completed in the present form.

Fall, 1985 H. C. Myung

CONTENTS

1

FLEXIBLE MALCEV-ADMISSIBLE ALGEBRAS

1.1. INTRODUCTION

Let A be an (nonassociative) algebra with multiplication denoted by juxtaposition xy over a field F of characteristic $\neq 2$. Associated with A are an anticommutative algebra A^- and a commutative algebra A^+ which are defined on the same vector space as A but with multiplications respectively given by

$$\text{Lie product:} \quad [x,y] = xy - yx ,$$

$$\text{Jordan product:} \quad x \circ y = \frac{1}{2}(xy + yx) .$$

Most celebrated anticommutative and commutative algebras are Lie and Jordan algebras. An algebra L with multiplication denoted by $[x,y]$ is called a *Lie algebra* if L satisfies the anticommutative law

$$[x,x] = 0 ,$$

and the Jacobi identity

$$(1.1) \qquad [[x,y],z] + [[y,z],x] + [[z,x],y] = 0$$

for all $x,y,z \in L$. A commutative algebra J with multiplication $x \circ y$ is termed a *Jordan algebra* if J satisfies the Jordan identity

$$(1.2) \qquad [(x \circ x) \circ y] \circ x = (x \circ x) \circ (y \circ x)$$

for all $x,y \in J$. Jordan algebras, named by Albert in 1946, were first introduced by physicist Pascual Jordan to attempt to introduce an infinite-dimensional algebraic setting for quantum mechanics essentially different from the standard setting of hermitian matrices. A half century after the inception of Jordan theory by Jordan, von Neumann and Wigner [1], there have been remarkable successes in mathematical studies of Jordan algebras. For bibliographies and a survey on Jordan algebras, the interested reader is referred to McCrimmon [2,3], Osborn [3], and Tomber [1]. On the other hand, applications of Jordan theory in physics seem doomed especially because of the lack of infinite-dimensional exceptional algebras (see McCrimmon [3]).

One of the most remarkable events in the history of all mathematics was the discovery of Lie groups and Lie algebras in the late 19th century. Starting from the works of Lie, Killing, and E. Cartan, the theory of Lie groups and Lie algebras has developed into a tremendous spectrum of mathematics. Their applications embrace virtually all areas of mathematics, including engineering and physics, from classical to quantum and relativistic. Today, as is well known, symmetry principles based on Lie theory are a fundamental tool and a main source of the development in theoretical physics. A good account of bibliographies, survey and exposition of this can be found in Tomber [1], Hawkins [1], Kac [1], and Howe [1].

Soon after World War II, it was Albert [2] who suggested the study of broader classes of nonassociative algebras which generalize Lie, Jordan and alternative algebras. The following three algebras first appeared in the Albert's 1948 paper [2] and will be instrumental for the investigations in this monograph.

Definition 1.1. An algebra A over a field F is called *flexible* if it satisfies the flexible law

$$(xy)x = x(yx)$$

for all $x,y \in A$. A *Jordan-admissible algebra* A over a field F of characteristic $\neq 2$ is an algebra whose attached plus algebra A^+ is a

Jordan algebra. An algebra A over F is said to be *Lie-admissible* if the attached minus algebra A^- is a Lie algebra. □

Throughout this monograph, otherwise stated, *all base fields are assumed to be of characteristic ≠ 2* . While the flexible law is a natural generalization of the anticommutative and commutative laws, it has proved to be a useful identity for the study of other nonassociative algebras (see Tomber [1]).

Flexible Jordan-admissible algebras were later called *noncommutative Jordan algebras* by Schafer [1] who showed that, under the presence of flexibility, relation (1.2) is equivalent to the identity $(x^2y)x = x^2(yx)$. The classification of simple noncommutative Jordan algebras with descending chain condition on inner ideals has essentially been completed by McCrimmon [1].

Since a brief introduction of Lie-admissible algebras in the Albert 1948 paper, the first paper in flexible Lie-admissible algebras was published in 1957 by Weiner [2], a student of Albert. There were three more works on Lie-admissible algebras by Schafer [2], Laufer and Tomber [1], and Oehmke [3], until Lie-admissible algebras came to the attention of physicist Santilli in 1967. The two papers by Schafer [2] and Oehmke [3] dealt with nodal noncommutative Jordan algebras which are Lie-admissible and are related to simple Lie algebras of characteristic $p > 0$. The work of Laufer and Tomber [1] was the first paper which gave a partial solution to Albert's original problem of the determination of flexible Lie-admissible algebras A with A^- semisimple. Two decades later, this problem was completely solved independently by Okubo and Myung [3] and Benkart and Osborn [1], when the ground field is algebraically closed of characteristic 0 .

Santilli in a paper of 1967 [1] recognized that Lie-admissible algebras had implications for physics. He followed this paper with several others (Santilli [2,3,4]). He pointed out that Lie-admissible algebras arise in a natural way in Newtonian mechanics via a generalization of Hamilton's equations for the representation of forces nonderivable from a potential. More recently, Okubo [6,10] suggested possible applications of flexible Lie-admissible algebras to quantum mechanics in a different view from Santilli.

Santilli's works on Lie-admissible algebras were not known to most mathematicians until the first joint workshop of physicists and mathematicians on Lie-admissible formulations in 1978. Since that time, there has been a major breakthrough in the theory of finite-dimensional Lie-admissible algebra over an algebraically closed field of characteristic 0, although there is no general theory of Lie-admissible algebras, nor will such theory exist in the near future. A grass-root tool has been the representation theory of finite-dimensional simple Lie algebras of characteristic 0. At the same time, there is a growing interest in applications of Lie-admissible algebras in physics, from classical to quantum, relativistic and gravitational (see Myung, Okubo and Santilli [1], Myung [10], and Santilli [5]).

Definition 1.2. An algebra M with multiplication $[x,y]$ over a field F of arbitrary characteristic is called a *Malcev algebra* if it satisfies the anticommutative law $[x,x] = 0$ and the Malcev identity

$$(1.3) \qquad [[x,y],[x,z]] = [[[x,y],z],x] + [[[y,z],x],x] + [[[z,x],x],y]$$

for all $x,y,z \in M$. An algebra A over F is said to be *Malcev-admissible* if the attached minus algebra A^- is a Malcev algebra. □

The identity (1.3) was first noted by Malcev [1] in 1955 when he considered the attached minus algebra A^- for an alternative algebra A. The original source of Malcev algebras therefore stems from Malcev-admissible algebras. A Malcev algebra, so named by Sagle [1] in 1961, was first called a Moufang-Lie algebra by Malcev [1]. In 1958, Kleinfeld [1] published a brief paper on Moufang-Lie rings. The first comprehensive study of Malcev algebras was undertaken by Sagle [1,2].

Let M be a Malcev algebra with product denoted by $[\ ,\ ]$. Define the *Jacobian* $J(x,y,z)$ in M by

$$J(x,y,z) = [[x,y],z] + [[y,z],x] + [[z,x],y] .$$

Proposition 1.1. In a Malcev algebra M, Malcev identity (1.3) is equivalent to the identity

$$J(x,y,[x,z]) = [J(x,y,z),x] \tag{1.4}$$

for all $x,y,z \in M$.

Proof. Assuming (1.3) holds, we have $J(x,y,[x,z]) = [[x,y],[x,z]] + [[y,[x,z]],x] + [[[x,z],x],y] = [[[x,y],z],x] + [[[y,z],x],x] + [[[z,x],x],y] + [[y,[x,z]],x] + [[[x,z],x],y] = [J(x,y,z),x]$. The converse follows from expanding (1.4). □

Since $J(x,y,z) = 0$ for all x,y,z in any Lie algebra, Proposition 1.1 implies that a Lie algebra is a Malcev algebra. In the next section, we will show that any alternative algebra is flexible Malcev-admissible. The basic example of simple, non-Lie Malcev algebra is the 7-dimensional attached minus algebra C_0^- of trace zero elements in an octonion algebra C of characteristic $\neq 2,3$. The algebra C_0^- was first noted by Sagle [2] who showed that any finite-dimensional simple non-Lie, Malcev algebra M over an algebraically closed field F of characteristic 0 is isomorphic to C_0^-, when M has an element x such that the *adjoint mapping* ad_x defined by

$$ad_x : y \to ad_x(y) = [x,y], \ y \in M , \tag{1.5}$$

is not nilpotent. Loos [1] later obtained the same result without this restriction. Kuzmin [1] extended this classification for central simple, non-Lie, Malcev algebras over F of characteristic $\neq 2,3$. Due to a result of Filippov [1], we now know that C_0^- is the only central simple non-Lie, Malcev algebra of arbitrary dimension over F of characteristic $\neq 2,3$.

Following similar techniques employed for the study of Lie-admissible algebras, Myung [5,8,9] has extended many known results to Malcev-admissible algebras. It seems also more desirable to study Malcev-admissible algebras, since these algebras include such algebras as octonion and para-octonion algebras which have been excluded from the study of Lie-admissible algebras. Furthermore, octonion algebras have been basic algebraic models in octonionic quantum mechanics developed by Günaydin and Gürsey [1,2]. Therefore, for physical applications, the theory of Malcev-admissible algebras might give rise to broader algebraic models to unify the known approaches.

The majority of investigations in this monograph are based on those results published since 1978. Many of these results were published in physics journals. As for Lie-admissible algebras, the main tool is the representation theory of finite-dimensional simple Lie algebras of characteristic 0.

In this chapter, we investigate some general results for flexible Malcev-admissible algebras of arbitrary dimension. However, the remainder of this monograph deals exclusively with finite-dimensional algebras.

1.2. Basic results

Let A be an algebra with multiplication denoted by xy over a field F of arbitrary characteristic. Define the associator (x,y,z) and $S(x,y,z)$ in A by

$$(x,y,z) = (xy)z - x(yz) ,$$

$$S(x,y,z) = (x,y,z) + (y,z,x) + (z,x,y)$$

for $x,y,z \in A$. In any algebra A, by direct expansion, we have the identity

$$S(x,y,z) - S(x,z,y) = J(x,y,z) , \tag{1.6}$$

where $J(x,y,z)$ is the Jacobian in A^-. The flexible law $(xy)x = x(yx)$ can be linearized to the relation

$$(x,y,z) = -(z,y,x) \tag{1.7}$$

for all $x,y,z \in A$. Hence, if A is flexible, then it satisfies the identity $S(x,y,z) = -S(x,z,y)$ and so by (1.6)

$$2S(x,y,z) = J(x,y,z) . \tag{1.8}$$

Lemma 1.2. Let A be a flexible algebra over the field F.

(i) A is Lie-admissible if and only if A satisfies the identity

$$2S(x,y,z) = 0 .$$

In particular, any flexible algebra of characteristic 2 is Lie-admissible.

(ii) A is Malcev-admissible if and only if A satisfies the identity

$$2S(x,y,[x,z]) = 2[S(x,y,z),x] .$$

Any flexible algebra of characteristic 2 is Malcev-admissible.

Proof. The results are immediate from (1.8) and Proposition 1.1. □

An algebra A over the field F is called an *alternative algebra* if it satisfies the left and right alternative laws

$$(x,x,y) = (y,x,x) = 0$$

for all $x,y \in A$. Linearizing this implies that the associator (x,y,z) in an alternative algebra A is skew symmetric in x,y,z. In particular, $(x,x,y) = 0$ gives the flexible law, which in turn implies $S(x,y,z) = - S(x,z,y) = 3(x,y,z)$ for all x,y,z in an alternative algebra A, and hence by (1.6)

$$J(x,y,z) = 6(x,y,z)$$

for all $x,y,z \in A$. Thus, we have

Lemma 1.3. An alternative algebra A over an arbitrary field is Lie-admissible if and only if A is either associative or of characteristic 2 or 3. □

Proposition 1.4. Any alternative algebra A is flexible Malcev-admissible.

Proof. Consider the function $f(w,x,y,z) = (wx,y,z) - x(w,y,z) - (x,y,z)w$ in A. This function is called the Kleinfeld function and is shown to be skew symmetric in four variables and vanishes whenever any pair of variables are equal (Kleinfeld [2,p.128]). In particular, $f(x,z,y,x) = f(z,x,y,x) = 0$ implies $(x,y,xz) = (x,y,z)x$ and $(x,y,zx) = x(x,y,z)$, since (x,y,z) is skew symmetric. Hence $[(x,y,z),x] = (x,y,[x,z])$, and since $S(x,y,z) = 3(x,y,z)$, we have by Lemma 1.2 that A is flexible Malcev-admissible. □

Definition 1.3. For an algebra A over an arbitrary field F, denote by $\mathrm{Hom}_F\, A = \mathrm{Hom}\, A$ the associate algebra of linear transformations on A over F. An element $d \in \mathrm{Hom}_F\, A$ is called a *derivation* of A if d satisfies the relation

$$d(xy) = xd(y) + (d(x))y$$

for all $x,y \in A$. Denote by $\mathrm{Der}\, A$ the set of derivations of A. It is readily seen that $\mathrm{Der}\, A$ is a Lie subalgebra of $(\mathrm{Hom}\, A)^-$. $\mathrm{Der}\, A$ is called the *derivation algebra* of A. For each $x \in A$, let L_x and R_x denote the *left* and *right multiplication* by x in A; i.e.,

$$L_x(y) = xy \;,\; R_x(y) = yx \;,\; y \in A \;.$$

If the characteristic of F is not two, then the left (= right) multiplication by x in the algebra A^+ will be denoted by $t_x = \frac{1}{2}(L_x + R_x)$. □

The following result is instrumental for our investigation.

Lemma 1.5. (i) An algebra A is flexible if and only if $\mathrm{ad}_A \subseteq \mathrm{Der}\, A^+$.

(ii) A is flexible Lie-admissible if and only if $\mathrm{ad}_A \subseteq \mathrm{Der}\, A$.

Proof. (i) Assume A is flexible. The flexible law is equivalent to the identity $(x,y,z) + (z,y,x) = 0$ which implies the identity $(x,y,z) + (z,y,x) + (x,z,y) + (y,z,x) = (y,x,z) + (z,x,y)$. By direct expansion, this last relation is equivalent to the identity

$$(1.9) \qquad [x,y \circ z] = [x,y] \circ z + y \circ [x,z]$$

which implies that $\mathrm{ad}_A \subseteq \mathrm{Der}\, A^+$. Letting $x = z$, (1.9) gives the flexible law.

(ii) Note that Lie-admissibility of A is equivalent to the relation that $\mathrm{ad}_A \subseteq \mathrm{Der}\, A^-$. Thus, if A is flexible Lie-admissible, then each ad_x is a derivation of A^- and of A^+ by (i). Since $xy = \frac{1}{2}[x,y] + x \circ y$, this implies that ad_x is a derivation of A, i.e.,

(1.10) $$[x,yz] = [x,y]z + y[x,z]$$

for all $x,y,z \in A$. When $x = z$, (1.10) gives the flexible law. □

For a subset S of an algebra A, we denote by $C_A^-(S)$ the *Lie centralizer* of S in A defined by

$$C_A^-(S) = \{x \in A \mid [S,x] = 0\} .$$

Corollary 1.6. Let A be a flexible algebra and S be a subset of A. Then $C_A^-(S)$ is a subalgebra of A^+. If, in addition, A is Lie-admissible then $C_A^-(S)$ is a subalgebra of A.

Proof. For $x,y \in C_A^-(S)$ and $s \in S$, by Lemma 1.5 (i) $\mathrm{ad}_s(x \circ y) = [s,x \circ y] = x \circ [s,y] + [s,x] \circ y = 0$ and hence $x \circ y \in C_A^-(S)$. Similarly, the second part follows from Lemma 1.5 (ii). □

Theorem 1.7. Let A be a flexible algebra such that A^- is a direct sum of simple ideals A_i^-. Then each A_i is an ideal of A, so that A is the direct sum of simple flexible algebras A_i.

Proof. We first show that each A_k is a subalgebra of A^+, and hence a subalgebra of A. Let $B_k^- = \Sigma_{i\neq k} A_i^-$, so that $A^- = A_k^- \oplus B_k^-$ and $A_k \subseteq C_A^-(B_k)$, since $[A_k,B_k] = 0$. In light of Lemma 1.5 (i) it suffices to verify that $A_k = C_A^-(B_k)$. For $x \in C_A^-(B_k)$, write $x = a_k + \Sigma_{i\neq k}\, a_i$ for some $a_j \in A_j$. Then we have $0 = [x,B_k] = \Sigma_{i\neq k}[a_i,A_i]$ since $[A_i,A_j] = 0$ for $i \neq j$. Since the sum is direct, this gives $[a_i,A_i] = 0$ for all

$i \neq k$, and hence $a_i = 0$ for all $i \neq k$, since A_i^- is simple. This shows that $x = a_k \in A_k$ and $A_k = C_A^-(B_k)$.

Since each A_k is an ideal of A^- , it suffices to show that A_k is an ideal of A^+ . If $a \in A_k$ and $b \in B_k$, we can write $b = \Sigma_{i\neq k}[x_i,y_i]$ for $x_i,y_i \in A_i$, using the fact that each A_i^- is simple and $B_k = \Sigma_{i\neq k}[A_i,A_i]$. By Lemma 1.5(i), we have $a \circ b = \Sigma\, a \circ [x_i,y_i] = \Sigma[a\circ x_i,y_i] - \Sigma[a,y_i] \circ x_i$ $= \Sigma[a \circ x_i,y_i] \in B_k$, since B_k^- is an ideal of A^- . Hence, A_k is an ideal of A^+ and so is an ideal of A . □

<u>Lemma 1.8</u>. Let A be an algebra over an arbitrary field F . For any $d \in \mathrm{Der}\, A$ and $\alpha,\beta \in F$,

$$(d - (\alpha + \beta)I)^n(xy) = \sum_{i=0}^{n} \binom{n}{i}(d - \alpha I)^i(x)(d - \beta I)^{n-i}(y)$$

for all $x,y \in A$, where n is a positive integer and I is the identity mapping on A .

<u>Proof</u>. We show this by induction on n . If $n = 1$, $(d - (\alpha + \beta)I)(xy)$ $= (d(x))y - \alpha(xy) + xd(y) - \beta(xy) = x(d - \beta I)(y) + [(d - \alpha I)(x)]y$. Assume that the relation holds for $n - 1$. Using the relation for $n = 1$, we have

$$\begin{aligned}(d - (\alpha + \beta)I)^n(xy) &= \Sigma\binom{n-1}{i}(d - \alpha I)^{i+1}(x)(d - \beta I)^{n-(i+1)}(y)\\ &\quad + \Sigma\binom{n-1}{i}(d - \alpha I)^i(x)(d - \beta I)^{n-i}(y)\\ &= \Sigma\binom{n}{i}(d - \alpha I)^i(x)(d - \beta I)^{n-i}(y) ,\end{aligned}$$

since $\binom{n-1}{i-1} + \binom{n-1}{i} = \binom{n}{i}$ for $n > i$. □

<u>Corollary 1.9</u>. For $\alpha \in F$ and $d \in \mathrm{Der}\, A$, let

$$A_\alpha(d) = \{ x \in A \mid (d - \alpha I)^n(x) = 0 \text{ for some } n > 0 \} .$$

Then $A_\alpha(d)A_\beta(d) \subseteq A_{\alpha+\beta}(d)$ for all $\alpha,\beta \in F$. If A is flexible, then $A_\alpha(ad_x) \circ A_\beta(ad_x) \subseteq A_{\alpha+\beta}(ad_x)$ for all $x \in A$ and $\alpha,\beta \in F$.

Proof. Assume $(d - \alpha I)^m(x) = (d - \beta I)^n(y) = 0$ for $x,y \in A$. It follows from Lemma 1.8 that $(d - (\alpha + \beta)I)^{m+n}(xy) = 0$, and hence $xy \in A_{\alpha+\beta}(d)$. The second part follows from the fact that if A is flexible, then $ad_A \subseteq Der\, A^+$ (Lemma 1.6(i)). □

Definition 1.4. Let A be an algebra over an arbitrary field F . For $x \in A$, define $x^1 = x$ and $x^{m+1} = x^m x$ inductively for positive integers m . An element $x \in A$ is said to be *nth power-associative* if $x^p x^q = x^{p+q}$ for all positive integers p,q such that $p + q = n$. If x is nth power-associative for all positive integers n , x is called *power-associative.* If every element of A is power-associative, then A is termed *power-associative.* In case every element of A is nth power-associative, A itself is called nth power-associative. An element a of A is said to be *nil* if it is power-associative and $a^k = 0$ for some $k > 0$. If every element of A is nil, A is said to be *nil.* □

Two important power-associativities are *third power-associativity*

(1.11) $$x^2x = xx^2, \quad \text{or} \quad [x^2,x] = 0$$

and *fourth power-associativity*

(1.12) $$x^2x^2 = (x^2x)x = x(x^2x) = x(xx^2) = (xx^2)x$$

for all $x \in A$. All flexible algebras, including commutative algebras, are third power-associative. Note that if A is commutative, (1.12) reduces to the identity $x^2x^2 = x^3x$. In fact, we show that this is the case for any third power-associative algebra.

Lemma 1.10. Assume that A is a third power-associative algebra over a field F of characteristic $\neq 2$. Then

(i) A satisfies the identity $[x^3,x] = 0$ and hence fourth power-associativity (1.12) is equivalent to the identity

$$x^2x^2 = x^3x \tag{1.13}$$

(ii) A satisfies the relation

$$(x^2x)x - x^2x^2 = ((x \circ x) \circ x) \circ x - (x \circ x) \circ (x \circ x) \tag{1.14}$$

for all $x \in A$. Hence, A is fourth power-associative if and only if A^+ is.

Proof (i) We first note that third power-associativity (1.11) can be linearized to the identity

$$[x \circ y,z] + [y \circ z,x] + [z \circ x,y] = 0 . \tag{1.15}$$

(Albert [1]). Letting $y = x$ and $z = x^2$ in (1.15) yields $[x^3,x] = 0$ by (1.11). Hence (1.12) reduces to (1.13).

(ii) Using $[x^3,x] = 0$, we compute $(x^2x)x - x^2x^2 = \frac{1}{2}[(x^2x)x + x(x^2x)] - x^2 \circ x^2 = (x^2x) \circ x - x^2 \circ x^2 = \frac{1}{2}(x^2x + xx^2) \circ x - x^2 \circ x^2 = (x^2 \circ x) \circ x - x^2 \circ x^2 = ((x \circ x) \circ x) \circ x - (x \circ x) \circ (x \circ x)$. Thus, A is fourth power-associative if and only if A^+ is fourth power-associative. □

Under an appropriate restriction on the characteristic, power-associativity of A is implied by third and fourth power-associativities of A. The following result is useful.

Lemma 1.11. (i) Any algebra A of characteristic 0 is power-associative if and only if it is third and fourth power-associative; that is, A satisfies (1.11) and (1.13).

(ii) Any commutative algebra A of characteristic $\neq 2,3,5$ is power-associative if and only if it is fourth power-associative. □

A proof of Lemma 1.11(i) can be found in Albert [1] or Osborn [2] while Lemma 1.11(ii) is proved in Albert [1]. There exist counter-examples to show that each restriction on the characteristic in Lemma 1.11 is in fact necessary (Albert [1]).

Power-associativity of A is also implied by other identities imposed on A . It is well known that Jordan and alternative algebras of arbitrary characteristic are power-associative (Jacobson [3] and Schafer [3]). Recall that an algebra A over F of arbitrary characteristic is called a non-commutative Jordan algebra, introduced by Schafer [1], if A is flexible and satisfies the identity $(x^2y)x = x^2(yx)$. Schafer [1] has shown that an algebra A of characteristic $\neq 2$ is noncommutative Jordan if and only if it is flexible Jordan-admissible and that any noncommutative Jordan algebra of characteristic $\neq 2$ is power-associative. Extending this result, Leadley and Ritchie [1] proved that a noncommutative Jordan algebra over a field F of characteristic 2 is power-associative, when F contains at least 3 elements.

1.3. CARTAN DECOMPOSITIONS OF A^-

In the study of Malcev-admissible algebras A , our basic technique is to utilize the known structure of the attached minus algebra A^- and to see what restrictions the structure of A^- imposes on the multiplication of A . Cartan subalgebras and Cartan decompositions are main tools in the traditional theory of Lie algebras. Cartan theory also applies to Malcev algebras though there exists only one, non-Lie, simple Malcev algebra.

Definition 1.5. Let M be a Malcev algebra over an arbitrary field F. Let H be a nilpotent subalgebra of M. For a function $\alpha : H \to F$, we define $M_\alpha(ad_H) \equiv M_\alpha$ as the subspace

$$M_\alpha = \{ x \in M \mid (ad_h - \alpha(h)I)^n(x) = 0 \text{ for some } n = n(h), h \in H \}$$

where $n(h)$ is a positive integer depending on $h \in H$. If $M_\alpha \neq 0$, α is called a *root* of H in M and M_α is called the *root space* corresponding to root α. In particular, M_0 is termed the *Fitting null component* of M relative to H. A nilpotent subalgebra H of M is called a *Cartan subalgebra* of M if $M_0(ad_H) = H$. A Cartan subalgebra H is said to be *split* over F if the eigenvalues of ad_h belong to F for all $h \in H$. If there exists a vector space direct sum $M = \sum_\alpha M_\alpha$ where α runs over all roots, then this sum is called a *Cartan decomposition* of M relative to H. □

Sagle [5] has introduced Cartan subalgebras, more generally pre-Cartan subalgebras, and Cartan decompositions for any anticommutative algebras. Assume that M is a finite-dimensional Malcev algebra over a field F of characteristic $\neq 2$. If F has at least $\dim_F M - 1$ elements, M has a Cartan subalgebra (Carlsson [1] and Malek [2]). This is well known for Lie algebras (Barnes [1]). As for Lie algebras, if H is a Cartan subalgebra of M which is split over F, then M has a Cartan decomposition $M = M_\alpha + M_\beta + \cdots + M_\delta$ which satisfies the properties:

$$(1.16) \qquad [M_\alpha, M_\beta] \subseteq M_{\alpha+\beta} \ (\alpha \neq \beta), \ [M_\alpha, M_\alpha] \subseteq M_{2\alpha} + M_{-\alpha}$$

for all roots α, β (Sagle [2], Carlsson [1], and Kuzmin [1]). In fact, relation (1.16) holds for arbitrary dimension (Sagle [2]). In particular, if M is a Lie algebra, then (1.16) reduces to

$$[M_\alpha, M_\beta] \subseteq M_{\alpha+\beta}$$

for all roots α, β .

Let A be a flexible Malcev-admissible algebra over F (not necessarily finite-dimensional). Assume that A^- has a Cartan subalgebra H which permits a Cartan decomposition

$$A^- = \sum_\alpha A_\alpha \; . \tag{1.17}$$

Lemma 1.12. Assume that A is a flexible Malcev-admissible algebra over a field F of characteristic $\neq 2$.

(i) Any Cartan subalgebra H of A^- is a subalgebra of A .

(ii) If H is a Cartan subalgebra of A^- which permits the decomposition (1.17), then

$$\begin{aligned} &A_\alpha \circ A_\beta \subseteq A_{\alpha+\beta} \quad \text{for all roots } \alpha, \beta, \\ &A_\alpha A_\beta \subseteq A_{\alpha+\beta} \quad \text{for } \alpha \neq \beta \text{ ,} \quad A_\alpha A_\alpha \subseteq A_{2\alpha} + A_{-\alpha} \text{ .} \end{aligned} \tag{1.18}$$

Proof. (i) Since $H = A_0^-(\mathrm{ad}_H) = \bigcap_{h \in H} A_0^-(\mathrm{ad}_h)$, it follows from Lemma 1.5 and Corollary 1.9 that H is a subalgebra of A^+ . Hence H is a subalgebra of A , since H is a subalgebra of A^- . Noting $xy = \frac{1}{2}[x,y] + x \circ y$, relation (1.18) is an immediate consequence of Lemma 1.5, Corollary 1.9, and (1.16). □

Lemma 1.13. Let A be a flexible algebra over F of characteristic $\neq 2$.

(i) If h is a third and fourth power-associative element of A and $x \in A$ is a common eigenvector of ad_h and ad_{h^2} , then $[x,h^3] = [x,h^4] = 0$ imply $[x,h^2] = 0$.

(ii) If h is an nth power-associative element of A for $n = 3,4,5$ and x is a common eigenvector of ad_h, ad_{h^2}, R_h, and R_{h^2}, then $[x,h^4] = [x,h^5] = 0$ imply $[x,h^3] = 0$.

Proof. (i) Since A is flexible, relation (1.9) holds for A and can be expressed as

$$(1.19) \qquad ad_{y \circ z} = t_z ad_y + t_y ad_z$$

for all $x,y,z \in A$, which shows

$$(1.20) \qquad ad_{h^2} = 2ad_h t_h = 2t_h ad_h ,$$

since flexibility is equivalent to the relation $R_x L_x = L_x R_x$. Letting $y = h$ and $z = h^2$ in (1.19) yields $ad_{h^3} = t_{h^2} ad_h + t_h ad_{h^2}$, which implies by (1.20)

$$(1.21) \qquad ad_{h^3} = [2(t_h)^2 + t_{h^2}]ad_h .$$

From (1.19) and (1.20), we obtain

$$(1.22) \qquad \begin{aligned} ad_{h^4} &= 2t_{h^2} ad_{h^2} = 4t_{h^2} t_h ad_h = 4ad_h t_h t_{h^2} \\ &= t_{h^3} ad_h + t_h ad_{h^3} . \end{aligned}$$

Assume $ad_h(x) = [h,x] = \lambda x$ and $[h^2,x] = \mu x$ for $\lambda,\mu \in F$. If $\lambda = 0$, then it follows from (1.20) that $\mu = 0$ or $[h^2,x] = 0$. Suppose $\lambda \neq 0$. Since $ad_{h^3}(x) = 0$ by the assumption, (1.21) gives $t_{h^2}(x) = -2t_h^2(x)$. Hence, in light of (1.22), we have $0 = [h^4,x] = 4ad_h t_h t_{h^2}(x) = -8\lambda t_h^3(x)$, since $ad_h t_h = t_h ad_h$ by flexibility. Thus, $t_h^3(x) = 0$ and by (1.20) $t_h(x) = \mu(2\lambda)^{-1}x$; hence $\mu = 0$ or $[h^2,x] = 0$.

(ii) Assume that $[h,x] = \lambda x$, $[h^2,x] = \mu x$, $xh = \nu x$ and $xh^2 = \omega x$ for

$\lambda,\mu,\nu,\omega \in F$. If $\lambda = 0$, then we use (1.21) to conclude $[h^3,x] = 0$. Suppose then $\lambda \neq 0$. If $\mu = 0$, then by (1.20) $t_h(x) = 0$, and hence by (1.21) $[h^3,x] = t_{h^2}ad_h(x) = \lambda\omega x$, since $xh^2 = h^2x$. Substituting $z = h^2$ and $y = h^3$ in (1.19) gives

$$(1.23) \qquad ad_{h^5} = t_{h^3}ad_{h^2} + t_{h^2}ad_{h^3} .$$

Using (1.23) and $[h^3,x] = \lambda\omega x$, we derive $0 = t_{h^2}[h^3,x] = \lambda\omega t_{h^2}(x) = \lambda\omega^2 x$, since $\mu = 0$. Thus, $\omega = 0$ and $[h^3,x] = 0$.

Assume now $\mu \neq 0$. By the assumption and (1.22),

$$0 = [h^4,x] = 2t_{h^2}[h^2,x] = 2\mu t_{h^2}(x)$$

and hence $t_{h^2}(x) = 0$. Since $\lambda \neq 0$, this gives $t_{h^2}[h,x] = 0$. Therefore, in light of (1.21), we have $[h^3,x] = 2t_h^2[h,x] = 2\lambda t_h^2(x) = 2\lambda\bar{\nu}^2 x$ for $\bar{\nu} = \nu - \frac{1}{2}\lambda$, since $h \circ x = \frac{1}{2}hx + xh = \frac{1}{2}(\nu - \lambda)x + \frac{1}{2}\nu x = (\nu - \frac{1}{2}\lambda)x$. Using $t_{h^2}(x) = 0$, this and (1.23) imply

$$\begin{aligned} 0 = [h^5,x] &= t_{h^3}[h^2,x] + t_{h^2}[h^3,x] \\ &= \mu t_{h^3}(x) + 2\lambda\bar{\nu}^2 t_{h^2}(x) \\ &= \mu t_{h^3}(x) , \end{aligned}$$

to give $t_{h^3}(x) = 0$ since $\mu \neq 0$. Therefore, it follows from (1.22) that $0 = [h^4,x] = t_{h^3}[h,x] + t_h[h^3,x] = \lambda t_{h^3}(x) + 2\lambda\bar{\nu}^2 t_h(x) = 2\lambda\bar{\nu}^3 x$, since $t_h(x) = \bar{\nu}x$. Hence, $\bar{\nu} = 0$ and $[h^3,x] = 0$. □

<u>Theorem 1.14.</u> Let A be a flexible Malcev-admissible algebra with multiplication denoted by xy (not necessarily finite-dimensional) over a field F of characteristic $\neq 2$ and let H be an abelian Cartan subalgebra (i.e., $[H,H] = 0$) of A^- . Assume that A^- has a Cartan decomposition

relative to H such that each ad_h $(h \in H)$ diagonally acts on the root space A_α for all roots α ; i.e., $[h,x] = \alpha(h)x$ for all $x \in A_\alpha$ and $h \in H$.

(i) If $h \in H$ and $x \in A_\alpha$ for $\alpha \neq 0$, then hx and xh are multiples of x .

(ii) If $x \in A_\alpha$, $y \in A_\beta$ and $\alpha \neq -\beta$ for $\alpha, \beta \neq 0$, then xy is a multiple of $[x,y]$.

(iii) If $HH = 0$, then $xy = \frac{1}{2}[x,y]$ for all $x,y \in A$, and hence A is a Malcev algebra isomorphic to A^- .

(iv) If the center of A^- is zero and H is a nil algebra under the product xy with property that there exists a positive integer n such that $h^n = 0$ for all $h \in H$, then $HH = 0$, and hence A is a Malcev algebra.

Proof. By the assumptions, we note that each root α is linear on H .

(i) Let α be a nonzero root of H . For $h,h' \in H$ and $x \in A$, since $ad_x \in \operatorname{Der} A^+$ and H is a subalgebra of A^+ (Lemma 1.12), we have

$$\alpha(h \circ h')x = [h \circ h',x] = [h,x] \circ h' + h \circ [h',x]$$

$$= \alpha(h)x \circ h' + \alpha(h')h \circ x . \tag{1.24}$$

We consider first the special case where $h = h'$ and $h \notin \ker \alpha$, the kernel of α . Then $\alpha(h)^2x = \alpha(h)[h,x] = \alpha(h)(hx - xh)$, while $\alpha(h^2)x = \alpha(h)(hx + xh)$ by (1.24). This implies that hx is a multiple of $[h,x]$. If $h \circ h = 0$, then by (1.24) $x \circ h = 0$ and hence

$$hx = \frac{1}{2}\alpha(h)x = \frac{1}{2}[h,x] . \tag{1.25}$$

Assume then that $h \notin \ker \alpha$ and $h' \in \ker \alpha$. Thus, by (1.24), $\alpha(h \circ h')x = \alpha(h)x \circ h'$ and so $x \circ h'$ is a multiple of x, to show that hx and xh are multiples of x for all $h \in H$ and $x \in A_\alpha$, $\alpha \neq 0$.

(ii) If $x \in A_\alpha$ and $y \in A_\beta$ for $\alpha \neq 0$, $\beta \neq 0$ with $\alpha \neq -\beta$, then

$$(1.26) \qquad [h \circ x,y] = h \circ [x,y] + [h,y] \circ x = \beta(h)y \circ x + h \circ [x,y]$$

for $h \in H$. By (i), the left side and the second term of right side of (1.26) are multiples of $[x,y]$. Since $\beta \neq 0$, we see that $x \circ y$ is a multiple of $[x,y]$. Hence xy is a multiple of $[x,y]$.

(iii) Assume $HH = 0$. Let $h \notin \ker \alpha$ and $h' \in \ker \alpha$ for a non-zero root α. By (1.24) $x \circ h' = 0$, and since $\alpha(h') = 0$, $h' \circ x = 0 = \frac{1}{2}\alpha(h')x = \frac{1}{2}[h',x]$. By the linearity of α, we have that (1.25) holds for all $h \in H$, $x \in A_\alpha$. Since $H \circ A_\alpha = 0$ and $A_\alpha \circ A_\beta \subseteq A_{\alpha+\beta}$, it follows from (1.26) that $\beta(h)\, x \circ y = 0$ for all $h \in H$, $x \in A_\alpha$, and $y \in A_\beta$, $\alpha \neq 0$, $\beta \neq 0$. Hence $xy = -xy = \frac{1}{2}[x,y]$, and this with (1.24) implies that A is a Malcev algebra isomorphic to A^-.

(iv) We show that $h^2 = 0$ for all $h \in H$. Suppose $n \geq 3$, and let m be the least integer such that $3m \geq n$. For any element $h \in H$, let $g = h^m$. Then $g^3 = g^4 = 0$. By hypothesis, each element of A_α for $\alpha \neq 0$ is a common eigenvector of ad_g and ad_{g^2}. Since H is abelian, it follows from Lemma 1.13(i) that g^2 is in the center of A^-. Hence $g^2 = h^{2m} = 0$. If $2m > 4$, then let p be the least integer with $3p \geq 2m$. Then $m > p$, since if $p \geq m$, then $3p \geq 3m > 2m$ and $p = m$, hence $3(m - 1) = 2m + (m - 3) \geq 2m$ ($2m > 4$ gives $m \geq 3$), contrary to the minimality of p. Since $h^{2m} = 0$, the argument just used implies $h^{2p} = 0$. By repeated applications of this, we have either $h^4 = 0$ or $h^2 = 0$. By part (i), every element of A_α for $\alpha \neq 0$ is an eigenvector of R_h for all $h \in H$. Thus, Lemma 1.13 (ii) implies $h^3 = 0$ for all

$h \in H$, since A^- has center 0 . By the argument above, we have $h^2 = 0$ for all $h \in H$, and since H is abelian, $0 = (h_1 + h_2)^2 = h_1^2 + 2h_1h_2 + h_2^2 = 2h_1h_2$ for $h_1, h_2 \in H$ to show that $HH = 0$. Thus by part (iii), A is a Malcev algebra. □

Several different versions of Theorem 1.14 have been proved. Theorem 1.14 was first proved by Myung [2] when A is finite-dimensional and $\dim A_\alpha = 1$ for $\alpha \neq 0$. Benkart [1] proved the present form for Lie-admissible algebras. A proof of Theorem 1.14 for the finite-dimensional case has been given by Malek [1]. The present form of Theorem 1.14 is due to Ko and Myung [1].

There are some important classes of Lie and Malcev algebras which satisfy the hypotheses of Theorem 1.14. Finite-dimensional split semisimple Lie and Malcev algebras over F of characteristic 0 are the best known algebras satisfying these hypotheses (Jacobson [2] and Kuzmin [1]). For the case of characteristic $p > 0$, two well known classes of algebras satisfying the conditions of Theorem 1.14 are the classical Lie algebras of Seligman [1] and generalized Witt algebras (Seligman [1] and Ree [1]).

Among infinite-dimensional Lie algebras of characteristic 0 which satisfy the conditions of Theorem 1.14 are the Virasoro algebra which arises in relativistic string dual model theory (Sherk [1]) and the Kac-Moody algebras which are best understood infinite-dimensional Lie algebras and are currently under active studies. To elucidate an application of Theorem 1.14 and for convenience, we give definitions of a Kac-Moody algebra and of a classical Lie algebra. Comprehensive bibliographies and exposition of these can be found in Kac [1] for Kac-Moody algebras and in Seligman [1] for classical Lie algebras. An application of Theorem 1.14 to the Virasoro and generalized Witt algebras will be discussed in the next section.

Let F be a field of characteristic 0. Assume that $C = (a_{ij})$ is an $n \times n$ matrix of rank ℓ over F and satisfies the conditions: $a_{ii} = 2$ for $i = 1, \cdots, n$; a_{ij} are non-positive integers for $i \neq j$; $a_{ij} = 0$ implies $a_{ji} = 0$. A matrix C satisfying these conditions is called a *generalized Cartan matrix*. Consider a triple $(H,\Pi,\hat{\Pi})$, called a *realization* of C, where H is a vector space over F, $\Pi = \{\alpha_1,\cdots,\alpha_n\} \subseteq H^*$ and $\hat{\Pi} = \{\hat{\alpha}_1,\cdots,\hat{\alpha}_n\} \subseteq H$ are indexed linearly independent subsets in the dual space H^* and H, respectively, which satisfy the condition, $\langle\alpha_i,\hat{\alpha}_j\rangle = a_{ij}$ $(i,j = 1,\cdots,n)$ and $n - \ell = \dim H - n$, where $\langle,\rangle$ is a pairing between H and H^*, with values in F. Note that $\dim_F H = 2n - \ell$ from the last condition.

We call Π and $\hat{\Pi}$ the *root basis* and the *coroot basis*, respectively, while elements of Π and $\hat{\Pi}$ are termed *simple roots* and *simple coroots*. As in finite-dimensional cases, the root lattice Q is defined by

$$Q = \sum_{i=1}^{n} Z\alpha_i, \quad Q_+ = \sum_{i=1}^{n} Z_+\alpha_i$$

where Z_+ is the set of nonnegative integers.

For a given realization of C, $(H,\Pi,\hat{\Pi})$, one can construct a Lie algebra, denoted by $\tilde{g}(C)$, with generators e_i, f_i $(i = 1,\cdots,n)$ and H, and satisfying the defining relations

$$[e_i,f_j] = \delta_{ij}\hat{\alpha}_i, \quad i,j = 1,\cdots,n,$$

$$[h,h'] = 0 \quad \text{for} \quad h,h' \in H,$$

$$[h,e_i] = \langle\alpha_i,h\rangle e_i, \quad h \in H,$$

$$[h,f_i] = -\langle\alpha_i,h\rangle f_i, \quad i = 1,\cdots,n, \quad h \in H.$$

It is shown that the Lie algebra $\tilde{g}(C)$ depends only on C (Kac [1,p.3]).

There is a unique maximal ideal I of $\tilde{g}(C)$ such that $I \cap H = 0$. The quotient algebra $g(C) = \tilde{g}(C)/I$ is called a *Kac-Moody algebra*. Note that the natural mapping $H \to g(C)$ is an embedding. The abelian subalgebra H of dimension $2n - \ell$ is called a *Cartan subalgebra* of $g(C)$ where C is called the *Cartan matrix* of $g(C)$. The degree n of C is the *rank* of $g(C)$. The Cartan decomposition of $g(C)$ relative to H is given by

$$g(C) = \sum_{\alpha \in Q} \oplus g_\alpha(C) ,$$

where $g_\alpha(C) = \{x \in g(C) \mid [h,x] = \alpha(h)x \text{ for all } h \in H\}$ is the root space corresponding to α , $g_0(C) = H$, and $\dim g_\alpha(C)$, called the *multiplicity* of α , is finite.

The center Z of a Kac-Moody algebra $g(C)$ is contained in H and is given by $Z = \{h \in H \mid \langle\alpha_i, h\rangle = 0 \text{ for all } i = 1,\cdots,n\}$ (Kac [1,p.9]). There is a better understood class of infinite-dimensional Kac-Moody algebras over the complex field, called (non-twisted or twisted) *affine Lie algebras*. The realization of these algebras is much simpler than non-affine infinite-dimensional Kac-Moody algebras (Kac [1,pp.73-102]). Every non-twisted affine Lie algebra has one-dimensional center while there exist twisted affine Lie algebras without center (Kac [1, Chapters 7 and 8]).

Following Seligman [1], a finite-dimensional Lie algebra L over a field F of characteristic $\neq 2,3$ is called *classical* if (1) the center of L is 0 , (2) $[L,L] = L$, (3) L has an abelian Cartan subalgebra H , relative to which, (a) L has a Cartan decomposition where each ad_h $(h \in H)$ diagonally acts on each root space, (b) if $\alpha \neq 0$ is a root, then $\dim [L_\alpha, L_{-\alpha}] = 1$, and (c) if α, β are roots and $\beta \neq 0$, then not all $\alpha + k\beta$ are roots.

If A is a finite-dimensional (power-associative) nilalgebra over

F , then $x^{\dim A + 1} = 0$ for all $x \in A$. In fact, if $x^{\dim A + 1} \neq 0$ then it is readily seen that $x, x^2, \cdots, x^{\dim A + 1}$ are linearly independent. For a nilalgebra A , if there exists a positive integer n such that $x^n = 0$ for all $x \in A$, then the least such integer is called the *nil-index* of A . Thus, if A is a finite-dimensional nilalgebra, then A is of nil-index $\leq \dim A + 1$. The following result is now an immediate consequence of Theorem 1.14.

Corollary 1.15. Assume that A is a flexible Lie-admissible algebra such that A^- is either isomorphic to a classical Lie algebra of characteristic $\neq 2,3$, or to a Kac-Moody algebra without center. If the Cartan subalgebra of A^- is a nil subalgebra of A , then A is a Lie algebra ismorphic to A^- . □

When A^- has center 0 , there are other conditions on A and A^- which imply that the algebra A or a Cartan subalgebra of A^- is nil under the product of A . Let A be a power-associative algebra over F . An ideal I of A is called *nil* if every element of I is nilpotent. If I and J are nil ideals of A , then $I + J$ is a nil ideal of A . In fact, let $a \in I$ and $b \in J$ with $a^m = 0$. Then $(a + b)^m = a^m + c = c \in J$ since J is an ideal of A , and hence $(a + b)^{mn} = c^n = 0$ for some $n > 0$ to imply that $I + J$ is nil. This implies that A contains a unique maximal nil ideal $N(A)$, called the *nilradical* of A . If $N(A) = 0$, then A is called *nil semisimple*.

Corollary 1.16. Let A be a finite-dimensional flexible power-associative Malcev-admissible algebra over F of characteristic 0 such that A^- is semisimple. Then A is a Malcev algebra isomoprhic to A^- .

Proof. We note first that A^- satisfies the conditions in

Theorem 1.14 and that A^- is a direct sum $A^- = A_1^- + \cdots + A_n^-$ of simple ideals A_i^- of A^- (Jacobson [2] and Kuzmin [1]). Hence, by Theorem 1.7, A is a direct sum of simple ideals A_i of A. If A_i is not a nil ideal of A then, by a result of Oehmke [1] or Osborn [2,p.293], A_i would contain an identity element. This is impossible since the center of A^- is zero. Hence, each A_i is a nilalgebra, and it follows from Theorem 1.14 that A is a Malcev algebra. □

The results above do not provide new algebras. However, there exist flexible Lie-admissible algebras A such that A is not Lie but A^- is semisimple. In view of Theorem 1.14 and Corollary 1.16, such algebras A cannot be power-associative, nor can A^- have a Cartan subalgebra which is nil in A.

Example 1.1. Let F be a field of characteristic 0 or prime to $n + 1$ where n is a positive integer and $n \geq 2$. Let A be the vector space $s\ell(n+1)$ of $(n+1) \times (n+1)$ trace zero matrices over F. Hence A^- is the Lie algebra $s\ell(n+1)$ of type $A_n (n \geq 2)$. Define a multiplication "*" on A by

$$(1.27) \qquad x * y = \frac{1}{2}[x,y] + \gamma\, x \# y$$

for some nonzero fixed scalar $\gamma \in F$, where $x \# y$ is defined by

$$(1.28) \qquad x \# y = xy + yx - \frac{2}{n+1}(\operatorname{tr} xy)I .$$

Here xy is the usual matrix product, tr denotes the trace, and I is the identity matrix. Clearly, $x \# y$ is a commutative product and hence $(A,*)^-$ is the simple Lie algebra $s\ell(n+1)$; but $(A,*)$ is not Lie, since $x * x = x \# x \neq 0$ for some $x \in A$. Since $x \circ y = \frac{1}{2}(x * y + y * x) = \gamma\, x \# y$, by a direct computation, it is readily seen that ad_x is a derivation of $(A,*)^+$ for all $x \in A$, and hence A is flexible. The algebra $(A,*)$ satisfies third power-associativity $(x * x) * x = x * (x * x)$, but not fourth power-associativity $(x * x) * (x * x) = [(x * x) * x] * x$; for, fourth power-associativity would imply the identity

$$(\text{tr } x^2)x^2 - (\text{tr } x^2)x - \frac{1}{n+1}(\text{tr } x^2)^2 I = 0$$

for all $x \in A$, but this is impossible since, for example, the diagonal matrix $x = \text{diag}\{1,2,-3,0,\cdots,0\}$ does not satisfy this identity. Hence, $(A,*)$ is not power-associative. □

The algebra defined by (1.27) will play an important role throughout our investigation and arise later from the classification of finite-dimensional flexible Lie-admissible algebras A over an algebraically closed field of characteristic 0 such that A^- is simple (also see Okubo and Myung [3] and Benkart and Osborn [1]). In the remainder of this section, we give some examples to show that the hypotheses that the center of A^- is 0 and the Cartan subalgebra H is nil in A are necessary in Theorem 1.14.

Example 1.2. Let A be a 3-dimensional algebra with basis x,y,h over F of characteristic $\neq 2$. Let A have multiplication given by

$$xh = x, \ yh = \frac{1}{2}(\alpha + 1)y, \ hy = \frac{1}{2}(1 - \alpha)y, \ h^2 = h ,$$

and all other products are 0, where $\alpha \neq 0, 1$ is a scalar in F. It is easy to check that A is flexible Lie-admissible and that A^- is given by

$$[x,y] = 0, \ [x,h] = x, \ [y,h] = \alpha y ,$$

so that A^- is a solvable Lie algebra. Note that $H = Fh$ is a Cartan subalgebra of A^- = Fh + Fx + Fy , where the root spaces are A_1 = Fx and A_α = Fy corresponding to roots 1 and α . The center of A^- is zero while H is not nil in A . This example shows that the algebra A^- in Theorem 1.14 need not be semisimple. □

Example 1.3. Let A be a 4-dimensional algebra with basis x,y,h,c over F of characteristic ≠ 2 and let A have multiplication given by

$$xy = \frac{1}{2}h + c,\ yx = -\frac{1}{2}h + c,\ xh = -hx = \frac{1}{2}x\ ,$$

$$yh = -hy = -\frac{1}{2}y\ ,\ h^2 = -c\ ,$$

where all other products are 0 . It is easily seen that A is flexible Lie-admissible and A^- = Fx + Fy + H is the Cartan decomposition relative to the Cartan subalgebra H = Fh + Fc which is nil of nil-index 3 . Also, Fc is the center of A^- and A^- is a reductive Lie algebra (the center of A^- is the solvable radical of A^-). But A is not Lie and, in fact, A is a nil algebra of nil-index 3 . This shows that the condition of center A^- = 0 is necessary in Theorem 1.14(iv). □

Example 1.4. Let A be a 4-dimensional algebra with basis e_1, e_2, e_3, e_4 over the field F , with multiplication defined by

$$e_1e_2 = e_1, e_1e_3 = -e_3e_1 = \frac{1}{2}e_4\ , e_2e_4 = e_4, e_3e_2 = e_3, e_2^2 = e_2\ ,$$

where all other products are 0 . By direct computation, it is seen that A is flexible Malcev-admissible and A^- is a non-Lie, solvable Malcev algebra such that $A^- = H + A_1 + A_{-1}$ is the Cartan decomposition relative to the Cartan subalgebra $H = Fe_2$, where $A_1 = Fe_4$ and $A_{-1} = Fe_1 + Fe_3$ are the root spaces corresponding to roots 1 and -1 . Thus, A^- has center 0 and satisfies the conditions of Theorem 1.14; but A is not a Malcev algebra. □

When the characteristic of F is 0, Kuzmin [2] has proven that the Malcev algebra A^- given in Example 1.4 is the only non-Lie, Malcev algebra of dimension 4.

Example 1.5. Let A be a 5-dimensional algebra with basis e_1, e_2, e_3, e_4, e_5 over the field F and let A be defined by

$$e_1e_2 = e_5 + \frac{1}{2}e_4, e_2e_1 = e_5 - \frac{1}{2}e_4, e_1e_4 = -e_4e_1 = \frac{1}{2}e_1,$$

$$e_2e_4 = -e_4e_2 = -\frac{1}{2}e_2, e_3e_4 = -e_4e_3 = \frac{1}{2}e_3, e_4^2 = -e_5,$$

where all other products are 0. One can verify that A is a flexible Malcev-admissible algebra such that A^- is not Lie and $A^- = s\ell(2) + R$ where $s\ell(2) = Fe_1 + Fe_2 + Fe_4$ is the 3-dimensional simple Lie algebra and $R = Fe_3 + Fe_5$ is the radical of A^-. Also, $H = Fe_4 + Fe_5$ is a Cartan subalgebra of A^- and $A_1 = Fe_2, A_{-1} = Fe_1 + Fe_3$ are the root spaces of A^- relative to H. The center of A^- is Fe_5 and H is a nil-algebra of nil-index 3. Note that A is not a Malcev algebra, while A^- satisfies the conditions of Theorem 1.14. □

When the characteristic of F is zero, it is shown by Kuzmin [2] that the Malcev algebra given in Example 1.5 is the only non-Lie, non-solvable Malcev algebra of dimension 5. Myung [5] classified all flexible Malcev-admissible algebras A of dimension 5 over an algebraically closed field of characteristic 0 such that A^- is isomorphic to the Malcev algebra of Example 1.5. Thus, the algebra A in Example 1.5 is a special case of this classification(Chapter 5).

Though Theorem 1.14 does not apply to algebras A such that A^- has nonzero center, there are some algebras of interest where A^- has one-dimensional center; for examples, matrix algebras, quadratic algebras, including quaternion and octonion algebras, and, as noted earlier, non-twisted affine Kac-Moody algebras. Included in these algebras are also reductive Lie or Malcev algebras with one-dimensional center. The case where A^- is reductive will be treated in Chapter 4. As shown by examples above, although algebras A satisfying the conditions of Theorem 1.14 are not Lie or Malcev algebras, when the center of A^- is one-dimensional, it is possible to determine these algebras under the assumption that the Cartan subalgebra H of A^- is nil in A and is of nil-index ≤ 4. This will be treated in Section 1.5.

1.4. Generalized Witt algebras

In this section, we make an application of Theorem 1.14 to a class of non-classical Lie algebras of characteristic $\neq 2$.

Definition 1.6. Let F be an arbitrary field and G be an additive subgroup of F. Denote by $G^{(m)} = G \times \cdots \times G$ (m copies) the direct product. Assume that W is the vector space over F with bases $\{e_\alpha^i\}$ where $i = 1,\cdots,m$ and $\alpha \in G^{(m)}$, and define a multiplication in W by

$$[e_\alpha^i, e_\beta^j] = \beta_i e_{\alpha+\beta}^j - \alpha_j e_{\alpha+\beta}^i \tag{1.29}$$

where $\alpha = (\alpha_1,\cdots,\alpha_m)$ and $\beta = (\beta_1,\cdots \beta_m)$. Then W becomes a Lie algebra under the multiplication (1.29) and is called a *generalized Witt algebra* (Seligman [1] and Ree [1]). □

Two special cases of the construction (1.29) are noteworthy. First when F has characteristic 0 and $G = \{0, \pm 1, \pm 2, \cdots\}$, the generalized Witt algebra obtained by taking $m = 1$ has the specialized multiplication

$$[e_j, e_k] = (k - j)e_{j+k} \tag{1.30}$$

where we have set $e_j^1 = e_j$. The resulting Lie algebra is called the *Virasoro algebra*, which plays a crucial role in dual strings theory. For the second special case, we take F to be a field of characteristic $p > 0$ and G to be the integers modulo p . Then the generalized Witt algebra obtained by (1.29) is the *Jacobson-Witt algebra* (Jacobson [1]).

We note that the elements $e_0^1, \cdots, e_0^m$ form a basis of an abelian Cartan subalgebra H of W and the linear span of $e_\alpha^1, \cdots, e_\alpha^m$ determines a root space for H . Thus, if $m > 1$, then W is not classical. Since $[e_0^i, e_\alpha^j] = \alpha_i e_\alpha^j$ by (1.29), H acts diagonally on each root space and hence W satisfies the conditions of Theorem 1.14. The Virasoro algebra can also be realized as a Z-graded subalgebra of the derivation algebra of an affine Kac-Moody algebra with one-dimensional center. This realization can be found in Kac [1,p.74]. The Virasoro algebra is often considered as the Lie algebra of regular vector fields on the multiplicative group of nonzero complex numbers (Kac [1,p.75]).

Theorem 1.17. Let A be a flexible algebra with product xy over a field F of characteristic $\neq 2$ such that A^- is isomorphic to a generalized Witt algebra. Then $xy = \frac{1}{2}[x,y]$ for all $x,y \in A$, so that A is a Lie algebra.

Proof. In light of Theorem 1.14 and the foregoing remarks, it suffices to show that $e_0^i e_0^j = 0$ for the basis $e_0^1, \cdots, e_0^m$ of the Cartan subalgebra H . We first prove

$$(1.31) \qquad e_0^i e_0^j = \delta_{ij} c_i e_0^i$$

where $c_i \in F$ and δ_{ij} is the Kronecker delta. Since ad_x $(x \in A)$ is a derivation of A , we have the equation

$$[e_0^i e_\alpha^j , e_{-\alpha}^k] = e_0^i [e_\alpha^j , e_{-\alpha}^k] + [e_0^i , e_{-\alpha}^k] e_\alpha^j \, . \tag{1.32}$$

If we let $j = k$ and choose α so that $\alpha_i = 0$ but $\alpha_j \neq 0$, then the right side of (1.32) is simply $-2\alpha_j e_0^i e_0^j$. By Theorem 1.14(i), the left side of (1.32) is a multiple of e_0^j. But since e_0^i and e_0^j commute, $e_0^i e_0^j = e_0^j e_0^i$ and the latter is a multiple of e_0^i. Thus, for $i \neq j$, $e_0^i e_0^j = 0$. Assume that $i = k$ in (1.32) and that α is chosen so that $\alpha_i = 0$, but $\alpha_j \neq 0$. The second term on the right side of (1.32) vanishes and the first term is $-\alpha_j e_0^i e_0^i$ whereas the left side of (1.32) is a multiple of $[e_\alpha^j , e_\alpha^i] = -\alpha_j e_0^i$ by Theorem 1.14. Therefore, $e_0^i e_0^i$ is a multiple of e_0^i, giving the desired relation (1.31).

Consider the relation

$$[e_0^i e_\beta^i , e_\alpha^i] = [e_0^i , e_\alpha^i] e_\beta^i + e_0^i [e_\beta^i , e_\alpha^i] \, . \tag{1.33}$$

If $\beta = 0$, then we have from (1.33) that $c_i \alpha_i e_\alpha^i = \alpha_i (e_\alpha^i e_0^i + e_0^i e_\alpha^i)$ which, together with $[e_0^i , e_\alpha^i] = \alpha_i e_\alpha^i$, implies that $e_0^i e_\alpha^i = \frac{1}{2}(c_i + \alpha_i) e_\alpha^i$ and $e_\alpha^i e_0^i = \frac{1}{2}(c_i - \alpha_i) e_\alpha^i$ for all $\alpha \neq 0$. If $\beta \neq -\alpha$ in (1.33), then

$$\frac{1}{2}(c_i - \alpha_i)(2\alpha_i) e_0^i = \alpha_i e_\alpha^i e_{-\alpha}^i + 2\alpha_i c_i e_0^i$$

and hence $e_\alpha^i e_{-\alpha}^i = -(c_i + \alpha_i) e_0^i$, which implies

$$\begin{aligned} -(c_i + \alpha_i)\alpha_i e_\alpha^i &= -(c_i + \alpha_i)[e_0^i , e_\alpha^i] \\ &= e_\alpha^i [e_{-\alpha}^i , e_\alpha^i] \\ &= e_\alpha^i [e_{-\alpha}^i , e_\alpha^i] \\ &= e_\alpha^i (2\alpha_i e_0^i) = 2\alpha_i \frac{1}{2}(c_i - \alpha_i) e_\alpha^i \, . \end{aligned}$$

This gives $c_i = 0$ for all i, and it follows from (1.31) that $HH = 0$. By Theorem 1.14(iii), $xy = \frac{1}{2}[x,y]$ for all $x,y \in A$ and hence A is a Lie algebra. □

The special case of Theorem 1.17 where A^- is isomorphic to a Jacobson-Witt algebra of characteristic > 0 has been proved by a direct computation in Tomber [2]. The following result is immediate from Theorem 1.17.

Corollary 1.18. Let A be a flexible algebra over a field F of characteristic 0 such that A^- is isomorphic to the Virasoro algebra. Then A is a Lie algebra. □

For any algebra A with multiplication xy , we have the relation $xy = \frac{1}{2}[x,y] + x \circ y$ where $[x,y] = xy - yx$ and $x \circ y = \frac{1}{2}(xy + yx)$. Hence, given a structure on the attached minus algebra A^- , the determination of multiplication xy in A is equivalent to that of the commutative Jordan product $x \circ y$. More generally, let "$\circ$" be a commutative product defined on the vector space A and define a new multiplication denoted by $x * y$ on the vector space A by the relation

$$x * y = \frac{1}{2}[x,y] + x \circ y \qquad (1.34)$$

for all $x,y \in A$. Denote the resulting algebra by $(A,*)$. Then $[x,y] = x * y - y * x$ and hence $(A,*)^- = A^-$. This shows that there is a vast possibility for defining multiplications "$*$" on the vector space A so that $(A,*)^-$ is isomorphic to A^- with a prescribed anticommutative product. Therefore, in the study of Lie- or Malcev-admissible algebras, the main effort is to determine the commutative products $x \circ y$, using the given structure of A^- and other conditions imposed on A . For example, the conditions imposed on A^- and flexibility of A in Theorems 1.14 and 1.17 force the commutative product $x \circ y$ to be identically zero. On the other hand, we show later that if A^- is a finite-dimensional simple Lie algebra of type A_n $(n \geq 2)$ over an algebraically closed field of characteristic 0 , then flexibility of A determines the commutative product

$x \circ y$ as the product $x \# y$ given by (1.28).

There exists another form of commutative product that has arisen in an earlier work of Weiner [1] and that was later identified by Benkart [1,2] with commutative products defined by linear forms, which play an important role in the study of Lie-admissible algebras satisfying third power-associativity. This commutative product is defined by

$$(1.35) \qquad x \circ y = \tau(x)y + \tau(y)x ,$$

where τ is a linear form on A with values in the base field F. Hence, if the product in (1.34) is specified to (1.35), the product defined by (1.34) is expressed as

$$(1.36) \qquad x * y = \frac{1}{2}[x,y] + \tau(x)y + \tau(y)x$$

which is power-associative, since $x^m = 2^{m-1}\tau(x)^{m-1}x$ for all $x \in A$. But "$*$" is not in general flexible, as we show in

Lemma 1.19. Let A be an algebra over F with an anticommutative product denoted by $[x,y]$, and let τ be a linear form on A. Then the product "$*$" on the vector space A defined by relation (1.36) is flexible if and only if $\tau([A,A]) = 0$.

Proof. Note that $x \circ y = \frac{1}{2}(x * y + y * x) = \tau(x)y + \tau(y)x$. In light of Lemma 1.5(i), it suffices to show that $[z,x \circ y] = x \circ [z,y] + [z,x] \circ y$ holds for all $x,y,z \in A$ if and only if $\tau([A,A]) = 0$. If x,y,z are in A, then we compute from (1.36)

$$[z,x \circ y] = \tau(x)[z,y] + \tau(y)[z,x] ,$$

$$x \circ [z,y] + [z,x] \circ y = \tau([z,x])y + \tau(y)[z,x] + \tau(x)[z,y] + \tau([z,y])x ,$$

which imply that if $(A,*)$ is flexible, then $\tau([z,x])y + \tau([z,y])x = 0$ for all $x,y,z \in A$; hence $\tau([A,A]) = 0$. Conversely, if $\tau([A,A]) = 0$, it is clear that A is flexible. □

Let L be the subalgebra of the Virasoro algebra generated by the basis $\{e_i \mid i = 0,1,2,\cdots\}$. As an application of Theorem 1.14 and Lemma 1.19 we have

Theorem 1.20. Let A be a flexible algebra with product $x * y$ over a field F of characteristic 0 such that A^- is isomorphic to the subalgebra L of the Virasoro algebra. Then there is a linear form τ on A with $\tau(e_i) = 0$ for all $i > 0$ such that the multiplication $x * y$ in A is given by (1.36). Conversely, any such product determines a flexible Lie-admissible algebra A such that $A^- \cong L$.

Proof. As in Lemma 1.19, let $x \circ y = \frac{1}{2}(x * y + y * x)$. The element e_0 spans a one-dimensional Cartan subalgebra H of A^- . By Theorem 1.14, e_je_k and so $e_j \circ e_k$ are multiples of $[e_j,e_k] = (k - j)e_{j+k}$ for all $j, k \geq 0$. Hence we can write $e_j \circ e_k = \gamma_{jk}e_{j+k}$, where $\gamma_{jk} = \gamma_{kj}$. Since A is flexible, we can use the relation $[e_j,e_0 \circ e_k] = [e_j,e_0] \circ e_k + e_0 \circ [e_j,e_k]$ to compute

$$\gamma_{0k}(k - j)e_{j+k} = - j\gamma_{jk}e_{j+k} + (k - j)\gamma_{0,j+k}e_{j+k} . \tag{1.37}$$

If $k = 0$, then equation (1.37) implies $\gamma_{0j} = \frac{1}{2}\gamma_{00}$ for all $j \neq 0$. Substituting this into (1.37) gives $\gamma_{jk} = 0$ for all $j, k > 0$. We define a linear form τ on A by $\tau(e_0) = \frac{1}{2}\gamma_{00}$ and $\tau(e_i) = 0$ for all $i > 0$. It is clear that $x \circ y = \tau(x)y + \tau(y)x$ for all $x,y \in A$ and $\tau([A,A]) = 0$. In light of Lemma 1.19, this proves the theorem. □

A systematic investigation of the product of the form (1.36) will be discussed in Chapter 2. The results in this section are based on the work of Benkart [1].

1.5. Flexible Malcev-admissible nilalgebras

If an algebra A is a nilalgebra, then, since A is its own nil radical, the structure theory of A is excluded from the traditional approach to the stucture theory of nonassociative algebras which heavily relies on the Peirce decompositions relative to idempotents. Besides Lie and Malcev algebras, very little has been known for the structure of nilalgebras. For an algebra A, denote by A^n the linear span of all products of any n elements of A in all possible associations. If $A^n = 0$ for some positive integer n, then A is said to be *nilpotent*. A nilalgebra is not necessarily nilpotent, as seen by anticommutative algebra (nilalgebras of nil-index 2). On the other hand, for many of the well known algebras, nilalgebras are necessarily nilpotent; for example, alternative and Jordan nilalgebras (Jacobson [3] and Schafer [3]). The nilpotence of commutative nilalgebras of finite dimension was conjectured originally by A. A. Albert. This long-standing conjecture was disproved by Suttles [1] in 1972 who gave a counterexample of a 5-dimensional commutative nilalgebra which is solvable but not nilpotent. This dimension is generally the best possible, since Gerstenhaber and Myung [1] have shown that every commutative nilalgebra of dimension 4 is nilpotent. All such algebras were determined in that paper. Commutative nilalgebras of dimension ≤ 3 are associative and their classification is well known (Kruse and Price [1, Chapter VI]).

It is a long-standing open problem whether there exist simple commutative nilalgebras of finite dimension. If such algebras do not exist, then, by virtue of a result of Block [2], any finite-dimensional simple, flexible nilalgebra over F of characteristic $\neq 2$ would be anticommutative. Using the structure of the Lie algebra A^-, there has been an attempt to investigate the structure of flexible nilalgebras of finite dimension (Myung [3,4]). The exposition in this section is based on the work of Gerstenhaber and Myung [1] and Myung [3,4].

For an algebra A over an arbitrary field F, a bilinear mapping $\phi : A \times A \to F$ is often called an F-valued *2-cochain* of A. If ϕ is a 2-cochain of A, define the skew symmetric 2-cochain ϕ^- of A by

$$\phi^-(x,y) = \phi(x,y) - \phi(y,x)$$

for all $x,y \in A$. If the characteristic of F is not 2, then the symmetric 2-cochain ϕ^+ of A can be defined by

$$\phi^+(x,y) = \frac{1}{2}[\phi(x,y) + \phi(y,x)] .$$

Then, we have $\phi = \frac{1}{2}\phi^- + \phi^+$.

Assume that A is Lie-admissible over F. Following Myung [7], an F-valued 2-cochain of A is called a *2-cocycle* of A if the 2-cochain ϕ^- is a 2-cocycle of the Lie algebra A^-, that is, ϕ^- satisfies

$$(1.38) \qquad \phi^-([x,y],z) + \phi^-([y,z],x) + \phi^-([z,x],y) = 0$$

for all $x,y,z \in A$. Similarly, if A is a Malcev-admissible algebra over F, then an F-valued 2-cochain of A is a *2-cocycle* of A if ϕ^- satisfies the relation

$$(1.39) \qquad \phi^-([x,y],[x,z]) = \phi^-([[x,y],z],x) + \phi^-([[y,z],x],x) + \phi^-([[z,x],x],y)$$

for all $x,y,z \in A$ (Myung [8]). Let c be a nonzero element in the Lie center $Z(A^-) = \{x \in A \mid [x,A] = 0\}$, and let A_0 be a subspace of A of codimension one, so that

$$A = A_0 \oplus Fc$$

is a vector space direct sum. If $x,y \in A_0$, then denote by $x * y$ and $\phi(x,y)c$ the projections of xy onto A_0 and Fc, respectively. Thus, we have

$$(1.40) \qquad xy = x * y + \phi(x,y)c, \; x,y \in A_0$$

where ϕ is an F-valued 2-cochain of the algebra $(A_0,*)$ with multiplication $x * y$. If we denote $[x,y]^* = x * y - y * x$, then by (1.40)

$$(1.41) \qquad [x + \lambda c, y + \mu c] = [x,y] = [x,y]^* + \phi^-(x,y)c$$

for $x,y \in A_0$ and $\lambda,\mu \in F$. It is readily seen from (1.41) that A is Lie- or Malcev-admissible if and only if $(A_0,*)$ is Lie- or Malcev-admissible and ϕ is a 2-cocycle of $(A_0,*)$.

Conversely, assume that there are given a Lie- or Malcev-admissible algebra $(A_0,*)$ over F and an F-valued 2-cocycle ϕ of $(A_0,*)$. Let Fc be a one-dimensional space over F, and let $A = A_0 \oplus Fc$ be the vector space direct sum. Assume that $cx = xc$ and c^2 are unique elements of A bilinearly depending on x and c. We define a multiplication in A by

$$(1.42) \qquad (x + \lambda c)(y + \mu c) = x * y + \mu cx + \lambda cy + \phi(x,y)c + \lambda\mu c^2$$

for $\lambda,\mu \in F$. Since the Lie product $[\ ,\ ]$ in A^- is given by (1.41), we see that the algebra A defined by (1.42) is Lie- or Malcev-admissible also. Relation (1.41) is the multiplication used for the construction of the non-twisted affine Kac-Moody algebras which are one-dimensional central extensions of infinite dimensional Lie algebras, the so-called loop algebras. The case where $cx = xc = \nu x$ and $c^2 \in Fc$ for a scalar $\nu \in F$ in (1.42) has arisen from the classification of finite-dimensional Lie- or Malcev-admissible algebras A over an algebraically closed field of characteristic 0 such that A^- is reductive (Okubo and Myung [1,3], Benkart and Osborn [1], and Myung [5]). This case will be investigated in Chapter 4.

Theorem 1.21. Let A be a flexible Malcev-admissible algebra with multiplication denoted by xy over a field F of characteristic $\neq 2$ such that A^- has a one-dimensional center Fc. Assume that A^- has an abelian Cartan subalgebra H which is nil in A and is of nil-index ≤ 4, and that A^- has a Cartan decomposition relative to H where each ad_h $(h \in H)$ diagonally acts on the root space A_α for all roots α. Then the multiplication xy in A is given by

$$xy = \frac{1}{2}[x,y] + \phi(x,y)c \tag{1.43}$$

for all $x,y \in A$, where ϕ is an F-valued symmetric 2-cochain of A such that $\phi(A_\alpha,A_\beta) = 0$ for all roots α, β with $\alpha \neq -\beta$, $A_0 = H$, and $\phi(c,A) = 0$. Moreover, ϕ satisfies the relation

$$\phi([x,y],z) + \phi([z,y],x) = 0 \tag{1.44}$$

for all $x,y,z \in A$. In this case A is a nilalgebra of nil-index ≤ 3, and A is a Malcev algebra if and only if $\phi = 0$. Conversely, for any prescribed Malcev algebra product [,] on A^- and a symmetric 2-cochain ϕ of A satisfying (1.44), if c is a fixed element of the center of A^-, then relation (1.43) defines a flexible Malcev-admissible product on A.

Proof. Since $h^4 = h^5 = 0$ for all $h \in H$, by Lemma 1.13(ii), h^3

belongs to center Fc, and this with Lemma 1.13(i) implies that h^3 is in Fc for all $h \in H$. Since H is abelian, we have $h_1h_2 = \frac{1}{2}[(h_1 + h_2)^2 - h_1^2 - h_2^2]$ and hence $HH \subseteq Fc$. Let $h_1h_2 = \phi(h_1,h_2)c$, where ϕ is an F-valued symmetric 2-cochain of H. By Corollary 1.6, center Fc is a subalgebra of A, and hence $c^2 = 0$. For each $h \in H$, $Fh + Fc$ is a nil subalgebra of A of nil-index ≤ 3. It is easily seen that $Fh + Fc$ is associative and hence $hc = ch = 0$ for all $h \in H$. Let $x \in A_\alpha$ for any nonzero root α. If $h, h' \in H$, then since $HH \subseteq Fc$,

$$0 = [h \circ h',x] = [h,x] \circ h' + h \circ [h',x] = \alpha(h)x \circ h' + \alpha(h')h \circ x ,$$

which implies that $A_\alpha \circ H = 0$ and $hx = - xh = \frac{1}{2}\alpha(h)x = \frac{1}{2}[h,x]$ for $h \in H$, $x \in A_\alpha$, $\alpha \neq 0$. In particular, we have $cA = Ac = 0$.

Let α,β be nonzero roots. For $x \in A_\alpha$ and $y \in A_\beta$, it follows that

$$(1.45) \qquad 0 = [h \circ x,y] = \beta(h)x \circ y + h \circ [x,y]$$

for all $h \in H$. If $\alpha + \beta \neq 0$, then (1.45) gives $x \circ y = 0$, since $\beta \neq 0$ and $H \circ A_\gamma = 0$ for $\gamma \neq 0$. Hence

$$xy = - yx = \frac{1}{2}[x,y] ,$$

$x \in A_\alpha$, $y \in A_\beta$ for any roots α,β with $\alpha + \beta \neq 0$. If $x \in A_\alpha$ and $y \in A_{-\alpha}$, then, since $\alpha \neq 0$ and $h \circ [x,y] \in Fc$, by (1.45) we have $x \circ y \in Fc$. Thus we can denote $x \circ y = \phi(x,y)c$ where $\phi(x,y) = \phi(y,x) \in F$ is uniquely determined by $x \in A_\alpha$ and $y \in A_{-\alpha}$. We extend ϕ bilinearly to A by defining $\phi(A_\alpha,A_\beta) = 0$ for all roots α,β with $\alpha + \beta \neq 0$. We have relation (1.43), since $xy = \frac{1}{2}[x,y] + x \circ y$ and $x \circ y = \phi(x,y)c$. The flexible law, which is equivalent to the identity $[x,y \circ z] = [x,y] \circ z + y \circ [x,z]$, gives relation (1.44). Since $Ac = cA = 0$, it follows from

(1.43) that A is a nilalgebra of nil-index ≤ 3. The converse is straightforward. □

A 2-cochain ϕ of an algebra A over F is called *invariant* (or *associative*) if ϕ satisfies the relation

$$(1.45) \qquad \phi(xy,z) = \phi(x,yz)$$

for all $x,y,z \in A$. Relation (1.44) shows that the symmetric 2-cochain ϕ given by (1.43) is an invariant bilinear form of the Malcev algebra A^-. If A is finite-dimensional, the *Killing form* $K(x,y)$ of A^- defined by

$$(1.46) \qquad K(x,y) = \operatorname{tr} ad_x \, ad_y$$

is the best known of such bilinear forms (Sagle [1]). Assume that ϕ is a 2-cochain of A. By a linearization, it is readily seen that (1.44) is equivalent to the relation

$$(1.47) \qquad \phi([x,y],x) = 0 \, , \quad \text{or} \quad \phi(xy,x) = \phi(yx,x).$$

Examples 1.3 and 1.5 are special cases of the construction given by (1.43) and (1.44).

Let A be the same algebra as in Theorem 1.21. Let H_0 be a subspace of H complementary to Fc, and let $A_0 = H_0 + \sum_{\alpha \neq 0} A_\alpha$. Then $A = A_0 \oplus Fc$. For $x,y \in A_0$, denote by $x * y$ and $\phi_0(x,y)c$ the projections of xy onto A_0 and Fc, respectively. It is readily seen that $x * y$ is an anticommutative product on A_0 and is equal to $\frac{1}{2}[x,y]^*$, where $[x,y]^*$ denotes the projection of $[x,y]$ onto A_0. Thus $(A_0,*)$ is a Malcev algebra and ϕ_0^- is a 2-cocycle of $(A_0,*)$ such that $\phi_0^+ = \phi$ on A_0. Hence, ϕ_0^+ satisfies (1.44), and (1.43) is reformulated as

$$(1.48) \qquad (x + \lambda c)(y + \mu c) = \frac{1}{2}[x,y]^* + \phi_0(x,y)c = x * y + \phi_0(x,y)c$$

for $x,y \in A_0$ and $\lambda,\mu \in F$. The converse of these remarks is

Corollary 1.22. Let A_0 be a Malcev algebra with multiplication denoted by $x * y$ over a field F of characteristic $\neq 2$ and let Fc be a one-dimensional space over F . Assume that ϕ_0 is a 2-cochain of A such that ϕ_0^- is a 2-cocycle of $(A_0,*)$ and ϕ_0^+ satisfies (1.44) or (1.47) under the product $x * y$. Then, the vector space $A = A_0 \oplus Fc$ with multiplication defined by (1.48) is a flexible Malcev-admissible nilalgebra of nil-index ≤ 3 such that $cA = Ac = 0$. □

The proof of Corollary 1.22 is straightforward. An element $c \neq 0$ of A satisfying $Ac = cA = 0$ is called an *absolute zero divisor* of A . Let A_0 be a finite-dimensional Lie or Malcev algebra. If K_0 is the Killing form of A_0 and ψ_0 is a 2-cocycle of A_0 , then the 2-cochain $\phi_0 = \frac{1}{2}\psi_0 + K_0$ satisfies the conditions of Corollary 1.22. When Theorem 1.21 is applied to a Kac-Moody algebra with one-dimensional center, we have

Corollary 1.23. Let A be a flexible Lie-admissible algebra such that A^- is isomorphic to a Kac-Moody algebra with one-dimensional center and with Cartan subalgebra H . If H is nil in A and of nil-index ≤ 4 then the multiplication of A is determined by (1.43) or (1.48). □

When the Cartan subalgebra H is just nil in A , Theorem 1.21 is an open problem. An algebra A over a field F is called *strictly power-associative* if the scalar extension $A_K = K \otimes_F A$ of A is power-associative for any extension field K of F . It is well known that a flexible power-associative algebra of characteristic $\neq 2,3,5$ is strictly power-associative (Albert [1]). In the following, we show that a flexible Malcev-admissible nilalgebra A of dimension ≤ 4 such that A^- is nilpotent is nilpotent such that $A^4 = 0$. The restriction of dimension ≤ 4

is generally best possible, since Suttles [1] has given a counter-example of dimension 5 in the commutative case.

Example 1.6. Let A be a 5-dimensional commutative algebra with basis $\{x_1,x_2,x_3,x_4,x_5\}$ over a field F of characteristic $\neq 2$. Let A have multiplication

$$x_1x_2 = x_3,\ x_1x_3 = x_4,\ x_1x_5 = -x_3,\ x_2x_3 = x_5,\ x_2x_4 = x_3\ ,$$

and all other products are zero. Clearly, $A^2A^2 = 0$ and so A is solvable, but not nilpotent. For any element $x = \sum_{i=1}^{5} \alpha_i x_i$ of A , we have $x^2 = \beta_3x_3 + \beta_4x_4 + \beta_5x_5$, where $\beta_3 = 2(\alpha_1\alpha_2 - \alpha_1\alpha_5 + \alpha_2\alpha_4)$, $\beta_4 = 2\alpha_1\alpha_3$ and $\beta_5 = 2\alpha_2\alpha_3$. We compute $x^3 = \alpha_1\beta_3x_4 + \alpha_2\beta_3x_5 + \alpha_2\beta_4x_3 - \alpha_1\beta_5x_3$ $= \alpha_1\beta_3x_4 + \alpha_2\beta_3x_5$ and $x^3x = 0$. Thus, A is a (power-associative) nilalgebra of nil-index 4 . □

As noted earlier, Example 1.6 is also the first counter-example to the conjecture of Albert that any commutative nilalgebra of finte dimension is nilpotent. The following are instrumental for the result mentioned above.

Lemma 1.24. Let A be a finite-dimensional, flexible, strictly power-associative nilalgebra over a field F of characteristic $\neq 2$.

(i) If x is an element of A such that ad_x is nilpotent, then R_x and L_x are nilpotent on A .

(ii) If S is a subalgebra of codimension one of A such that ad_x is nilpotent on A for all $x \in S$, then S is an ideal of A . In particular, every commutative subalgebra of codimension one of A is an ideal of A .

Proof. (i) Consider the commutative algebra A^+ . If the characteristic is 0 , it is shown that the left multiplication t_x in A^+ is nilpotent on A (Gerstenhaber [1]). If the characteristic is greater than 2, then we adjoin an identity element 1 to A^+ to obtain a commutative algebra $(A^+)'$ of degree one. Oehmke [4] proves that t_x is nilpotent on $(A^+)'$ and so on A^+ for all $x \in A$. Thus, in any case, t_x is nilpotent for all $x \in A$. Using the flexible law $R_xL_x = L_xR_x$, we have that if ad_x is nilpotent, then $R_x = \frac{1}{2}ad_x + t_x$ and $L_x = t_x - \frac{1}{2}ad_x$ are nilpotent too, since $t_x ad_x = ad_x t_x$.

(ii) Let S be a codimension one subalgebra of A. Let a be an element of A but not in S. Suppose that S is not an ideal of A. Since a and S span A, we may assume that there exists an element x of S such that $ax \equiv \lambda a \pmod{S}$ for some $\lambda \neq 0$ in F. Since S is a subalgebra of A, we have $aR_x^n \equiv \lambda^n a \pmod{S}$ and $0 \equiv \lambda^n a \pmod{S}$ for some n, since R_x is nilpotent, by part (i). This forces $\lambda = 0$, a contradiction, and so $ax \in S$ for all $x \in S$. Similarly, we have $xa \in S$ for all $x \in S$, and hence S is an ideal of A. □

The nility of A and nilpotence of ad_x are necessary for Lemma 1.24. Let L be the 3-dimensional solvable Lie algebra with multiplication $[x,y] = x$, $[x,z] = [y,z] = 0$. Then $Fy + Fz$ is an abelian subalgebra of L but not an ideal of A, where we note that ad_y is not nilpotent. Let A_0 be a Lie or Malcev algebra over F with product $x * y$. Assume that the direct sum $A = A_0 + Fc$ has the multiplication defined by relation (1.42), where $cx = xc = \nu x$ for all $x \in A_0$, $\nu \neq 0$ in F, $\phi = 0$, and $c^2 = c$. If B is an abelian subalgebra of A_0 of codimension one, then $B + Fc$ is a commutative subalgebra of A of codimension one but not an ideal of A. Note that A is not a nilalgebra.

<u>Theorem 1.25</u>. Let A be a finite-dimensional, flexible, strictly power-associative algebra over a field F of characteristic $\neq 2$. Assume

that A is a noncommutative nilaglebra such that A^- is nilpotent and A^- contains an abelian subalgebra of codimension one. Then the Lie center Z of A^- is an ideal of A and contains absolute zero divisors of A . Moreover, the set of abolute zero divisors of A equals $[A,A] \cap Z$.

Proof. Let B be a codimension one, abelian subalgebra of A^- . Since A^- is nilpotent, all ad_x are nilpotent. Applying Lemma 1.24 to A^- implies that A is an ideal of A^- . We first show that B is a subalgebra of A . Let $A = B + Fh$ be a vector space direct sum. Then $[A,A] = [B,h] \neq 0$ since B is an ideal of A^- . For $x,y \in B$, let $x \circ y \equiv \alpha h \pmod{B}$. If $g \neq 0$ is in $[A,A]$, then we let $g = [b,h]$ for $b \in B$. Since ad_b is a derivation of A^+ and B is abelian, applying ad_b to $x \circ y \equiv \alpha h \pmod{B}$ yields $0 = \alpha[b,h] = \alpha g$ and $\alpha = 0$. Hence B is a subalgebra of A^+ and is a subalgebra of A , which combines with Lemma 1.24 (ii) to show that B is an ideal of A .

Since ad_h induces a nilpotent linear transformation on B , B can be decomposed as a direct sum

$$B = M_1 \oplus M_2 \oplus \cdots \oplus M_r$$

of cyclic subspaces M_i in B relative to ad_h such that $n_1 \geq \cdots \geq n_r$ where $n_i = \dim M_i$ and n_1 is the nil-index of ad_h in B . Let $x_{i,1}, \cdots, x_{i,n_i}$ denote a cyclic basis of M_i , so that $[h,x_{i,k-1}] = x_{i,k}$ and $[h,x_{i,n_i}] = 0$, $k = 2, 3, \cdots, n_i$. Since B is abelian and $[h,B] \neq 0$, the Lie center Z of A^- is contained in B , and hence Z is the centralizer of h in B . Therefore, if we let $x_1 = x_{1,n_1}, \cdots, x_r = x_{r,n_r}$, then $x_1, \cdots, x_r$ form a basis of Z . Since B is an ideal of A , $hx_i = x_ih \in B$ and $[h,h \circ x_i] = h \circ [h,x_i] = 0$. Hence $h \circ x_i \in Z$ and this gives

(1.49) $$hx_i = x_i h \in Z,\ i = 1, 2, \cdots, r .$$

Let p be such that $n_1 \geq \cdots \geq n_p \geq 2$, and $n_i = 1$ if $i > p$. Since $[h,B] \neq 0$ and so $n_1 \geq 2$, $p \geq 1$. If $x \in B$ and $i \leq p$, then

$$0 = [x_{i,n_i-1},\ x \circ h] = x \circ [x_{i,n_i-1}, h] = - x \circ x_i .$$

Since $[x_i,B] = 0$, this gives

(1.50) $$Bx_i = x_i B = 0,\ i = 1, 2, \cdots, p .$$

If $j > p$, then by (1.49) $0 = [x_j \circ h, x_{i,k}] = x_j \circ [h, x_{i,k}] = x_j \circ x_{i,k+1}$ for $i = 1, \cdots, p$ and $1 \leq k \leq n_i - 1$. Therefore

(1.51) $$x_j x_{i,k} = x_{i,k} x_j = 0,\ 1 \leq i \leq p,\ 2 \leq k \leq n_i,\ p < j .$$

If $i \leq p$ and $j > p$, then $[h, x_j \circ x_{i,1}] = x_j \circ [h, x_{i,1}] = x_j \circ x_{i,2} = 0$ by (1.51). Since Z is the centralizer of h in B, it follows that $x_j x_{i,1} = x_{i,1} x_j \in Z$ for $j > p$ and $1 \leq i \leq p$. Therefore, by (1.49)-(1.51), we see that Z is an ideal of A.

Since B is abelian in A, $[A,A] = [h,B]$ and hence $[A,A] \cap Z$ is spanned by $x_1, x_2, \cdots, x_p$. We show that every element z of $[A,A] \cap Z$ is an absolute zero divisor of A. In view of (1.50), it suffices to show $zh = 0$. Let $h^2 \equiv \lambda h \pmod{B}$ for $\lambda \in F$ and let $z = [b,h]$ for $b \in B$. Then $[b,h^2] = h \circ [b,h] + [b,h] \circ h = 2h \circ z = 2hz$, while $[b,h^2] = \lambda[b,h] = \lambda z$. Hence $2zh = \lambda z$, and since R_h is nilpotent by Lemma 1.24, $zh = 0$. □

"Strict" power-associativity is needed only to show that all t_x are nilpotent on A. The condition that A is not commutative is essential in Theorem 1.25. It can be readily seen that the algebra in Example 1.6 has no nonzero absolute zero divisors. If A^- is not nilpotent, then Theorem

1.25 is not valid, as shown by a non-nilpotent solvable Lie algebra.

Theorem 1.26. Let A be a flexible nilalgebra over a field F of characteristic $\neq 2$ such that A^- is a nilpotent Malcev algebra. If $\dim A \le 4$, then A is also nilpotent such that all products of any 4 elements in A are zero.

Proof. We first treat the case where A is commutative. If $\dim A \le 2$, it is easily seen that A is associative and $A^3 = 0$. Assume $\dim A = 3$. Thus the nil-index of A is less than 5. If it is 4, then A is spanned by x, x^2, x^3 for some element x of A with $x^3 \neq 0$, and hence A is associative such that $A^4 = 0$. If the nil-index of A is 2, then $A^2 = 0$ since $xy = \frac{1}{2}[(x+y)^2 - x^2 - y^2] = 0$. We show that if $\dim A = 3$ or 4 and the nil-index of A is 3, then A is associative and $A^3 = 0$. Linearizing the identity $x^3 = x^2x = 0$ yields $(xy)z + (yz)x + (zx)y = 0$ for all $x,y,z \in A$, which is equivalent to

$$R_yR_z + R_zR_y = -R_{yz} \tag{1.52}$$

for all $y,z \in A$. Setting $y = z = x$ gives

$$2R_x^2 = -R_{x^2}. \tag{1.53}$$

Letting $y = x$, $z = x^2$ in (1.52), and noting that $x^3 = 0$, we have $R_xR_{x^2} + R_{x^2}R_x = 0$, which with (1.53) implies that $R_x^3 = 0$. Choose any $x \in A$ with $x^2 \neq 0$, and let B be the linear span of x, x^2, so that $\dim B = 2$. Then B is stable under R_x which therefore operates on the one- or two-dimensional quotient space A/B. As R_x is nilpotent, we have $R_x^2(A/B) = 0$, so $(yx)x \in B$ for all $y \in A$, i.e., $(yx)x = \alpha x + \beta x^2$ for some $\alpha,\beta \in F$. Since $R_x^3 = 0$, multiplying by x shows that $\alpha = 0$, after which using (1.53) and the fact that R_y is also nilpotent shows $\beta = 0$ also. Thus

$yx^2 = 0$ by (1.53), and since every product is a linear combination of squares, this shows that the product of any three elements is zero. In particular, A is associative.

We now assume that $\dim A = 4$ and the nil-index of A is 4. Let x be any element of A with $x^3 \neq 0$ and let S be the linear span of x, x^2, x^3. We first claim that $A^2 = Fx^2 + Fx^3$. It suffices to show that $y^2 \in Fx^2 + Fx^3$ for all $y \in A$, and we may further assume $y \notin S$ and $y^2 \neq 0$, otherwise the matter is trivial. Let T be the linear span of y, y^2, y^3. Then $\dim T = 2$ or 3 and hence $\dim (S \cap T) = \dim T - 1$, since $\dim S = 3$ and $\dim (S + T) = 4$. Furthermore, $S \cap T$ is a proper subalgebra of S, hence must be contained in $Fx^2 + Fx^3$. In fact, if S_0 is any proper subalgebra of S and u is an element of S_0, then $u = \alpha x + \beta x^2 + \gamma x^3$ for $\alpha,\beta,\gamma \in F$, and $u^3 = 0$ gives $\alpha = 0$, since $x^4 = 0$. By a similar argument, $S \cap T$ contains y^2, and so $y^2 \in Fx^2 + Fx^3$, as asserted. It follows from this that $A^2A^2 = 0$. For arbitrary y, we have $yx^2 \in A^2$, hence $yx^2 = \alpha x^2 + \beta x^3$ for some $\alpha, \beta \in F$. We claim $\alpha = 0$. Otherwise, setting $z = \alpha^{-1}(y - \beta x)$ gives $zx^2 = x^2$. Using the fact that $A^2A^2 = 0$ and computing $[(z + x^2)^2(z + x^2)](z + x^2) = 0$, we find that the left side of this equation is $2x^2$, a contradiction. Hence $yx^2 = \beta x^3$. If now we replace x by $x + x^2$, thereby replacing x^2 by $x^2 + 2x^3$ but leaving x^3 unchanged, it follows that yx^3 is also a multiple of x^3. In fact, $yx^3 = 0$, for if $yx^3 = \lambda x^3$, then computing $[(y + x^3)^2(y + x^3)](y + x^3) = 0$, one finds that the left side of the equation is $2\lambda^3x^3$, hence $\lambda = 0$. Replacing the original y by $y - \beta x$, for which we have $(y - \beta x)x^2 = 0$, one can choose $y \in A$, so that $y \notin S$ and $yx^2 = yx^3 = 0$, so $yA^2 = 0$. Thus, the product of any four of the elements x, x^2, x^3, y vanishes. Since these span A, it follows that the product of any 4 elements of A is zero, so $A^4 = 0$.

We assume next that A is not commutative. Since A^- is nilpotent, we must have $\dim A = 3$ or 4. It is shown that any nilpotent Malcev algebra of dimension ≤ 4 is a Lie algebra (Gainov [1]). Suppose $\dim A = 3$. There is a unique non-abelian nilpotent Lie algebra of dimension 3, which has multiplication $[x,y] = z$ with all other Lie products zero for a basis x,y,z (Bourbaki [1,p.120]). Hence Fz is the center Z of A^-, and by Theorem 1.25 z is an absolute zero divisor of A. From this and flexibility, one gets $[x \circ y,x] = x \circ [y,x] = -xz = 0$, and similarly $[x \circ y,y] = [x^2,x] = [x^2,y] = 0$. It follows from this that $A^2 \subseteq Fz$ and hence $A^3 = 0$. In this case, A is associative. Assume $\dim A = 4$. Then $\dim Z = 1$ or 2. If $\dim Z = 1$, then there is a basis x,y,z,h of A for which A^- has multiplication $[h,x] = y, [h,y] = z$, and all other Lie products are zero (Bourbaki [1,p.120]). Letting $Z = Fz$ and $B = Fx + Fy + Fz$, it follows from Lemma 1.24 and Theorem 1.25 that B and Z are ideals of A and z is an absolute zero divisor of A. Since B is abelian and $xh \in B$, we get $0 = [x \circ h,x] = x \circ [h,x] = x \circ y$, hence $xy = yx = 0$. The equation $[h,yh] = [h,y]h = zh = 0$ implies $yh = -hx = \lambda z$ for some $\lambda \in F$, which with $[xh,h] = [x,h]h = -yh$ gives $xh = \lambda y + \beta z$ for $\beta \in F$. We use $[h,x^2] = xy + yx = 0$ to obtain $x^2 = \alpha z$ for some $\alpha \in F$. The equation $[h,h^2] = [h^2,y] = 0$ gives $h^2 = \gamma z$ for $\gamma \in F$, while $0 = [h^2,x] = hy + yh = (2\lambda + 1)z$ implies $\lambda = -\frac{1}{2}$. Therefore, the multiplication in A is given by

$$x^2 = \alpha z,\ xh = -\frac{1}{2}y + \beta z,\ hx = \frac{1}{2}y + \beta z,$$

(1.54)

$$yh = -hy = -\frac{1}{2}z,\quad h^2 = \gamma z$$

for $\alpha,\beta,\gamma \in F$, and all other products are zero. We see from this that $A^2 \subseteq Fy + Fz$, $A^3 \subseteq Fz$, and $A^4 = 0$. In this case, A is a nilalgebra of nil-index 3. Conversely, it is easily checked that the algebra A defined by (1.54) is a flexible Lie-admissible nilalgebra such that A^- is nilpotent with one-dimensional center.

Finally, assume that $\dim A = 4$ and $\dim Z = 2$. There is a basis x,y,z,h of A such that A^- has multiplication given by $[h,x] = y$ and all other Lie products are zero (Bourbaki [1]). Letting $B = Fx + Fy + Fz$ and $Z = Fy + Fz$, by Lemma 1.24 and Theorem 1.25, B and Z are ideals of A and Z is the center of A^-. Hence, y is an absolute zero divisor of A. From $[h,x^2] = 2xy = 0 = [h,h^2]$, we have $x^2 = \alpha y + \beta z$ and $h^2 = \sigma y + \tau z$ for $\alpha,\beta,\sigma,\tau \in F$. The equation $[h,xh] = [h,x]h = yh = 0$ yields $xh = \delta y + \lambda z$ and $hx = (\delta + 1)y + \lambda z$, while setting $zx = xz = \gamma y + \gamma' z$ gives $(xz)x = \gamma' xz$, hence $\gamma' = 0$ since R_x is nilpotent by Lemma 1.24. Similarly, $hz = zh = \nu y$ for some $\nu \in F$. Since Z is a nilalgebra of dimension 2, $z^3 = 0$ and this implies $z^2 = \mu y$ for some $\mu \in F$. This together with the fact that $x \in B$ shows that $0 = x^2x^2 = (\alpha y + \beta z)^2 = \beta^2 z^2 = \beta^2 \mu y$, and hence $\beta^2\mu = 0$. Since h belongs to the subalgebra $Fy + Fz + Fh$ of A (Lemma 1.24), $h^2h^2 = 0$ to give $\mu\tau^2 = 0$. Similarly, one gets $\mu\lambda^2 = 0$ from $(xh)^4 = 0$. Therefore, we have shown that A is given by

$$x^2 = \alpha y + \beta z,\ xz = zx = \gamma y,\ xh = \delta y + \lambda z\ ,$$

(1.55)

$$hx = (\delta + 1)y + \lambda z,\ zh = hz = \nu y,\ z^2 = \mu y,\ h^2 = \sigma y + \tau z\ ,$$

and all other products are zero, and $\alpha,\beta,\gamma,\delta,\lambda,\mu,\nu,\sigma,\tau \in F$ with $\mu\beta^2 = \mu\lambda^2 = \mu\tau^2 = 0$. In this case, A is a nilalgebra of nil-index 4 if $\mu = 0$, and of nil-index 3 if $\mu \neq 0$. It follows from (1.55) that $A^3 \subseteq Fy$, $A^2A^2 \subseteq F(\mu y)$ and hence $A^2A^2 = 0$, since if $\mu \neq 0$, then $\beta = \lambda = \tau = 0$ and $A^2 \subseteq Fy$. Thus, $A^4 = 0$. Conversely, it is easy to see that the algebra A defined by (1.55) is a flexible Lie-admissible nilalgebra, where A^- is nilpotent with two-dimensional center. □

The proof of Theorem 1.26 for the commutative case is given by Gerstenhaber and Myung [1], while the noncommutative case is drawn from Myung [3]. It is not known whether or not there exists a simple, flexible Lie-admissible nilalgebra A such that A^- is nilpotent. We have resolved this for dimension ≤ 4 and for the algebra A described in Theorem 1.25. As shown by Example 1.6, Theorem 1.26 does not hold for an arbitrary dimension. We however conjecture that the algebra A described in Theorem 1.25 is nilpotent.

As a final result, we give a condition that a finite-dimensional flexible power-associative Malcev-admissible algebra A is a nilalgebra, in terms of a Cartan subalgebra of A^-. A basis $\{u_i\}$ of a power-associative algebra A is called a *nil-basis* if each basis element u_i is nilpotent. Shestakov [1] has shown that a finite-dimensional Jordan algebra with a nil-basis of characteristic ≠ 2 is a nilalgebra and hence nilpotent. Shestakov proves this result by utilizing the classification of simple Jordan algebras. Included in Shestakov [1] is that any finite-dimensional power-associative algebra A with a nil-basis over a field of characteristic 0 is a nilalgebra. The proof of this in Shestakov [1] is based on the fact that a finite-dimensional simple commutative (non-nil) power-associative algebra over an algebraically closed field of characteristic 0 is a Jordan algebra. A more elementary proof of this, which does not invoke the classification of simple algebras, may be contained in the following generalized version of the result of Shestakov.

Lemma 1.27. Let A be a finite-dimensional power-associative algebra over a field F of characteristic 0 or greater than 5 and dim A . Then A is a nilalgebra if and only if A has a nil-basis.

Proof. Since A is nil if and only if A^+ is nil, one can assume that A is commutative. Suppose that A is not a nilalgebra. If N denotes the nilradical of A , then $A/N \neq 0$ is a direct sum of simple algebras (Oehmke [1]) and has a nil-basis, and hence each simple summand of A/N has a nil-basis. Thus we can further assume that A is simple and hence has a unit element 1 (Oehmke [1]).

Assume that the characteristic of F is zero, and let $\{u_1, \cdots, u_n\}$ be a nil-basis. We write $1 = \sum_{i=1}^{n} \alpha_i u_i$ for $\alpha_i \in F$. In terms of the right multiplication R_x in A , we have $I = \sum_i \alpha_i R_{u_i}$, where I is the identity mapping on A . By a result of Gerstenhaber [1], each R_{u_i} is a nilpotent linear transformation on A and hence the trace of all R_{u_i} is zero, but $tr(I) = n \neq 0$, a contradiction. This proves that A is a nilalgebra.

Assume now that the characteristic of F is greater than 5 and dim A . At this point, we use the known classification of finite-dimensional simple commutative power-associative algebras over an algebraically closed field of characteristic > 5 . This classification has been completed by a number of authors and appropriate references can be found in Oehmke [1,2]. Since the characteristic of F is greater than 5 , A is strictly power-associative (Albert [1]). Any scalar extension of A has also a nil-basis. Therefore we can assume that F is algebraically closed. Suppose that A is not a nilalgebra. As before, assume that A is a simple commutative algebra over F . If the degree of A is not two, then it is known that A is a Jordan algebra. By a result of Shestakov [1] noted above, A is a nilalgebra (so nilpotent), and this is a contradiction. When the degree of A is two, A is a Jordan algebra, except for the case when A is the algebra described by Oehmke [2]. In the latter case, A must contain a subalgebra of dimension p^k for some positive integer k where p is the characteristic of F (this subalgebra is a commutative, associative, differentiably simple algebra over F (Oehmke [2])). This is absurd since $p > \dim A$. Thus, A must be nil in this case also. □

Theorem 1.28. Let A be a finite-dimensional flexible power-associative Malcev-admissible algebra over a field F of characteristic 0 or greater than 5 and $\dim A$. Then A is a nilalgebra if and only if A^- has a Cartan subalgebra with a nil-basis.

Proof. The assumption on the characteristic of F guarantees the existence of a Cartan subalgebra of A^- (Barnes [1] and Malek [2]). Let H be a Cartan subalgebra of A^- which has a nil-basis. In light of Lemma 1.27, it suffices to show that A has a nil-basis. Since the characteristic of F is 0 or greater than 5, A is strictly power-associative. As before, we can assume that F is algebraically closed, since the scalar extension of H is also a Cartan subalgebra. Let $A^- = H + A_\alpha + A_\beta + \cdots + A_\delta$ be the Cartan decomposition of A^- relative to H. If A_α^n denotes the linear span of Jordan products $x_1 \circ x_2 \circ \cdots \circ x_n$ of any elements $x_1, \cdots, x_n$ of A_α in any association, then by relation (1.18) (Section 1.3) we have $A_\alpha^n \subseteq A_{n\alpha}$. Hence $A_\alpha^n = A_{n\alpha} = 0$ for some $n > 0$. This in particular implies that A_α has a nil-basis for all nonzero roots α and hence A has a nil-basis. □

Theorem 1.28 has been proved for Lie-admissible algebras under stronger assumptions that the Cartan subalgebra H is nil in A and the characteristic of A is zero (Myung [1]).

2

POWER-ASSOCIATIVE MALCEV-ADMISSIBLE ALGEBRAS

2.1. INTRODUCTION

The main effort in Chapter 1 was to determine flexible algebras A according to the given structures of the attached minus algebras A^- when A^- is a Lie or Malcev algebra. In this chapter, we determine all power-associative algebras A when A^- is isomorphic to the attached minus algebra of the $n \times n$ matrix algebra, an octonion algebra, or to a finite-dimensional split semisimple Malcev algebra. This problem has been motivated by some recent works in particle physics. Eder [1,2] used a power-associative product defined on the real associative enveloping algebra of spin $\frac{1}{2}$, 1, or $\frac{3}{2}$ matrices. The assumption of power-associativity in this work was necessary to have a well-defined notion of the exponential of a spin matrix and to measure a deviation from the standard spin theory. The power-associative product used in Eder's spin theory satisfies the additional hypothesis that the Lie product under the new product remains unchanged. It appears that extensions of Eder's work are likely to require other power-associative products defined on $n \times n$ matrices for arbitrary n. In a different point of view, Okubo [6,10] has attempted to use flexible Lie-admissible algebras A to generalize the Heisenberg equation whose solutions require the underlying algebra A to be power-associative. Power-associative products defined on octonions might suggest similar extensions of the octonionic quantum mechanics. The results in this chapter also generalize earlier works by Weiner [1,2], Okubo and Myung [2,3], Benkart and Osborn [1], and include the para-octonion and pseudo-octonion algebras introduced by Okubo [5] and Okubo and Myung [2].

Let A be an algebra with multiplication denoted by xy over a field F and let "$*$" be another multiplication defined on the same vector space as A such that $[x,y]^* = x * y - y * x = [x,y] = xy - yx$.

If A is Lie- or Malcev-admissible, then so is $(A,*)$. Therefore, another motivation for investigating the algebra $(A,*)$ is to produce further examples of Lie- or Malcev-admissible algebras that might be of interest in physics, and to provide results that might be useful in studying Malcev-admissible algebras. Our investigation begins with determining third power-associative algebras $(A,*)$ such that $[x,y]^* = [x,y]$ when A is the $n \times n$ matrix algebra $M(n,F)$, an octonion algebra, or a finite-dimensional split simple Lie algebra (Sections 2.3, 2.4 and 2.5). For the same algebra A , power-associative products "$*$" defined on A satisfying $[x,y]^* = [x,y]$ will be determined by investigating fourth power-associative product "$*$" on A such that $[x,y]^* = [x,y]$.

Assume that A is an algebra with product denoted by xy over F and let $(A,*)$ be an algebra with multiplication "$*$" defined on the vector space A such that $[x,y]^* = [x,y]$. Hence, $x * y$ is given by

$$x * y = \frac{1}{2}[x,y] + x \circ y \tag{2.1}$$

where $x \circ y = \frac{1}{2}(x * y + y * x)$. Conversely, for any commutative product "$\circ$" defined on A , the algebra $(A,*)$ defined by (2.1) has the property that

$$[x,y]^* = [x,y] \tag{2.2}$$

for all $x,y \in A$. Thus, our major efforts are devoted to determining commutative products defined on A which satisfy certain given constraints.

Assume that $(A,*)$ is third power-associative and satisfies (2.2). Since $x * x = x \circ x$, third power-associativity $(x * x) * x = x * (x * x)$ is equivalent to

$$[x,x \circ x] = 0 , \tag{2.3}$$

which has a partial linearization

(2.4) $$[y,x \circ x] + 2[x,x \circ y] = 0$$

and which has a full linearization

(2.5) $$[x,y \circ z] + [y,z \circ x] + [z,x \circ y] = 0$$

for all $x,y,z \in A$. If the characteristic of F is not 2 or 3, then (2.3)-(2.5) are equivalent. In light of Lemma 1.10, if $(A,*)$ is third power-associative then $(A,*)$ is fourth power-associative if and only if $(A,\circ) = (A,*)^+$ is. Thus, by Lemma 1.11(i), the problem of finding all power-associative Malcev-admissible algebras of characteristic 0 such that A^- is a certain specified type of Malcev algebra is to find all commutative products $x \circ y$ on a given Malcev algebra A^- with product "[,]" which satisfies relation (2.3) or (2.5), and fourth power-associativity

(2.6) $$(x \circ x) \circ (x \circ x) = x \circ (x \circ (x \circ x))$$

for all $x,y,z \in A$. Power-associative Malcev-admissible products on A then must be given by (2.1), where "$\circ$" is one of these commutative products.

One of the main tools to be employed is the Peirce decompositions in various types of algebras. We review briefly the facts needed. An element u in an associative algebra A is called *idempotent* if $u^2 = u$, and two idempotents u, v are *orthogonal* if $uv = 0 = vu$. An idempotent u induces the vector space direct sum decomposition $A = A_{11} + A_{10} + A_{01} + A_{00}$ where

(2.7) $$A_{ij} = \{x \in A \mid ux = ix, xu = jx\}$$

for $i, j = 0, 1$. If $u_1, u_2, \cdots, u_n$ is a set of pairwise orthogonal idempotents whose sum is the identity element, then there is a vector space direct sum $A = \sum_{i,j=1}^{n} A_{ij}$ where

$$(2.8) \qquad A_{ij} = \{x \in A \mid u_k x = \delta_{ki} x \text{ and } x u_k = \delta_{jk} x,\ k = 1, \cdots, n\} .$$

Here δ_{jk} denotes the Kronecker delta. In cases (2.7) or (2.8), we have the multiplication properties

$$(2.9) \qquad A_{ij} A_{k\ell} \subseteq \delta_{jk} A_{i\ell} .$$

If $(A, \circ)$ is a commutative algebra of characteristic $\neq 2,3$ and satisfies fourth power-associativity (2.6), then an idempotent $u = u \circ u$ induces a vector space direct sum $A = A_1 + A_{\frac{1}{2}} + A_0$ where

$$(2.10) \qquad A_\lambda = A_\lambda(u) = \{x \in A \mid u \circ x = \lambda x\}$$

for $\lambda = 0, \frac{1}{2}$, or 1 . The multiplication between the A_λ is given by

$$(2.11) \qquad A_\lambda \circ A_\lambda \subseteq A_\lambda,\ A_\lambda \circ A_{1-\lambda} = 0,\ A_{\frac{1}{2}} \circ A_{\frac{1}{2}} \subseteq A_1 + A_0,\ A_{\frac{1}{2}} \circ A_\lambda \subseteq A_{\frac{1}{2}} + A_{1-\lambda}$$

for $\lambda = 0, 1$ (Osborn [2, §6], for example).

When A is an algebra with product xy over F of characteristic $\neq 2,3$ and has an identity element e , the commutative product $x \circ y$ on A defined by

$$(2.12) \qquad x \circ y = \tau(y)x + \tau(x)y + \beta(xy + yx) + \sigma(x,y)e$$

plays a main role in our investigation, where τ is a linear form of A into F , $\sigma : A \times A \to F$ is a symmetric bilinear form, and β is a fixed scalar in F . In fact, the product defined by (2.12) satisfies relation (2.3), as seen by

<u>Lemma 2.1</u>. Let A be a third power-associative algebra with product xy and with identity element e . Let "$\circ$" be a commutative product on

A defined by (2.12). Then the algebra $(A,*)$ defined by (2.1) is third power-associative and satisfies (2.2).

Proof. Since $x * y = \frac{1}{2}[x,y] + x \circ y$, $[x,y]^* = [x,y]$. Hence, it suffices to prove that $[x,x \circ x] = 0$ for all $x \in A$. This follows from the calculation

$$[x,x \circ x] = [x,2\tau(x)x + 2\beta x^2 + \sigma(x,x)e]$$

$$= 2\tau(x)[x,x] + 2\beta[x,x^2] + \sigma(x,x)[x,e] = 0$$

since A is third power-associative. □

Proving the converse of Lemma 2.1 is the main topic in Sections 2.3 and 2.4 when A is an $n \times n$ matrix algebra or an octonion algebra.

2.2. Para-octonion and pseudo-octonion algebras

There are two interesting algebras of dimension 8 which have arisen from different sources. The first type of these algebras is a pseudo-octonion algebra first introduced by Okubo [5] who also conceived its relevance with the SU(3) particle physics. This algebra is a flexible Lie-admissible algebra whose minus algebra is isomorphic to the Lie algebra $\mathfrak{sl}(3,F)$, or to $\mathfrak{su}(3)$ for the real number field. The second type is a para-octonion algebra introduced by Okubo and Myung [2] which has arisen from the construction of division algebras. This algebra is not Lie-admissible but a flexible Malcev-admissible algebra whose minus algebra is isomorphic to the minus algebra of an octonion algebra. However, these two algebras share a common property that there exists a nondegenerate symmetric bilinear form $(\, , \,)$ satisfying the relation

$$(xy)x = x(yx) = (x,x)y \tag{2.13}$$

for all x, y. The para- and pseudo-octonion algebras have been re-investigated many times in conjunction with the study of algebras satisfying identity (2.13) (Okubo [3,4,7,8,9,11,12], Benkart and Osborn [2,3], Okubo and Osborn [1,2], for example).

Definition 2.1. A bilinear form $(\, , \,)$ or a quadratic form N on a nonassociative algebra A is said to *permit composition* if

$$(xy,xy) = (x,x)(y,y) \quad \text{or} \quad N(xy) = N(x)N(y) \tag{2.14}$$

for all $x,y \in A$, and the form is *invariant* if

$$(xy,z) = (x,yz) \tag{2.15}$$

for all $x,y,z \in A$. A quadratic form N on A over a field F of characteristic $\neq 2$ is called *nondegenerate* if the associated (symmetric) bilinear form $(\, , \,)$ defined by $(x,y) = \frac{1}{2}[N(x + y) - N(x) - N(y)]$ is nondegenerate. An algebra A over a field F of characteristic $\neq 2$ is termed a *composition algebra* if it possesses a nondegenerate quadratic form which permits composition. A composition algebra A is said to be *unital* if A has an identity element. □

It is well known that a unital composition algebra A has dimension 1, 2, 4 or 8, and must be one of the algebras: $F1$, a quadratic algebra (of dimension 2), a (generalized) quaternion algebra, or an (Cayley-Dickson) octonion algebra. Following Okubo [8,9], each of these algebras will be called a *Hurwitz algebra*. On the other hand, the classification of general composition algebras is not known although they still have dimension 1, 2,

4, or 8 (Okubo and Osborn [1]). The composition algebras of dimension 2 has been classified by Petersson [1] as one of the algebra $(A,*)$ defined by: (i) a unital composition algebra A of dimension 2 with involution $x \to \bar{x}$ and norm N, (ii) $x * y = \bar{x}y$, (iii) $x * y = x\bar{y}$, (iv) $x * y = u\bar{x}\bar{y}$ for some element $u \in A$ with $N(u) = 1$. Here xy denotes the product in A. The last three algebras are not associative, but Lie-admissible. This can be easily seen from the fact that A is an associative commutative algebra. Let Q be a quaternion algebra with product xy and let $x \to \bar{x}$ and N be the standard involution and norm of Q, respectively. Assume that a, b, c are elements of Q satisfying $N(a)N(b)N(c) = 1$. When F is algebraically closed, Shapiro [1] has classified any composition algebra of dimension 4 as one of the algebras $(Q,*)$ defined by: (i) Q, (ii) $x * y = axbyc$, (iii) $x * y = axb\bar{y}c$, (iv) $x * y = a\bar{x}byc$, (v) $x * y = a\bar{x}b\bar{y}c$. The classification for dimension 8 remains unsolved. Okubo [1,2] has shown that any flexible composition algebra over an arbitrary field of characteristic $\neq 2$ is one of Hurwitz algebras, para-Hurwitz or pseudo-octonion algebras.

Discussion of para-octonion and pseudo-octonion algebras in detail involves a lengthy exposition and will be too far removed from the main topics of our investigation. The information needed for the present discussion is essentially the construction and multiplication tables of these algebras. Let A be a Hurwitz algebra with product xy over a field F of characteristic $\neq 2$. Let N be the quadratic form on A and denote by $x \to \bar{x}$ the standard involution of A. Then A satisfies a quadratic equation

$$x^2 - 2T(x)x + N(x)e = 0 \tag{2.16}$$

for all $x \in A$, where e is an identity element of A, N permits composition, so $N(e) = 1$, and the trace $T : A \to F$ is a linear form with $T(e) = 1$ (Schafer [3], for example). The involution $x \to \bar{x}$ is given by

$$\bar{x} = 2T(x)e - x \; . \tag{2.17}$$

Definition 2.2. Let A be a Hurwitz algebra over F with product xy and with standard involution $x \to \bar{x}$ given by (2.17). The algebra $(A,*)$ with multiplication "$*$" defined by

$$x * y = \bar{x}\bar{y} \tag{2.18}$$

on the vector space A is called a *para-Hurwitz algebra*. Also, $(A,*)$ is called a *para-quadratic*, *para-quaternion*, or *para-octonion algebra* if A is a quadratic, quaternion, or octonion algebra. □

Consider the vector space direct sum

$$A = Fe + A_0 \; , \tag{2.19}$$

where $A_0 = \{x \in A \mid T(x) = 0\}$. In light of (2.17) and (2.18), it follows that $x * y = xy$ for all $x,y \in A_0$, and $e * e = e$, $e * x = x * e = -x$ for $x \in A_0$. Thus, if A is an octonion algebra, then it is possible to choose a basis $e_1, e_2, \cdots, e_7$ of A_0, so that the multiplications in A and $(A,*)$ are given by Table 2.1, where α, β, γ are scalars in F with $\alpha\beta\gamma \neq 0$ (see Schafer [3] or Okubo and Osborn [1]).

TABLE 2.1

	e	e_1	e_2	e_3	e_4	e_5	e_6	e_7
e	e	εe_1	εe_2	εe_3	εe_4	εe_5	εe_6	εe_7
e_1	εe_1	$-\alpha e$	e_3	$-\alpha e_2$	e_5	$-\alpha e_4$	$-e_7$	αe_6
e_2	εe_2	$-e_3$	$-\beta e$	βe_1	e_6	e_7	$-\beta e_4$	$-\beta e_5$
e_3	εe_3	αe_2	$-\beta e_1$	$-\alpha\beta e$	e_7	$-\alpha e_6$	βe_5	$-\alpha\beta e_4$
e_4	εe_4	$-e_5$	$-e_6$	$-e_7$	$-\gamma e$	γe_1	γe_2	γe_3
e_5	εe_5	αe_4	$-e_7$	αe_6	$-\gamma e_1$	$-\alpha\gamma e$	$-\gamma e_3$	$\alpha\gamma e_2$
e_6	εe_6	e_7	βe_4	$-\beta e_5$	$-\gamma e_2$	γe_3	$-\beta\gamma e$	$-\beta\gamma e_1$
e_7	εe_7	$-\alpha e_6$	βe_5	$\alpha\beta e_4$	$-\gamma e_3$	$-\alpha\gamma e_2$	$\beta\gamma e_1$	$-\alpha\beta\gamma e$

In Table 2.1, ε indicates $+1$ or -1, and if $\varepsilon = 1$, then Table 2.1 is the multiplication table of an octonion algebra A, while if $\varepsilon = -1$, then it represents the multiplication table of a para-octonion algebra. The blocks through e, e_1 and e, e_1, e_2, e_3 give multiplication tables of a para-quadratic and para-quaternion algebra for $\varepsilon = -1$, respectively. It is clear that $[x,y]^* = [x,y]$ for all $x,y \in A$, and hence $(A,*)$ satisfies relation (2.2). Since $T(xy) = T(yx)$, A_0 is closed under product $[\ ,\ ]$ and A_0^- is isomorphic to $A^-/Fe = (A,*)^-/Fe$. Hence, A_0^- is isomorphic to one of algebras defined in Table 2.2.

TABLE 2.2

	e_1	e_2	e_3	e_4	e_5	e_6	e_7
e_1	0	e_3	$-\alpha e_2$	e_5	$-\alpha e_4$	$-e_7$	αe_6
e_2	$-e_3$	0	βe_1	e_6	e_7	$-\beta e_4$	$-\beta e_5$
e_3	αe_2	$-\beta e_1$	0	e_7	$-\alpha e_6$	βe_5	$-\alpha\beta e_4$
e_4	$-e_5$	$-e_6$	$-e_7$	0	γe_1	γe_2	γe_3
e_5	αe_4	$-e_7$	αe_6	$-\gamma e_1$	0	$-\gamma e_3$	$\alpha\gamma e_2$
e_6	e_7	βe_4	$-\beta e_5$	$-\gamma e_2$	γe_3	0	$-\beta\gamma e_1$
e_7	$-\alpha e_6$	βe_5	$\alpha\beta e_4$	$-\gamma e_3$	$-\alpha\gamma e_2$	$\beta\gamma e_1$	0

Denote by $M(\alpha,\beta,\gamma)$ the algebra defined by Table 2.2. Kuzmin [1] has shown that $M(\alpha,\beta,\gamma)$ with $\alpha\beta\gamma \neq 0$ is a 7-dimensional central simple, non-Lie, Malcev algebra over F and is isomorphic to $A^-/Fe \cong A_0$ for some octonion algebra A, and conversely, every finite-dimensional central simple, non-Lie, Malcev algebra over F of characteristic $\neq 2,3$ is isomorphic to $M(\alpha,\beta,\gamma)$ for some scalars α,β,γ with $\alpha\beta\gamma \neq 0$. Furthermore, two algebras defined by Table 2.2 are isomorphic if and only if their associated octonion algebras are isomorphic (Kuzmin [1]).

The construction of pseudo-octonion algebras is quite different from that of a para-octonion algebra. Following the original construction of Okubo [5], let M_3 be the vector space of 3×3 trace zero matrices over the complex number field K, and let μ be a complex number satisfying the equation

(2.20) $$3\mu(1 - \mu) = 1 .$$

We introduce a product "*" on M_3 defined by

(2.21) $$x * y = \mu xy + (1 - \mu)yx - \frac{1}{3}(\mathrm{tr}\ xy)I$$

for $x,y \in M_3$, where xy is the matrix product, tr denotes the trace and I is the identity matrix. Okubo [5] called the algebra $(M_3,*)$ defined by (2.21) the *pseudo-octonion algebra over the complex field.* A more general definition can be given as follows (Okubo and Osborn [1] or Okubo and Myung [2]).

Definition 2.3. Let F be a field of characteristic $\neq 2,3$ which contains a cube root of unity, $\omega \neq 1$. Let A be the set of 3×3 trace zero matrices over F. An algebra $(A,*)$ with multiplication $x * y$ defined by (2.21) is called a *pseudo-octonion algebra over* F, where μ is a scalar in F satisfying (2.20). □

Since $[x,y]^* = (2\mu - 1)[x,y]$, $(A,*)^-$ is isomorphic to $\mathfrak{sl}(3,F)$ and hence $(A,*)$ is Lie-admissible. Define a bilinear form $(\ , \ )$ on $(A,*)$ by

(2.22) $$(x,y) = \frac{1}{6}\mathrm{tr}\ xy ,$$

which is clearly nondegenerate and symmetric.

Theorem 2.2. Let $(A,*)$ be a para-Hurwitz algebra over a field F of characteristic $\neq 2$ or a pseudo-octonion algebra over F of characteristic $\neq 2,3$. Assume that if $(A,*)$ is para-Hurwitz, then $(\ , \ )$ denotes the bilinear form associated with the nondegenerate quadratic form N, and if $(A,*)$ is pseudo-octonion, then $(\ , \ )$ denotes the bilinear form defined by (2.22). Then, $(A,*)$ satisfies relation (2.13).

Proof. Assume first that $(A,*)$ is para-Hurwitz. Since $(yx)\bar{x} = y(x\bar{x}) = (x,x)y$ in a composition algebra A (Jacobson [4,p.442]), we have

$$(x * y) * x = (\overline{\bar{x}\bar{y}}) * x = (yx)\bar{x} = (x,x)y ,$$

and by left-right symmetry we also obtain $x * (y * x) = (x,x)y$.

Assume then that $(A,*)$ is a pseudo-octonion algebra with multiplication defined by (2.21) where $\mu \varepsilon F$ satisfies the equation (2.20). We compute from (2.21)

$$\begin{aligned}(x * y) * x = x * (y * x) &= \mu^2 xyx + (1 - \mu)^2 xyx + \mu(1 - \mu)(yx^2 + x^2y) \\ &\quad - \tfrac{1}{3}(\text{tr } xy)x - \tfrac{1}{3}\mu(\text{tr } xyx)I - \tfrac{1}{3}(1 - \mu)(\text{tr } xyx)I \\ &= [1 - 2\mu(1 - \mu)]xyx + \mu(1 - \mu)(yx^2 + x^2y) - \tfrac{1}{3}(\text{tr } xy)x - \tfrac{1}{3}(\text{tr } xyx)I\end{aligned}$$

$$(2.23) \qquad = \tfrac{1}{3}(xyx + yx^2 + x^2y) - \tfrac{1}{3}(\text{tr } xy)x - \tfrac{1}{3}(\text{tr } xyx)I ,$$

since $\mu^2 + (1 - \mu)^2 = 1 - 2\mu(1 - \mu) = \frac{1}{3}$ by (2.20). We next show that A satisfies the cubic identity

$$(2.24) \qquad x^3 - \tfrac{1}{2}(\text{tr } x^2)x - \tfrac{1}{3}(\text{tr } x^3)I = 0$$

for all $x \varepsilon A$. By a scalar extension argument, we may assume that F is algebraically closed. Since zxz^{-1} and x satisfy the same polynomial equation, we can let x be a triangular matrix with characteristic roots $\alpha,\beta,\gamma \varepsilon F$. Since $\text{tr } x = 0$,

$$(2.25) \qquad \alpha + \beta + \gamma = 0 .$$

By the Cayley-Hamilton theorem, x satisfies the equation $x^3 + (\alpha\beta + \beta\gamma + \alpha\gamma)x - (\alpha\beta\gamma)I = 0$, since $\text{tr } x = 0$. We use (2.25) to obtain

$0 = (\alpha + \beta + \gamma)^2 = \alpha^2 + \beta^2 + \gamma^2 + 2(\alpha\beta + \beta\gamma + \alpha\gamma)$, and hence $\alpha\beta + \beta\gamma + \alpha\gamma = -\frac{1}{2}\mathrm{tr}\ x^2$. Similarly, expanding $0 = (\alpha + \beta + \gamma)^3$ and using (2.25), we have $\alpha\beta\gamma = \frac{1}{3}\mathrm{tr}\ x^3$, as desired.

When (2.24) is linearized by replacing x by $x + \lambda y$ $(\lambda \in F)$, the coefficient of λ , which must vanish, leads to the identity

$$xyx + yx^2 + x^2y - (\mathrm{tr}\ xy)x - \frac{1}{2}(\mathrm{tr}\ x^2)y - (\mathrm{tr}\ xyx)I = 0$$

for all $x,y \in A$. Combining this with (2.23) yields relation (2.13). □

Lemma 2.3. Let A be an algebra with multiplication denoted by xy over a field F of characteristic $\neq 2$. Assume $(\, , \,)$ is a symmetric nondegenerate bilinear form on A . Then $(\, , \,)$ permits composition and is invariant if and only if $(\, , \,)$ satisfies relation (2.13).

Proof. Assume that $(\, , \,)$ permits composition and is invariant. We set $N(x) = (x,x)$ for $x \in A$. Replace y by $y + z$ to linearize the composition law $(xy,xy) = (x,x)(y,y)$ to

$$(xy,xz) = (yx,zx) = N(x)(y,z)$$

for all $x,y,z \in A$. Since $(\, , \,)$ is invariant, this yields

$$0 = ((xy)x - N(x)y,z) = (x(yx) - N(x)y,z)$$

for all $x,y,z \in A$, which gives the desired relation (2.13) by the nondegeneracy of $(\, , \,)$.

Assume that (2.13) holds for A . We linearize (2.13) to

$$x(yz) + z(yx) = (xy)z + (zy)x = 2(x,z)y \ . \tag{2.26}$$

Substituting xy for x and yz for z in this, we obtain

$$(xy)(yz) = 2(xy,z)y - z(yxy) = 2(x,yz)y - (yzy)x \ ,$$

which implies by (2.13) that $2(xy,z)y - (y,y)zx = 2(x,yz)y - (y,y)zx$ and hence $(xy,z)y = (x,yz)y$ for all $x,y,z \in A$. Thus $(\, , \,)$ is invariant. Finally, we have $(xy,xy) = ((xy)x,y) = (N(x)y,y) = N(x)(y,y) = N(x)N(y)$ to show that $(\, , \,)$ permits composition. □

Theorem 2.4. Let A be an algebra with multiplication denoted by xy over a field F of characteristic $\neq 2$ and let A have a symmetric nondegenerate bilinear form $(\, , \,)$ satisfying relation (2.13). Then

(i) A is simple.

(ii) A has no identity element if $\dim A \geq 2$.

(iii) If $\dim A = 1$, then A is isomorphic to the field F.

Proof. (i) Since $(\, , \,)$ is invariant by Lemma 2.3, the idea is to use the theorem of Dieudonné that if A has no nonzero ideal B with $B^2 = 0$, then A is a direct sum of simple ideals (Schafer [3,p.24], for example). Since $(\, , \,)$ permits composition, $\dim A = 1, 2, 4,$ or 8 (Okubo and Osborn [1]). Suppose that there is an ideal $B \neq 0$ of A with $B^2 = 0$. Let $a \neq 0$ be in B. By (2.26), which is a linearization of (2.13), we have $(aa)x + (xa)a = 2(a,x)a$, and $2(a,x)a = 0$ for all $x \in A$, since $xa \in B$. The nondegeneracy of $(\, , \,)$ forces $a = 0$, a contradiction. Thus A is a direct sum $A = A_1 \oplus \cdots \oplus A_n$ of simple ideals A_i. Since $A_iA_i = A_i$ and $(\, , \,)$ is invariant, $(A_i,A_j) = 0$ for $i \neq j$, and hence $(\, , \,)$ is nondegenerate on each A_i. Since $(x,x)(y,y) = (xy,xy) = 0$ for all $x \in A_i$, $y \in A_j$, $i \neq j$, we must have $n = 1$, and A is simple.

(ii) We can prove a slightly stronger result that if A has a one-sided identity element, then it is an identity element and $\dim A = 1$. Assume that A has a right identity element e, i.e., $xe = x$ for $x \in A$. By the composition law, we have $(x,x) = (x,x)(e,e)$, and since there is an element x such that $(x,x) \neq 0$, this gives $(e,e) = 1$. It follows from this and (2.13) that $x = (e,e)x = e(xe) = ex$ for all $x \in A$, and so e is an identity element of A. Thus relation (2.26) yields $(xy)e + (ey)x = 2(e,x)y$, so $xy + yx = 2(e,x)y$ which gives $xy + yx = 2(e,y)x$ by interchanging x and y. Hence, $(e,x)y = (e,y)x$ for all $x,y \in A$ and $x = (e,x)e$ for all $x \in A$, implying that $A = Fe$ and $\dim A = 1$.

(iii) Assume $\dim A = 1$ and let $A = Fe$, where $e^2 = \lambda e$. Since $(,)$ is nondegenerate, $(e,e) \neq 0$ and this gives $\lambda \neq 0$ by the composition law. Thus $\lambda^{-1}e$ is an identity element of A and A is a field. □

The following are immediate consequences of Theorems 2.2, 2.4, and Lemma 2.3.

<u>Corollary 2.5</u>. Para-Hurwitz algebras of dimension ≥ 2 and pseudo-octonion algebras are simple algebras without identity element. □

<u>Corollary 2.6</u>. Let A and $(,)$ satisfy the hypothesis of Theorem 2.4. If $(,)$ is invariant and permits composition, then A satisfies the properties (i) - (iii) of Theorem 2.4. □

In passing, we remark that a pseudo-octonion algebra M_3 over the complex number field K has a real form which is a real division algebra in the sense that the equations $ax = b$ and $ya = b$ have unique solutions for any elements $a \neq 0, b$. If $x \in M_3$, then denote by x^+ the Hermitian conjugate of x and let $\tilde{M}_3$ be the set of Hermitian matrices x in M_3; i.e., $x^+ = x$. It follows from (2.21) that $(x * y)^+ = x^+ * y^+$ for all $x,y \in \tilde{M}_3$, and hence $(\tilde{M}_3,*)$ with multiplication "*" defined by (2.21) is an 8-dimensional algebra over the real number field $\mathbb{R}$ which satisfies relation (2.13) and properties (i) and (ii) in Theorem 2.4. Since

$(x,x) \neq 0$ for $x \neq 0$ in $\tilde{M}_3$ (in fact, $(x,x) > 0$ for $x \neq 0$), it is readily seen that $(\tilde{M}_3,*)$ is a real division algebra whose complexification is $(M_3,*)$ (Okubo [5]). Since μ is a solution of equation (2.20), we have

$$\mu = \frac{1}{2} \pm \frac{\sqrt{3}}{6} i .$$

We use this to rewrite (2.21) as

$$\begin{aligned} x * y &= \left(\frac{1}{2} \pm \frac{\sqrt{3}}{6} i\right) xy + \left(\frac{1}{2} \mp \frac{\sqrt{3}}{6} i\right) yx - \frac{1}{3}(\mathrm{tr}\ xy)I \\ &= - i \left\{\mp \frac{\sqrt{3}}{6}[x,y] + \frac{1}{2} i \left(xy + yx - \frac{2}{3}(\mathrm{tr}\ xy)I\right)\right\} . \end{aligned}$$

Hence $(\tilde{M}_3,*)$ is isomorphic to the real algebra $S(\delta,\frac{1}{2})$ of 3×3 skew-Hermitian matrices in M_3 with multiplication "$*'$" defined by

$$x *' y = \delta[x,y] - \frac{1}{2} i[xy + yx - \frac{2}{3}(\mathrm{tr}\ xy)I]$$

for $x,y \in S(\delta,\frac{1}{2})$, when δ is specified to $\delta = \pm \frac{\sqrt{3}}{6}$. In fact, the mapping $x \to ix$ gives an isomorphism of $(\tilde{M}_3,*)$ to $S(\pm \frac{\sqrt{3}}{6},\frac{1}{2})$. The algebra $S(\delta,\frac{1}{2})$ has arisen in the classification of finite-dimensional real flexible division algebras by Benkart, Britten and Osborn [2]. For any real number $\delta \neq 0$, $S(\delta,\frac{1}{2})$ is a division algebra, called a *generalized pseudo-octonion algebra*, such that $S(\delta,\frac{1}{2})^-$ is isomorphic to the Lie algebra $\mathfrak{su}(3)$.

Para-Hurwitz and pseudo-octonion algebras are essential sources of algebras satisfying (2.13), since Okubo and Osborn [1] showed that if A is an algebra over F of characteristic $\neq 2,3$ with nondegenerate symmetric bilinear form satisfying (2.13) and if A has an idempotent, then A is an algebra with certain constants inserted into the multiplication tables of para-Hurwitz and pseudo-octonion algebras. They showed in the same paper that if F is either algebraically closed or real closed, then the hypothesis of the existence of an idempotent may be dropped. In a subsequent paper, Okubo and Osborn [2] treated the special case of the characteristic 3 and proved that if F is a field of characteristic 3 having no proper quadratic field extensions and if A has an idempotent and a nondegenerate symmetric bilinear form satisfying (2.13), then A is either a para-Hurwitz algebra or the pseudo-octonion algebra of characteristic 3 defined by the multiplication table:

Definition 2.4. An 8-dimensional algebra A with basis x_{ij}, $i, j \in Z_3 = \{0,1,2\}$ $(x_{00} = 0)$, over a field F of characteristic 3 defined by the multiplication table

$$x_{ij}x_{k\ell} = (jk - i\ell - 1)x_{i+k,j+\ell} \tag{2.27}$$

is called the *pseudo-octonion algebra of characteristic 3* where $x_{ij} = x_{k\ell}$ if and only if $i \equiv k$, $j \equiv \ell \pmod 3$. □

The bilinear form defined by $(x_{ij}, x_{k\ell}) = -\delta_{i(-k)}\delta_{j(-\ell)}$ is symmetric, nondegenerate, and satisfies relation (2.13) (Okubo and Osborn [2]). We note that the definition of a pseudo-octonion algebra given by (2.21) does not apply to the characteristic 3 case.

2.3. Power-Associative Products on Matrices

Denote by $M(n,F)$ the associative algebra of $n \times n$ matrices over a field F. Throughout the matrix multiplication will be denoted by juxtaposition xy and "$*$" will denote a multiplication defined on the vector space A such that $[x,y]^* = x * y - y * x = [x,y]$ for all

$x,y \in A$. We first determine all third power-associative algebra $(A,*)$ when $A = M(n,F)$ over a field F of characteristic $\neq 2,3$.

Lemma 2.7. If $A = M(n,F)$ and $(A,*)$ is third power-associative, then, for any idempotent $u^2 = u$ of the associative algebra A , $u \circ u \in Fu + FI$, where $x \circ y = \frac{1}{2}(x * y + y * x)$ and I is the identity matrix.

Proof. Let $A = A_{11} + A_{10} + A_{01} + A_{00}$ be the Peirce deomposition of A relative to u in A and let

$$u \circ u = u_{11} + u_{10} + u_{01} + u_{00}$$

for $u_{ij} \in A_{ij}$, $i,j = 0, 1$. Since $(A,*)$ is third power-associative, by (2.3) $0 = [u,u \circ u] = u_{10} - u_{01}$ and hence $u = u_{11} + u_{00}$. For each $z \in A$, in light of (2.4), we have

$$0 = [z,u \circ u] + 2[u,z \circ u] = [z,u_{11}] + 2[u,z \circ u] .$$

The first term of this lies in A_{11} while the second term is in $[u,A] \subseteq A_{10} + A_{01}$. Hence $[u_{11},A_{11}] = 0$, and by a similar argument $[u_{00},A_{00}] = 0$. Since $A = M(n,F)$, A_{11} is isomorphic to $M(k,F)$ and A_{00} is isomorphic to $M(n - k,F)$ where k is the matrix rank of u . Thus we conclude that u_{11} is a multiple of u and u_{00} is a multiple of $I-u$, and hence $u \circ u = u_{11} + u_{00}$ is in $Fu + F(I - u) = Fu + FI$. □

Lemma 2.8. Let B be an arbitrary algebra with identity element I over a field F of characteristic $\neq 2$ and with product denoted by xy . Let "$\circ$" be a commutative product defined on B such that $u \circ u \in Fu + FI$ for each idempotent u in B . Let $u_1, \cdots, u_n$ be a set of pairwise orthogonal idempotents of B whose sum is the identity element I of B , and, for $i \neq j$, let a be an element of B such that

$u_k a = \delta_{ki} a$, $au_k = \delta_{jk} a$, $k = 1, \cdots, n$ and $a^2 = 0$. If we let $U = Fu_1 + \cdots + Fu_n + Fa$, then there exist a linear form $\tau : U \to F$, a scalar $\beta \in F$, and symmetric bilinear form $\sigma : U \times U \to F$ such that $x \circ y$ is given by relation (2.12) with $e = I$ for all x, y in U. If $n = 2$, then β can be taken to be zero.

Proof. Let $U_0 = Fu_1 + \cdots + Fu_n$. We show first that there exist a linear form $\tau : U_0 \to F$, a scalar $\beta \in F$, and a symmetric bilinear form $\sigma : U_0 \times U_0 \to F$ such that (2.12) holds for $x, y \in U_0$. The result is clearly true by hypothesis when $n = 1$, and we suppose first that $n \geq 3$. Since $u_i + u_j$ is an idempotent of B for $i \neq j$, we have

$$(2.28) \quad u_i \circ u_j = \tfrac{1}{2}[u_i + u_j) \circ (u_i + u_j) - u_i \circ u_i - u_j \circ u_j] \in Fu_i + Fu_j + FI .$$

For each i, choose $j \neq i$ and define $\tau(u_i)$ to be the coefficient of u_j in the representation of $u_i \circ u_j$ as a linear combination of u_i, u_j, I. We note $\tau(u_i)$ is independent of the choice of j, since, for $k \neq i, j$, by (2.28) $u_i \circ (u_j + u_k) \in Fu_i + F(u_j + u_k) + FI$ and $u_i \circ (u_j + u_k) = u_i \circ u_j + u_i \circ u_k$. Then, for $i \neq j$, by (2.28) there exists a scalar $\sigma(u_i, u_j) = \sigma(u_j, u_i)$, so that $u_i \circ u_j = \tau(u_j) u_i + \tau(u_i) u_j + \sigma(u_i, u_j) I$. By hypothesis, there are scalars β_i and $\sigma(u_i, u_i)$ in F such that

$$u_i \circ u_i = 2[\tau(u_i) + \beta_i] u_i + \sigma(u_i, u_i) I .$$

Therefore, for $i \neq j$, we have

$$\begin{aligned}(u_i + u_j) \circ (u_i + u_j) &= u_i \circ u_i + 2u_i \circ u_j + u_j \circ u_j \\ &= 2[\tau(u_j) + \beta_i + \tau(u_i)] u_i + 2[\tau(u_i) + \beta_j + \tau(u_j)] u_j \\ &\quad + [\sigma(u_i, u_i) + 2\sigma(u_i, u_j) + \sigma(u_j, u_j)] I \in F(u_i + u_j) + FI ,\end{aligned}$$

which shows that $\beta_i = \beta_j$. We call this common scalar β and note that (2.12) holds when $x = u_i$ and $y = u_j$ for i, j not necessarily distinct. Extending τ and σ by linearity to all elements of U_0 , we have (2.12) for all $x,y \in U_0$ when $n \geq 3$.

If $n = 2$, then there is one degree of freedom in choosing τ, β, σ and it is possible to choose these so that $\beta = 0$. To see this, we let $\tau(u_i)$ and $\sigma(u_i,u_i)$ satisfy

$$u_i \circ u_i = 2\tau(u_i)u_i + \sigma(u_i,u_i)I$$

for $i = 1, 2$. Then there exist scalars $\gamma_1, \gamma_2, \sigma(u_i,u_2) \in F$ such that

$$u_1 \circ u_2 = [\tau(u_2) + \gamma_1]u_1 + [\tau(u_1) + \gamma_2]u_2 + \sigma(u_1,u_2)I \ , \tag{2.29}$$

and since $u_1 + u_2 = I$, this relation holds if $\gamma_1, \gamma_2, \sigma(u_1,u_2)$ are replaced by $\gamma_1 - \gamma, \gamma_2 - \gamma, \sigma(u_1,u_2) + \gamma$ for any $\gamma \in F$. On the other hand, the requirement that $u_1 \circ u_1 + 2u_1 \circ u_2 + u_2 \circ u_2 = I \circ I$ is a multiple of I implies that $\gamma_1 = \gamma_2$. Thus we may choose $\gamma_1 = \gamma_2 = 0$ in (2.29), so that (2.12) holds with $\beta = 0$.

We now extend τ, β, σ to U . For any scalar $\xi \in F$, we note that $u_i + \xi a, u_j - \xi a$, together with all u_k for $k \neq i, j$, form a set of orthogonal idempotents of B whose sum is I . Let U' be the linear span of these idempotents. We can apply the conclusions just obtained for U_0 to U' . By hypothesis, $(u_i + \xi a) \circ (u_i + \xi a)$ is a multiple of $u_i + \xi a$ and I , and since this holds for any value $\xi \in F$, $a \circ a$ and $u_i \circ a$ must be in $Fu_i + Fa + FI$. Thus there exist scalars $\tau(a)$, $\sigma(a,a)$, γ, ν, ρ , and $\sigma(u_i,a) = \sigma(a,u_i)$ such that

$$a \circ a = \gamma u_i + 2\tau(a)a + \sigma(a,a)I \ ,$$

$$u_i \circ a = \nu u_i + \rho a + \sigma(u_i,a)I \ .$$

Using (2.12) for U_0 and equating the coefficients of u_i and ξa in $(u_i + \xi a) \circ (u_i + \xi a)$, we find that $2\tau(u_i) + 2\beta + 2\xi\nu + \xi^2\gamma = 2\rho + 2\xi\tau(a)$ which must be true for all $\xi \in F$. Thus, $\gamma = 0$, $\nu = \tau(a)$, and $\rho = \tau(u_i) + \beta$ to give

$$(2.30)\qquad \begin{aligned} a \circ a &= 2\tau(a)a + \sigma(a,a)I , \\ u_i \circ a &= \tau(a)u_i + \tau(u_i)a + \beta a + \sigma(u_i,a)I . \end{aligned}$$

Hence relation (2.12) holds for $u_i \circ a$ and $a \circ a$. Similarly, (2.12) holds for $u_j \circ a$.

By the arguments above with U' in place of U_0, there exist τ', β', σ' specifying the product "$\circ$" in U'. Then $(u_i + \xi a)\circ(u_j - \xi a)$ and $(u_i + \xi a) \circ (u_i + \xi a)$ can be computed two different ways, one using (2.12) for U_0 and (2.30), and the other using τ', β', σ' for U'. Equating the coefficients of u_i, u_j, a in those two products, we obtain

$$(2.31)\qquad \begin{aligned} \tau'(u_j - \xi a) &= \tau(u_j) - \xi\tau(a) , \\ \tau'(u_i + \xi a) &= \tau(u_i) + \xi\tau(a) , \\ \beta' &= \beta . \end{aligned}$$

Hence

$$\begin{aligned} (u_i + \xi a)u_k &= [\tau(u_i) + \xi\tau(a)]u_k + \tau'(u_k)(u_i + \xi a) + \sigma'(u_i + \xi a,u_k)I \\ &= \tau(u_i)u_k + \tau(u_k)u_i + \sigma(u_i,u_k)I + \xi a \circ u_k \end{aligned}$$

for all $\xi \in F$, to show that

$$(2.32)\qquad \begin{aligned} \tau'(u_k) &= \tau(u_k) , \\ a \circ u_k &= \tau(u_k)a + \tau(a)u_k + \sigma(a,u_k)I \end{aligned}$$

for some scalar $\sigma(a,u_k) = \sigma(u_k,a)$. We have therefore established (2.12) for the basis elements of U and hence τ, σ can be extended linearly to all elements of U , as desired. □

Note that the subspace U in Lemma 2.8 is a subalgebra of B . Lemmas 2.7 and 2.8 are instrumental to determine all third power-associative product "$*$" on matrices such that $[x,y]^* = [x,y]$. For this, it is sufficient to establish the following result, which may be of interest in its own right.

<u>Theorem 2.9</u>. Let $A = M(n,F)$ over F of characteristic $\neq 2,3$, and let "$\circ$" be a commutative product on A such that $u \circ u \in Fu + FI$ for each idempotent u of A . Then there exist a linear form $\tau : A \to F$, a scalar $\beta \in F$, and a symmetric bilinear form $\sigma : A \times A \to F$ such that $x \circ y$ is given by relation (2.12) for all $x,y \in A$. If $n = 2$, then β can be taken to be zero.

<u>Proof</u>. We show that (2.12) holds for the basis of matrix units e_{ij}, i, j = 1, 2, $\cdots$, n , where $e_{ij}e_{k\ell} = \delta_{jk}e_{i\ell}$ and $A_{ij} = Fe_{ij}$, $i \neq j = 1, 2, \cdots, n$, are the Peirce spaces relative to idempotents $e_{11}, \cdots, e_{nn}$, By Lemma 2.8 there exist a scalar β , which can be taken to be zero for $n = 2$, a linear form τ , and a symmetric bilinear form σ defined on $Fe_{11} + \cdots + Fe_{nn} + Fe_{ij}$ for all $i \neq j$, so that (2.12) holds for elements in this subspace. In order to prove that (2.12) is satisfied by the remaining products involving matrix units, the idea is to replace U_0 and A_{ij} (= Fa) in Lemma 2.8 with $U_0' = Fu_1' + \cdots + Fu_n'$ where $u_i' = e_{ii} + \xi e_{ij}$, $u_j' = e_{jj} - \xi e_{ij}$, and $u_k' = e_{kk}$ for all $k \neq i, j$, together with a suitably chosen Peirce space $A_{k\ell}'$ relative to idempotents $u_1', \cdots, u_n'$. Lemma 2.8 and arguments used in that lemma show that there exist τ', β', σ' associated with $U_0' + A_{k\ell}'$ satisfying (2.12), and that

$$\tau'(e_{ii} + \xi e_{ij}) = \tau(e_{ii}) + \xi\tau(e_{ij}) ,$$

(2.33) $$\tau'(e_{jj} - \xi e_{ij}) = \tau(e_{jj}) - \xi\tau(e_{ij}) ,$$

$$\tau'(e_{kk}) = \tau(e_{kk}) , \quad \beta' = \beta ,$$

by (2.31) and (2.32).

First, let i, j, k, ℓ be distinct, and consider $U_0' + A'_{k\ell}$ where $A'_{k\ell} = Fe_{k\ell}$. There are two expressions for the product $(e_{ii} + \xi e_{ij}) \circ e_{k\ell}$, namely,

$$(e_{ii} + \xi e_{ij}) \circ e_{k\ell} = \tau'(e_{k\ell})(e_{ii} + \xi e_{ij}) + [\tau(e_{ii}) + \xi\tau(e_{ij})]e_{k\ell}$$
$$+ \sigma'(e_{ii} + \xi e_{ij}, e_{k\ell})I ,$$

$$(e_{ii} + \xi e_{ij}) \circ e_{k\ell} = \tau(e_{k\ell})e_{ii} + \tau(e_{ii})e_{k\ell} + \sigma(e_{ii}, e_{k\ell})I + \xi e_{ij} \circ e_{k\ell} ,$$

and they combine to show that $\tau'(e_{k\ell}) = \tau(e_{k\ell})$ and

$$e_{ij} \circ e_{k\ell} = \tau(e_{k\ell})e_{ij} + \tau(e_{ij})e_{k\ell} + \sigma(e_{ij}, e_{k\ell})I$$

for some scalar $\sigma(e_{ij}, e_{k\ell}) = \sigma(e_{k\ell}, e_{ij})$. Thus (2.12) holds for $e_{ij} \circ e_{k\ell}$.

Next, for i, j, k distinct, the subalgebra $U_0' + F(e_{jk} - \xi e_{ik})$ satisfies relation (2.12) since $e_{jk} - \xi e_{ik} \in A'_{jk}$. We compute $e_{kk} \circ (e_{jk} - \xi e_{ik})$ two different ways to obtain $\tau'(e_{jk} - \xi e_{ik}) = \tau(e_{jk}) - \xi\tau(e_{ik})$ by (2.31) which in turn implies

$$(e_{ii} + \xi e_{ij}) \circ (e_{jk} - \xi e_{ik}) = [\tau(e_{jk}) - \xi\tau(e_{ik})](e_{ii} + \xi e_{ij})$$
$$+ [\tau(e_{ii}) + \xi\tau(e_{ij})](e_{jk} - \xi e_{ik}) + \sigma'(e_{ii} + \xi e_{ij}, e_{jk} - \xi e_{ik})I .$$

On the other hand, this product is expressed as

$$(e_{ii} + \xi e_{ij}) \circ (e_{jk} - \xi e_{ik}) = \tau(e_{ij})e_{jk} + \tau(e_{jk})e_{ii} + \sigma(e_{ii},e_{jk})I$$

$$+ \xi e_{ij} \circ e_{jk} - \xi[\tau(e_{ii})e_{ik} + \tau(e_{ik})e_{ii} + \beta e_{ik} + \sigma(e_{ii},e_{ik})I] - \xi^2 e_{ij} \circ e_{ik} ,$$

which implies

$$e_{ij} \circ e_{jk} = \tau(e_{jk})e_{ij} + \tau(e_{ij})e_{jk} + \beta e_{ik} + \sigma(e_{ij},e_{jk})I ,$$

$$e_{ij} \circ e_{ik} = \tau(e_{ik})e_{ij} + \tau(e_{ij})e_{ik} + \sigma(e_{ij},e_{ik})I$$

for some scalars $\sigma(e_{ij},e_{jk})$ and $\sigma(e_{ij},e_{ik})$. This gives two more types of products for which (2.12) is satisfied, and by left-right symmetry the same must be true for products of the form $e_{ik} \circ e_{jk}$.

Therefore, the only remaining type of product yet to be determined is that of the form $e_{ij} \circ e_{ji}$. For this, let $b = e_{ji} - \xi e_{ii} + \xi e_{jj} - \xi^2 e_{ij}$. It is easily checked that $b \in A'_{ji}$ and hence $U'_0 + Fb$ is a subalgebra of A . Using (2.33), we compute $(e_{ii} + \xi e_{ij}) \circ b$ to derive the relation

$$\xi e_{ij} \circ e_{ji} = \tau'(b)(e_{ii} + \xi e_{ij}) + [\tau(e_{ii}) + \xi\tau(e_{ij})]b + \beta b - e_{ii} \circ b$$

$$+ \sigma'(e_{ii} + \xi e_{ij},b)I + \xi^2 e_{ij} \circ e_{ii} - \xi^2 e_{ij} \circ e_{jj} + \xi^3 e_{ij} \circ e_{ij} ,$$

where the coefficient of e_{ji} on the right side is $\xi\tau(e_{ij})$ since the coefficient of e_{ji} in $e_{ii} \circ b$ is β. But then if the roles of i and j are interchanged, the coefficient of e_{ij} must be $\xi\tau(e_{ji})$ and hence $\xi\tau(e_{ji}) = \xi\tau'(b) + \xi^2\tau(e_{ii}) - \xi^2\tau(e_{jj}) + \xi^3\tau(e_{ij})$, or

$$\tau'(b) = \tau(e_{ji}) - \xi\tau(e_{ii}) + \tau(e_{jj}) - \xi^2\tau(e_{ij}) .$$

Therefore, the terms in (2.34) that are linear in ξ give

$$e_{ij} \circ e_{ji} = \tau(e_{ji})e_{ij} + \tau(e_{ij})e_{ji} + \beta(e_{ii} + e_{jj}) + \sigma(e_{ij},e_{ji})I$$

for some scalar $\sigma(e_{ij},e_{ji})$. Extending τ, σ by linearity to A , the proof is complete. □

In view of Lemma 2.7, the following result is immediate from Theorem 2.9.

Corollary 2.10. Let $A = M(n,F)$ over a field F of characteristic $\neq 2,3$. Then any third power-associative product "$*$" on A such that $[x,y]^* = [x,y]$ for all $x,y \in A$ is given by

$$x * y = \frac{1}{2}[x,y] + \tau(y)x + \tau(x)y + \beta(xy + yx) + \sigma(x,y)I , \tag{2.35}$$

where τ is a linear form on A, σ is a symmetric bilinear form on A, and β is a scalar that can be taken to be zero for $n = 2$. □

When an algebra $(A,*)$ satisfies certain other identities, we can further specify τ and σ in (2.35). Recall that $(A,*)$ is called flexible if $(x * y) * x = x * (y * x)$ for all $x,y \in A$. Clearly, flexibility strengthens the third power identity. Substituting $x * y = \frac{1}{2}[x,y] + x \circ y$ into the flexible identity, we see that flexibility is equivalent to

$$[x \circ y,x] = x \circ [y,x] . \tag{2.36}$$

Corollary 2.11. Let $A = M(n,F)$ over F of characteristic zero, and let "$*$" be a flexible product on A satisfying $[x,y]^* = [x,y]$ for all $x,y \in A$. Then there exist $\beta, \gamma, \theta, \zeta \in F$ such that

$$\begin{aligned} x * y &= \frac{1}{2}[x,y] + \beta(x \# y) + \gamma K(x,y)I , \\ I * y &= y * I = \theta y , \\ I * I &= \zeta I , \end{aligned} \tag{2.37}$$

where x, y are in $s\ell(n)$, the matrices of trace zero, where $x \# y$ is defined by

$$x \# y = xy + yx - \frac{2}{n}(\mathrm{tr}\ xy)I \ , \tag{2.38}$$

where $K(\ ,\ )$ denotes the Killing form on $s\ell(n)$, and where tr is the trace.

Proof. Since "$*$" is third power-associative, by Corollary 2.10 $x \circ y$ is given by (2.12). Substituting (2.12) into (2.36) yields

$$\tau(x)[y,x] + \beta[xy + yx,x] = \tau([y,x])x + \tau(x)[y,x] + \beta x[y,x]$$
$$+ \beta[y,x]x + \sigma(x,[y,x])I \ ,$$

which implies

$$0 = \tau([y,x])x + \sigma(x,[y,x])I \tag{2.39}$$

for all $x,y \in A$. Thus we obtain $\tau([y,x]) = 0$ for all $x \in s\ell(n)$ and $y \in A$, and hence τ vanishes on $s\ell(n)$. It follows from this and (2.39) that $\sigma(x,[y,x]) = 0$ for all $x,y \in A$, which is linearized to the relation

$$\sigma(x,[y,z]) = \sigma([x,y],z) \tag{2.40}$$

for all $x,y,z \in A$. If F is algebraically closed, then by Humphreys [1,p.118], for example, $\sigma(\ ,\ )$ must be a multiple of the Killing form $K(\ ,\ )$ on $s\ell(n)$ and $K(x,y) = 2n(\mathrm{tr}\ xy)$. When F is not algebraically closed, the same result holds by a standard argument of a scalar extension. Hence if $\sigma(x,y) = \delta K(x,y)$, then by (2.35)

$$x * y = \frac{1}{2}[x,y] + \beta(x \# y) + \gamma K(x,y)I$$

for $x,y \in s\ell(n)$, where $\gamma = \delta + \frac{\beta}{n^2}$ and $x \# y$ is given by (2.38).

Relation (2.40) with $z = I$ implies that I is orthogonal to

$\mathcal{sl}(n)$ under σ . Thus, for $y \in \mathcal{sl}(n)$,

$$I * y = I \circ y = \tau(y)I + \tau(I)y + 2\beta y + \sigma(I,y)I = \theta y$$

where $\theta = \tau(I) + 2\beta$. Finally, $I * I = \zeta I$ for $\zeta = 2\tau(I) + 2\beta + \sigma(I,I)$, to give the desired multiplications. □

We now turn our attention to classifying those products "$*$" defined on $A = M(n,F)$ which are power-associative by determining the commutative products "$\circ$" on A that correspond to them. If $(A,*)$ is power-associative, then so is $(A,*)^+ = (A,\circ)$, and hence "$\circ$" satisfies both (2.12) and the fourth power identity (2.6). Conversely, if "$\circ$" satisfies (2.12) and (2.6), then $(A,*)$ is third and fourth power-associative by Lemma 2.1 and Lemma 1.10. In the remainder of this section, we determine all commutative products "$\circ$" satisfying both (2.12) and (2.6), and show that the corresponding products "$*$" are power-associative for characteristic $\neq 2,3$.

Lemma 2.12. Let $(A,\circ)$ be an arbitrary commutative algebra over a field F of characteristic $\neq 2,3$ and let e be an idempotent of $(A,\circ)$. Assume that $(A,\circ)$ satisfies (2.6). If $x \in A_{\frac{1}{2}}(e)$ and e generate a 2-dimensional subalgebra of A then $x^2 = x \circ x = 0$.

Proof. In view of (2.11), we have $x^2 = \xi e$, but $\xi^2 e = x^2 \circ x^2 = x \circ (x \circ x^2) = \frac{1}{2}\xi^2 e$ and hence $x^2 = 0$. □

Lemma 2.13. Let B be an algebra with product xy over a field F of characteristic $\neq 2,3$ and let I be the identity element of B . Let "$\circ$" be a commutative product on B defined by

$$x \circ y = \tau(y)x + \tau(x)y + \beta(xy + yx) + \sigma(x,y)I \tag{2.41}$$

where τ is a linear form on B, σ is a symmetric bilinear form on B, and β is a scalar. Let $I \circ I = \zeta I$ where

$$\zeta = 2\tau(I) + 2\beta + \sigma(I,I) . \tag{2.42}$$

If $\zeta \neq 0$, then $e = \zeta^{-1}I$ is an idempotent of $(B,\circ)$ and

$$e \circ x = \lambda x + \eta(x)e \quad \text{for each} \quad x \in B \tag{2.43}$$

where $\lambda = \tau(e) + 2\beta\zeta^{-1}$ and $\eta(x) = \tau(x) + \sigma(x,I)$. If $(B,\circ)$ satisfies (2.6), then $\lambda = 0, \frac{1}{2}$, or 1. If $\beta = 0$ then I can be taken to be a nonzero element of B.

<u>Proof</u>. When x is taken from the Peirce spaces of B relative to e, the result is immediate from relation (2.10). □

When B is an associative algebra, the associator $(x,y,z)^\circ = (x \circ y) \circ z - x \circ (y \circ z)$ in "$\circ$" defined by (2.41) can be computed as

$$\begin{aligned}(x,y,z)^\circ &= [\tau(x)\tau(y) + \sigma(x,y)(\tau(I) + 2\beta) + \beta\tau(xy + yx)]z \\ &- [\tau(y)\tau(z) + \sigma(y,z)(\tau(I) + 2\beta) + \beta\tau(yz + zy)]x \\ &+ [\beta\sigma(xy + yx,z) - \beta\sigma(yz + zy,x) + \sigma(x,y)\sigma(z,I) - \sigma(y,z)\sigma(x,I)]I \\ &+ \beta^2[y,[x,z]] ,\end{aligned} \tag{2.44}$$

$$\begin{aligned}(I,y,z)^\circ &= \eta(y)(\tau(I) + 2\beta)z \\ &- [\tau(y)\tau(z) + \sigma(y,z)(\tau(I) + 2\beta) + \beta\tau(yz + zy)]I \\ &+ [2\beta\sigma(y,z) - \beta\sigma(yz + zy,I) + \sigma(y,I)\sigma(z,I) - \sigma(y,z)\sigma(I,I)]I .\end{aligned} \tag{2.45}$$

When "$\circ$" is given by (2.41), the fourth power identity $(x \circ x,x,x)^\circ = 0$ is equivalent to

(2.46) $$2\beta(xx,x,x)^{\circ} + \sigma(x,x)(I,x,x)^{\circ} = 0 ,$$

and if $\beta = 0$, then this reduces to

(2.47) $$\sigma(x,x)(I,x,x)^{\circ} = 0$$

which holds for a nonzero element I, not necessarily an identity element of B . We first treat the case $\beta = 0$, and show the following general result.

Theorem 2.14. Let B be a vector space of dimension greater than one over a field F of characteristic $\neq 2,3$. Suppose a multiplication "$\circ$" is defined on B by

(2.48) $$x \circ y = \tau(x)y + \tau(y)x + \sigma(x,y)I$$

where τ is a linear form on B , σ is a symmetric bilinear form on B , and I is a nonzero element of B . Then the product "$\circ$" satisfies the fourth power identity (2.6) if and only if one of the following conditions holds:

(a) τ is arbitrary and σ is identically zero.

(b) σ is arbitrary and τ is determined by $\tau(x) = -\sigma(x,I)$.

(c) τ and σ satisfy the relation

(2.49) $$0 = \tau(x)\tau(y) - \sigma(I,x)\sigma(I,y) + \sigma(x,y)\sigma(I,I) .$$

If $\sigma(I,I) \neq 0$ in (2.49), then the forms $\tau(x)$ and $\sigma(I,x)$ can be defined arbitrarily subject only to $\tau(I) = 0$ and $\sigma(I,I) \neq 0$, and the remaining values of σ can be determined from (2.49). If $\sigma(I,I) = 0$, then σ can be arbitrarily defined subject to that condition, and τ is given by $\tau(x) = \sigma(I,x)$.

Proof. It follows from (2.48) that $I \circ I = \zeta I$ where $\zeta = 2\tau(I)$

$+ \sigma(I,I)$. If $\zeta \neq 0$, then $e = \zeta^{-1}I$ is an idempotent of $(B,\circ)$ and by (2.43) $e \circ x = \tau(e)x + \eta(x)e$ where $\eta(x) = \tau(x) + \sigma(x,I)$. Suppose the product "$\circ$" is fourth power-associative. By Lemma 2.13, there are four possibilities: either $\zeta = 0$, or $\zeta \neq 0$ and $\tau(e) = 0, \frac{1}{2}$, or 1 .

Assume $\zeta \neq 0$ and $\tau(e) = 0$. Let $x \in B$ not be a multiple of e , so that x and e span a 2-dimensional subalgebra by (2.43). Since $e \circ x = \eta(x)e$, $x - \eta(x)e$ belongs to the Peirce space $B_0(e)$ relative to e , which is a subalgebra by (2.11). Since $x - \eta(x)e$ belongs to the 2-dimensional subalgebra generated by e and x , $x - \eta(x)e$ must square to a multiple of itself, from which we see

$$(x - \eta(x)e) \circ (x - \eta(x)e) = 2\tau(x)x + \sigma(x,x)\zeta e - \eta(x)^2 e = 2\tau(x)(x - \eta(x)e) .$$

Therefore, $-2\tau(x)\eta(x) = \sigma(x,x)\zeta - \eta(x)^2$ and this implies

$$\tau(x)^2 - \sigma(x,e)^2\zeta^2 + \sigma(x,x)\zeta = 0 .$$

Since $e \circ e = 2\tau(e)e + \sigma(e,e)I = \sigma(e,e)\zeta e = e$ and hence $\sigma(I,I) = \zeta^2\sigma(e,e) = \zeta$, we have $\tau(x)^2 - \sigma(x,I)^2 + \sigma(x,x)\sigma(I,I) = 0$. Relation (2.49) is simply a linearized form of this. Since $\tau(I) = \zeta\tau(e) = 0$, relation (2.49) holds for all x, y including multiples of I . The 2-dimensional algebra with basis e, $y = x - \eta(x)e$ and with multiplication given by $e = e \circ e$, $e \circ y = 0$, $y \circ y = \lambda y$ is associative for any choice of $\lambda \in F$, in particular for the value $\lambda = 2\tau(x)$ that we have found. Conversely, if relation (2.49) is satisfied and x is an element not a multiple of e , then reversing the arguments above shows that $(x - \eta(x)e) \circ (x - \eta(x)) = 2\tau(x)(x - \eta(x)e)$ and hence x is contained in such a 2-dimensional associative subalgebra. Therefore, (2.49) is a necessary and sufficient condition for "$\circ$" to be power-associative. We have assumed that $\tau(e) = 0$, and that $\zeta \neq 0$, and these last conditions imply that $\tau(I) = 0$ and $\sigma(I,I) = \zeta \neq 0$, which is precisely the situation described in the first part of (c).

Let us consider next the case that $\tau(e) = \frac{1}{2}$ and $\zeta \neq 0$. Then for each x not a multiple of e, the vector $y = x - 2\eta(x)e$ has the property that $e \circ y = \frac{1}{2}y$. Hence by Lemma 2.12 $y \circ y = 0$ for all such y, so that

$$0 = (x - 2\eta(x)e) \circ (x - 2\eta(x)e) = 2\tau(x)x + \sigma(x,x)\zeta e - 2\eta(x)(\tau(x)x + \eta(x)e) - \eta(x)(x - 2\eta(x)e) .$$

Thus, $0 = \tau(x) - \eta(x) = \sigma(x,e)\zeta$ and $\sigma(x,x)\zeta = 0$ for all such x. Since $e = e \circ e = 2\tau(e)e + \sigma(e,e)\zeta e = e + \sigma(e,e)\zeta e$, $\sigma(e,e) = 0$. Hence the form σ is identically zero, and case (a) of Theorem 2.14 holds in this situation. For any vector space and any linear form τ defined on it, the product $x \circ y = \tau(y)x + \tau(x)y$ is power-associative since $x \circ x = 2\tau(x)x$, so that the conditions in (a) are necessary and sufficient.

Assume that $\zeta \neq 0$ and $\tau(e) = 1$, so that e acts as the identity element of $(B,\circ)$. Thus the subalgebra spanned by e and x is associative regardless of how the values of σ and τ are specified. But then $e \circ x = \tau(e)x + \tau(x)e + \sigma(x,e)I = x$ and $\tau(x) = -\sigma(x,I)$, so that case (b) holds under the hypothesis $\tau(e) = 1$.

We turn our attention to the remaining case that $\zeta = 0$, that is, $2\tau(I) + \sigma(I,I) = 0$ or $\sigma(I,I) = -2\tau(I)$. Since the fourth power-identity (2.6) is equivalent to relation (2.47), we have by (2.45) that (2.6) is equivalent to

$$\sigma(x,x)\eta(x)\tau(I) = 0,\ \sigma(x,x)[\sigma(x,x)\tau(I) + (\sigma(x,I) - \tau(x))\eta(x)] = 0 \tag{2.50}$$

for all x not multiples of I, since $\sigma(I,I) = -2\tau(I)$.

Consider the case where $\tau(I) \neq 0$. Then we deduce from the first equation of (2.50) that $\sigma(x,x)\eta(x) = 0$, which together with the second equation implies

$$\sigma(x,x) = 0 \tag{2.51}$$

for all x not multiples of I. Since $\tau(I) \neq 0$ and $2\tau(I) + \sigma(I,I) = 0$, $\sigma(I,I) \neq 0$. Let y be an element of B such that $\sigma(y,I) = 0$ and let $x = y + \lambda I$ for $\lambda \in F$. Then x is not a multiple of I and by (2.51) $\sigma(y,y) + \lambda^2\sigma(I,I) = 0$ for all $\lambda \in F$. Since $\sigma(I,I) \neq 0$, this is impossible and hence the case $\tau(I) \neq 0$ cannot occur whenever the dimension of B is greater than one.

We assume then that $\tau(I) = 0$, so that $\sigma(I,I) = 0$ and $\eta(I) = 0$ also. Then (2.50) implies

$$\sigma(x,x)\eta(x)[\sigma(x,I) - \tau(x)] = 0 \tag{2.52}$$

for all $x \in B$. We replace x with $x + I$ in (2.52) to obtain $[\sigma(x,x) + 2\sigma(x,I)]\eta(x)[\sigma(x,I) - \tau(x)] = 0$, which by (2.52) implies that $\sigma(x,I)\eta(x)[\sigma(x,I) - \tau(x)] = 0$ for all x. Therefore, $\sigma(x,I) = 0, \tau(x)$, or $-\tau(x)$ for each $x \in B$, and we deduce from this that the linear form $\sigma(\ ,I)$ on B vanishes on the kernel of τ. Thus $\sigma(\ ,I)$ is a scalar multiple of τ, that is, $\sigma(x,I) = \lambda\tau(x)$ for each $x \in B$. Substituting this in (2.52) yields

$$(\lambda^2 - 1)\sigma(x,x)\tau(x)^2 = 0 \tag{2.53}$$

for all x.

If $\lambda = 1$, then $\sigma(x,I) = \tau(x)$ and relation (2.49) clearly holds since $\sigma(I,I) = 0$. This case is described by the last sentence of the theorem. If $\lambda = -1$, then a restricted form of (b) holds in which $\tau(I) = \sigma(I,I) = 0$. Assume that $\lambda^2 \neq 1$, so that $\sigma(x,x)\tau(x)^2 = 0$ by (2.53). If τ is identically zero, then $\sigma(x,I) = \lambda\tau(x) = 0$ for all x in B, and otherwise σ can be arbitrarily defined. This is the description of a special case of (b). Assume $\tau \neq 0$ and let K be the kernel of τ. Since $\sigma(x,x)\tau(x)^2 = 0$ for all x, we have $\sigma(y,y) = 0$ for each $y \notin K$. Thus, for every $y \notin K$ and every $z \in K$,

$$0 = \sigma(y + \nu z, y + \nu z) = 2\nu\sigma(y,z) + \nu^2\sigma(z,z) ,$$

and since this holds for all ν in F, $\sigma(y,z) = \sigma(z,z) = 0$ for all $y \notin K$ and $z \in K$. Thus, σ is identically zero when $\lambda^2 \neq 1$ and $\tau \neq 0$. This gives case (a) and completes the proof. □

To extend Theorem 2.14 to the product "∘" defined by (2.41) when the given algebra is $M(n,F)$, the following lemmas are instrumental.

Lemma 2.15. Let $A = M(n,F)$ over a field F of characteristic $\neq 2$, and let $f : A \times A \to F$ be a symmetric bilinear form such that $f(u,u) = 0$ for every idempotent $u \in A$. Then f is identically zero.

Proof. If u and v are orthogonal idempotents, then since $u + v$ is an idempotent,

$$f(u,v) = \tfrac{1}{2}[f(u + v,u + v) - f(u,u) - f(v,v)] = 0 .$$

For each idempotent u of rank 1 where rank means the rank as a matrix, the Peirce space A_{00} relative to u is isomorphic to $M(n - 1,F)$ and so has a basis of idempotents, namely, e_{ii} for all i and $e_{ii} + e_{ij}$ for all $i \neq j$. Thus, $f(u,A_{00}) = 0$. If $a \in A_{10}$ and $\xi \in F$, then $u + \xi a$ is an idempotent and

$$0 = f(u + \xi a,u + \xi a) = f(u,u) + 2\xi f(u,a) + \xi^2 f(a,a)$$

for all $\xi \in F$. It follows from this that $f(u,a) = 0$, or $f(u,A_{10}) = 0$, and by symmetry $f(u,A_{01}) = 0$. Since the Peirce space A_{11} is Fu when u has rank 1, we also have $f(u,A_{11}) = 0$, implying $f(u,A) = 0$. But then f must vanish on A, since A has a basis of rank 1 idempotents. □

Lemma 2.16. Let B be an algebra with product xy and with identity element I over a field F of characteristic $\neq 2$. Assume that "$\circ$" is the product defined by (2.41) on B and satisfies the properties:

(d) β is an arbitrary nonzero scalar and τ, β, σ satisfy the relations

$$\tau(I) + 2\beta \neq 0 = \tau(x)\tau(y) + \beta\tau(xy + yx) + \sigma(x,y)(\tau(I) + 2\beta) \tag{2.54}$$

for all $x,y \in B$, where τ can be defined arbitrarily subject only to $\tau(I) + 2\beta \neq 0$, and σ can be determined from (2.54). Then the algebra B^+ is isomorphic to $(B,\circ)$ under the mapping $\phi : B^+ \to (B,\circ)$ defined by $\phi(x) = (2\beta)^{-1}(x - \tau(x)e)$, where $e = \xi^{-1}I$.

Proof. Setting $y = I$ in (2.54) and dividing out the factor $\tau(I) + 2\beta$ yields $\eta(x) = \tau(x) + \sigma(x,I) = 0$ for all x, so $\eta(I) = 0$, in particular. This implies that $\zeta = 2\tau(I) + 2\beta + \sigma(I,I) = \tau(I) + 2\beta \neq 0$, so that $e = \zeta^{-1}I$ is an idempotent of $(B,\circ)$. But then e is the identity element of $(B,\circ)$, since $e \circ x = \lambda x$ by (2.43) and $\lambda = \tau(e) + 2\beta\zeta^{-1} = (\tau(I) + 2\beta)\zeta^{-1} = 1$. Thus

$$\begin{aligned}
\phi(x) \circ \phi(y) &= \frac{1}{4\beta^2}[x - \tau(x)e] \circ [y - \tau(y)e] \\
&= \frac{1}{4\beta^2}[x \circ y - \tau(y)x - \tau(x)y + \tau(x)\tau(y)e] \\
&= \frac{1}{4\beta^2}[\beta(xy + yx) + \sigma(x,y)\zeta e + \tau(x)\tau(y)e] \\
&= \frac{1}{4\beta^2}[\beta(xy + yx) - \beta\tau(xy + yx)e] \\
&= (2\beta)^{-1}[\tfrac{1}{2}(xy + yx) - \tau(\tfrac{1}{2}(xy + yx))e] = \phi(\tfrac{1}{2}(xy + yx)) .
\end{aligned}$$

Hence, $B^+ \cong (B,\circ)$. □

Theorem 2.17. Let $A = M(n,F)$ over a field F of characteristic $\neq 2,3$. Then the commutative product "∘" on A defined by (2.41) is fourth power-associative if and only if one of the four sets of conditions (a), (b), (c) for $\beta = 0$, and (d) described in Theorem 2.14 and Lemma 2.16 holds, where I is the identity matrix, and τ, β, σ are determined subject to those restrictions described in (a), (b), (c), and (d).

Proof. When $\beta = 0$, we have shown in Theorem 2.14 that (2.6) is equivalent to one of conditions (a), (b), and (c). Since β can be taken to be zero when $n = 2$, we may assume in what follows that $\beta \neq 0$ and $n \geq 3$, and we show that (2.6) is equivalent to condition (d) of Lemma 2.16. Assume that the product "∘" satisfies the fourth power identity (2.6), which is equivalent to relation (2.46). By Lemma 2.13, $I \circ I = \zeta I$ and there are four possibilities: either $\zeta = 0$, or $\zeta \neq 0$ and $\lambda = 0$, $\frac{1}{2}$, or 1. We first dispose of the case $\zeta \neq 0$, $\lambda = \frac{1}{2}$. In this situation it follows from (2.43) that the element $x - 2\eta(x)e$ for each $x \in A$ lies in the Peirce space $A_{\frac{1}{2}}$ relative to the idempotent $e = \zeta^{-1}I$, and hence $A_{\frac{1}{2}}$ has codimension 1 in A . In view of (2.11), the product

$$[x - \eta(x)e] \circ [y - \eta(y)e] = x \circ y - \eta(x)y - \eta(y)x + 4\eta(x)\eta(y)e$$

belongs to $A_0 + A_1$; but then $A_0 + A_1 = Fe$. Thus

$$\tau(x)y + \tau(y)x + \beta(xy + yx) + \sigma(x,y)\zeta e - \eta(x)y - \eta(y)x$$

is a multiple of e . Since $n \geq 3$ and we can find $x,y \in A$ such that x, y, e, $xy + yx$ are linearly independent, β must be zero. This contradicts the assumption $\beta \neq 0$, and shows that the case $\lambda = \frac{1}{2}$ cannot occur when $\beta \neq 0$.

For the remaining three cases, let u and v be orthogonal idempotents of the associative algebra A and let ξ be a scalar. When $u + \xi v$ is substituted into (2.46), the terms that are linear in ξ give

$$(2.55)\qquad 0 = 2\beta(u,u,v)^{\circ} + \sigma(u,u)(I,u,v)^{\circ} + \sigma(u,u)(I,v,u)^{\circ} + 2\sigma(u,v)(I,u,u)^{\circ} .$$

When $u + v \neq I$, using (2.44) and (2.45), the coefficient of v in (2.55) yields the relation

$$(2.56)\qquad \begin{aligned} 0 = {} & 2\beta\tau(u)^2 + 4\beta^2\tau(u) + 2\beta\sigma(u,u)(\tau(I) + 2\beta) \\ & + \sigma(u,u)\eta(u)(\tau(I) + 2\beta) , \end{aligned}$$

which must hold for all idempotents u with rank $\leq n - 2$. If u has rank $n - 1$, then $I - u$ has rank 1 and (2.56) with $I - u$ replacing u can be expressed as

$$(2.57)\qquad \begin{aligned} 0 = {} & 2\beta\tau(u)^2 + 4\beta^2\tau(u) + 2\beta\tau(I)(\tau(I) + 2\beta) - 4\beta\tau(u)(\tau(I) + 2\beta) \\ & + 2\beta\sigma(I - u,I - u)(\tau(I) + 2\beta) + \sigma(I - u,I - u)\eta(I - u)(\tau(I) + 2\beta) . \end{aligned}$$

Assume now that $\zeta \neq 0$ and $\lambda = 0$. In this case, $\tau(I) + 2\beta = \zeta\lambda = 0$, and (2.56) and (2.57) combine to give

$$(2.58)\qquad \tau(u)^2 + 2\beta\tau(u) = 0$$

for all idempotents u . The bilinear form f on A defined by $f(x,y) = \tau(x)\tau(y) + \beta\tau(xy + yx)$ vanishes on all idempotents, and must be identically zero by Lemma 2.15. Therefore,

$$(2.59)\qquad \tau(x)\tau(y) + \beta\tau(xy + yx) = 0 \quad \text{for all} \quad x,y \in A .$$

If y lies in the kernel of τ , then $xy + yx$ does also by (2.59), and this shows that the kernel of τ is an ideal of A^+ , where A^+ is the

Jordan algebra with product given by $\frac{1}{2}(xy + yx)$. Since A is a simple associative algebra, A^+ is simple also (see Herstein [2,p.4], for example). Thus τ must vanish on A , but then $0 = \tau(I) + 2\beta = 2\beta$, to give a contradiction. Hence the case $\zeta \neq 0$ and $\lambda = 0$ cannot occur when $\beta \neq 0$.

Consider the case $\zeta \neq 0$ and $\lambda = 1$. In this situation, the idempotent e acts as the identity of $(A,\circ)$, and by (2.43) $\zeta(I) + 2\beta = \zeta\lambda = \zeta$ and $0 = \eta(x) = \tau(x) + \sigma(x,I)$ for all $x \in A$. Thus equation (2.57) can be simplified to

$$\begin{aligned} 0 &= 2\beta\tau(u)^2 + 4\beta^2\tau(u) + 2\beta\tau(I)\zeta - 4\beta\tau(u)\zeta + 2\beta\sigma(I-u,I-u)\zeta \\ &= 2\beta\tau(u)^2 + 4\beta^2\tau(u) + 2\beta\zeta\eta(I) - 4\beta\zeta\eta(u) + 2\beta\sigma(u,u)\zeta \\ &= 2\beta\tau(u)^2 + 4\beta^2\tau(u) + 2\beta\sigma(u,u)\zeta \ . \end{aligned}$$

This last relation together with (2.56) gives

$$0 = \tau(u)^2 + 2\beta\tau(u) + \sigma(u,u)\zeta$$

which holds for all idempotents u . Thus, as before, the bilinear form

$$f(x,y) = \tau(x)\tau(y) + \beta\tau(xy + yx) + \sigma(x,y)\zeta$$

vanishes on A , giving relation (2.54). Since $\tau(I) + 2\beta = \zeta \neq 0$, the conditions described by (d) in Lemma 2.16 are satisfied in this case.

To complete the proof, it remains to treat the case $\zeta = 0$, or $2\tau(I) + 2\beta + \sigma(I,I) = 0$ by (2.42). For any idempotent u of A , relation (2.46) reduces to

$$\sigma(u,u)(I,u,u)^\circ = 0 \ . \tag{2.60}$$

If $\sigma(u,u) = 0$ for all idempotents u , then σ is identically zero.

Thus, using (2.44), the coefficient of v in (2.55) gives $\tau(u)^2 + 2\beta\tau(u) = 0$ for all idempotents u, which is relation (2.58). As before, this leads to equation (2.59), which in turn implies $\tau = 0$, just as in the case $\lambda = 0$. But then $0 = 2\tau(I) + 2\beta + \sigma(I,I) = 2\beta$ gives a contradiction. Assume that $\sigma(v,v) \neq 0$ for some idempotent v of A. Then by (2.60) $0 = (I,v,v)^{\circ}$, and in light of (2.45) the coefficients of v and I yield

$$\eta(v)(\tau(I) + 2\beta) = 0 ,$$

$$\sigma(v,v)(\tau(I) + 2\beta) + \eta(v)[\sigma(v,I) - \tau(v)] - 2\beta\eta(v) = 0 ,$$

which imply that $\tau(I) + 2\beta = 0$. It follows from this, (2.56) and (2.57) that $\tau(u)^2 + 2\beta\tau(u) = 0$ for all idempotents u of A. By the same argument as above, that identity leads to a contradiction. Thus the case $\zeta = 0$ cannot occur when $\beta \neq 0$, and this completes the proof of Theorem 2.17. □

Definition 2.5. An algebra B over a field F of characteristic $\neq 2$ is called a *quadratic algebra* if there is a nonzero element e of B such that, for every element $x \in B$, the subspace $Fx + Fe$ is a subalgebra of B. □

If B is a quadratic algebra, then every element x of B satisfies a quadratic equation given by (2.16) for some scalars $T(x)$ and $N(x)$ in F. It is clear that if x is not a multiple of e, then $T(x)$ and $N(x)$ are uniquely determined by x. Theorems 2.9, 2.14, and 2.17 essentially determine all power-associative products "*" defined on $M(n,F)$ such that $[x,y]^* = [x,y]$. As a consequence of these results, we have

Theorem 2.18. Let $A = M(n,F)$ over a field F of characteristic $\neq 2,3$, and let "*" be a product defined on A satisfying the properties that $(A,\star)$ is power-associative and that $[x,y]^\star = [x,y]$. Then the product "$\circ$" defined by $x \circ y = \frac{1}{2}(x \star y + y \star x)$ satisfies one of four cases (a) - (d) of Theorem 2.17, and in each of these cases $(A,\circ)$ is a Jordan algebra. Thus, $(A,\star)$ is Jordan-admissible as well as Lie-admissible (Definition 1.1). In case (d) of Theorem 2.17, $(A,\circ)$ is isomorphic to A^+ , and in the other cases $(A,\circ)$ as well as $(A,\star)$ is a quadratic algebra. Conversely, if "$\star$" is a third power-associative product on A satisfying $[x,y]^\star = [x,y]$ and such that "$\circ$" satisfies the fourth power identity (2.6), then $(A,\star)$ is power-associative.

Proof. Assume that $(A,\star)$ is power-associative and that $[x,y]^\star = [x,y]$ for all $x,y \in A$. Then, by Corollary 2.10, the product "$\star$" is given by (2.35), while "$\circ$" is described by relation (2.41) by Theorem 2.9. Since $(A,\star)^+ = (A,\circ)$, $(A,\circ)$ is power-associative, so that the product "$\circ$" satisfies each of four conditions (a) - (d) of Theorem 2.17 by Theorems 2.14 and 2.17. We show that each of the four cases gives a Jordan algebra. If "$\circ$" satisfies case (d), then $(A,\circ) \cong A^+$ by Lemma 2.16, and hence $(A,\circ)$ is a Jordan algebra.

In the remaining cases, $\beta = 0$, and the product has the form $x \circ y = \tau(y)x + \tau(x)y + \sigma(x,y)I$. It suffices to show that $(A,\circ)$ satisfies the Jordan identity

$$\begin{aligned} 0 = (x \circ x,y,x)^\circ &= (2\tau(x)x + \sigma(x,x)I,y,x)^\circ \\ &= 2\tau(x)(x,y,x)^\circ + \sigma(x,x)(I,y,x)^\circ \\ &= \sigma(x,x)(I,y,x)^\circ \end{aligned}$$

for all $x,y \in A$. This relation holds for case (a) since the form σ is identically zero. For the other two cases we show that $(I,y,x)^\circ = 0$. In case (b), $I \circ y = \tau(I)y + (\tau(y) + \sigma(I,y))I = \tau(I)y$, hence

$$(I,y,x)^{\circ} = (I \circ y) \circ x - I \circ (y \circ x) = \tau(I)y \circ x - \tau(I)x \circ y = 0 \ .$$

For case (c), setting $x = y = I$ in (2.49) gives $\tau(I) = 0$, so that $I \circ y = \eta(y)I$ for $\eta(y) = \tau(y) + \sigma(I,y)$. Thus,

$$(I,y,x)^{\circ} = \eta(y)I \circ x - \eta(x \circ y)I = (\eta(x)\eta(y) - \eta(x \circ y))I \ ,$$

and we compute

$$\begin{aligned}
\eta(y)\eta(x) - \eta(x \circ y) &= [\tau(y) + \sigma(I,y)][\tau(x) + \sigma(I,x)] \\
&\quad - \eta(\tau(y)x + \tau(x)y + \sigma(x,y)I) \\
&= \tau(y)\tau(x) + \tau(y)\sigma(I,x) + \sigma(I,y)\tau(x) + \sigma(I,y)\sigma(I,x) \\
&\qquad - \tau(y)\eta(x) - \tau(x)\eta(y) - \sigma(x,y)\sigma(I,I) \\
&= -\tau(y)\tau(x) + \sigma(I,y)\sigma(I,x) - \sigma(x,y)\sigma(I,I) = 0
\end{aligned}$$

by (2.49). Hence $(I,y,x)^{\circ} = 0$ and $(A,\circ)$ is a Jordan algebra. In the cases (a) - (c), since $x * x = x \circ x = 2\tau(x)x + \sigma(x,x)I$ and $I * x = I \circ x = \tau(I)x + \eta(x)I$, $FI + Fx$ is a subalgebra of $(A,\circ)$ and $(A,*)$. Thus $(A,\circ)$ and $(A,*)$ are quadratic algebras in these cases.

Conversely, assume that $(A,*)$ is third power-associative such that $[x,y]^* = [x,y]$ and such that "$\circ$" satisfies (2.6). Thus, by Theorem 2.9 the product "$\circ$" is given by (2.12) or (2.41), and satisfies the conditions (a) - (d) of Theorem 2.17. Consequently, $(A,\circ)$ is a Jordan algebra and so is power-associative in each of these cases. For each $x \subset A$, the mth power of x in $(A,\circ)$ commutes with x in the matrix algebra A, since it lies in the subalgebra of A generated by x and I. This implies that the mth powers of x in $(A,\circ)$ and $(A,*)$ coincide, since $x * y = \frac{1}{2}[x,y] + x \circ y$. Thus, $(A,*)$ is power-associative. □

We note that in the cases (a) - (c) where $\beta = 0$, the results of Theorem 2.18 for the product "$\circ$" does not depend on the associative multiplication in A , except that in cases (b) and (c) the element I has been designated to play a special role. Thus, in view of Theorem 2.14, such results would hold for a fourth power-associative commutative product (2.48) defined on an arbitrary vector space, by specifying a linear form τ , a symmetric bilinear form σ , and a fixed nonzero element I , which satisfy the conditions described in Theorem 2.14. The discussion in this section is based on work of Benkart and Osborn [4].

2.4. Power-associative products on octonions

In this section we extend the results in Section 2.3 to octonion algebras and to finite-dimensional central simple alternative algebras. We begin with determining all third power-associative products "$*$" on $M(\alpha,\beta,\gamma)$ such that $[x,y]^* = [x,y]$, where $M(\alpha,\beta,\gamma)$ is a 7-dimensional central simple Malcev algebra given by Table 2.2.

Lemma 2.19. Let "$*$" be a third power-associative product on $M(\alpha,\beta,\gamma)$ over a field F of characteristic $\neq 2$ such that $[x,y]^* = [x,y]$, where $M(\alpha,\beta,\gamma)$ is an algebra with product $[x,y]$ defined by Table 2.2 for scalars α, β, γ with $\alpha\beta\gamma \neq 0$. Then the product $x * y$ is given by

$$(2.61) \qquad x * y = \frac{1}{2}[x,y] + \tau(y)x + \tau(x)y$$

for some linear form τ on $M(\alpha,\beta,\gamma)$.

Proof. Note that the third power identity for "$*$" is equivalent to the identity (2.3), $[x,x \circ x] = 0$, which is linearized to relation (2.5), where $x \circ y = \frac{1}{2}(x * y + y * x)$. Let

$$e_i \circ e_j = \sum_{k=1}^{7} c_{ij}^k e_k \ , \ c_{ij}^k \in F$$

where $c_{ij}^k = c_{ji}^k$ for $i, j, k = 1, 2, \cdots, 7$. Equating $[e_i,e_i \circ e_i] = 0$ gives $\sum_k c_{ii}^k [e_i,e_k] = 0$, which implies by Table 2.2

$$c_{ii}^k = 0 \ , \ 1 \leq i \neq k \leq 7 \ .$$

Using this and substituting $x = y = e_i$, $z = e_j$ in relation (2.5), we have $2 \sum_k c_{ij}^k [e_i,e_k] = c_{ii}^i[e_i,e_j]$, which gives by Table 2.2

(2.62) $$c_{ij}^j = \frac{1}{2} c_{ii}^i, \ c_{ij}^k = 0 \ \text{ for } i, j, k \text{ distinct.}$$

Thus, for a linear form τ on $M(\alpha,\beta,\gamma)$ defined by the rule $\tau(e_i) = \frac{1}{2}c_{ii}^i$, $i = 1, \cdots, 7$, we have $e_i \circ e_j = \tau(e_j)e_i + \tau(e_i)e_j$. Since $x * y = \frac{1}{2}[x,y] + x \circ y$, this gives the desired multiplication (2.61). □

If the product "*" given by (2.61) is flexible, then by Lemma 1.19 τ is identically zero, since $[M,M] = M$. Noting that a finite-dimensional central simple, non-Lie, Malcev algebra is isomorphic to an algebra given by Table 2.2 with $\alpha\beta\gamma \neq 0$, we have

<u>Corollary 2.20</u>. Let A be a finite-dimensional flexible algebra over a field F of characteristic $\neq 2,3$ such that A^- is a central simple, non-Lie, Malcev algebra over F . Then A is isomorphic to a 7-dimensional simple Malcev algebra given by Table 2.2 with $\alpha\beta\gamma \neq 0$. □

<u>Theorem 2.21</u>. Let A be an octonion or para-octonion algebra over a field F of characteristic $\neq 2$. Let "*" be a product defined on A such that $[x,y]^* = [x,y]$. Then the product "*" is third power-associative if and only if there exist a linear form τ on A and a symmetric

bilinear form σ on A such that

$$x * y = \tfrac{1}{2}[x,y] + \tau(x)y + \tau(y)x + \sigma(x,y)e \tag{2.63}$$

for all $x,y \in A$, where e is the identity element of A.

<u>Proof</u>. Write $A = A_0 \oplus Fe$ as in relation (2.19), where A_0 is the set of elements x with $T(x) = 0$. Then A_0 is closed under $[\ ,\ ]$ and A_0^- is isomorphic to an algebra defined by Table 2.2 with $\alpha\beta\gamma \neq 0$. Assume that "$*$" is third power-associative. It follows from (2.5) that Fe is a subalgebra of $(A,*)$. If x and y are elements of A, then let $x \tilde{*} y$ and $\sigma_0(x,y)e$ be the projections of $x * y$ onto A_0 and Fe, respectively. Thus

$$x * y = x \tilde{*} y + \sigma_0(x,y)e\ , \quad x,y \in A\ . \tag{2.64}$$

Since $e * x = x * e$ for all $x \in A$ and $[x,y]^* = [x,y] \in A_0$ for $x,y \in A_0$, σ_0 is a symmetric bilinear form on A. Thus, "$\tilde{*}$" defines a third power-associative product on A_0 such that $x \tilde{*} y - y \tilde{*} x = [x,y]$. Let $x \circ y = \frac{1}{2}(x * y + y * x)$ and $x \tilde{\circ} y = \frac{1}{2}(x \tilde{*} y + y \tilde{*} x)$. Then, by (2.64) $x \tilde{\circ} y$ is the projection of $x \circ y$ onto A_0. Thus, if we set

$$e_i \circ e_j = \sum_{k=0}^{7} c_{ij}^k e_k\ , \quad e_0 = e,$$

then $e_i \tilde{\circ} e_j = \sum_{k=1}^{7} c_{ij}^k e_k$. It follows from this and (2.62) that if $i \neq 0$ and $j \neq 0$, then $c_{ij}^k = 0$ for all distinct $1 \le i, j, k \le 7$. Since Fe is a subalgebra of $(A,\circ)$, $c_{00}^k = 0$ for $k \neq 0$. For $0 \neq i \neq k \neq 0$, $[e_i, e \circ e_i] = 0$ gives $c_{0i}^k = 0 = c_{i0}^k$. From $[e_i, e \circ e_j] = [e \circ e_i, e_j] + [e_i \circ e_j, e]$ by (2.5), we have $c_{0j}^j[e_i,e_j] = c_{0i}^i[e_i,e_j]$, and hence $c_{0i}^i = c_{0j}^j$ for all $i \neq 0$, $j \neq 0$. Therefore, by Lemma 2.19 third power-associativity of "$*$" leaves 44 free parameters

c_{ij}^{0} $(0 \le i \le j \le 7)$, $c_{0i}^{i} = c_{0j}^{j}$ $(1 \le i,j \le 7)$, $c_{ij}^{i} = \frac{1}{2}c_{ii}^{i} (1 \le i \ne j \le 7)$,

and all other parameters are zero.

These relations imply

(2.65)
$$e_i \circ e_j = \frac{1}{2}c_{jj}^{j}e_i + \frac{1}{2}c_{ii}^{i}e_j + c_{ij}^{0}e \ , \ i \ne 0, \ j \ne 0 \ ,$$
$$e \circ e_j = c_{01}^{1}e_j + c_{0j}^{0}e, \ j \ne 0 \ , \ e \circ e = c_{00}^{0}e \ .$$

For $x = \sum_{i=0}^{7} x_i e_i$ and $y = \sum_{i=0}^{7} y_i e_i$, we define

$$\tau(x) = x_0 c_{01}^{1} + \frac{1}{2}\sum_{j=1}^{7} x_j c_{jj}^{j} \ ,$$

$$\sigma(x,y) = \sum_{i,j=0}^{7} x_i y_j c_{ij}^{0} - \tau(x)y_0 - \tau(y)x_0 \ .$$

It is routine to check from (2.65) that τ and σ defined above satisfy the desired relation (2.63). The converse follows from Lemma 2.1. □

<u>Corollary 2.22</u>. Let A be a Hurwitz or para-Hurwitz algebra of dimension > 2 over a field F of characteristic ≠ 2 , and let "*" be a product defined on A such that $[x,y]^* = [x,y]$. Then the product "*" is third power-associative if and only if it is given by relation (2.63) for some linear form τ on A and symmetric bilinear form σ . The product "*" is flexible if and only if

(2.66) $$x * y = \frac{1}{2}[x,y] + \sigma(x,y)e, \ e * y = y * e = \theta y, \ e * e = \zeta e$$

for scalars $\theta, \zeta \in F$, $x,y \in A_0$, and $\sigma(x,[y,z]) = \sigma([x,y],z)$ for all $x,y,z \in A$. Assume that F has characteristic ≠ 2,3 . Then the product "*" is third and fourth power-associative if and only if it is power-associative if and only if τ and σ satisfy one of the three conditions (a), (b), and (c) in Theorem 2.14, with $I = e$. If $(A,*)$ is power-associative, then $(A,*)$ is Jordan-admissible as well as Malcev-admissible, and $(A,*)$ and $(A,\circ)$ are quadratic algebras.

Proof. Since any Hurwitz algebra is a subalgebra of an octonion algebra and contains the identity element e, the first part follows from Theorem 2.21. If $(A,*)$ is flexible, then the same calculation as in Corollary 2.11 shows that relation (2.66) holds for $(A,*)$. The last part follows from Theorems 2.14 and 2.18, and remarks following Theorem 2.18. □

Corollary 2.22 for a quaternion algebra was proved by Benkart [1] and the octonion case is a result of Myung [9]. We show that the principal results established for a matrix algebra and an octonion algebra hold for finite-dimensional central simple alternative algebras.

Theorem 2.23. Let A be a finite-dimensional central simple alternative algebra over a field F of characteristic $\neq 2,3$, and let "$*$" be a product on A such that $[x,y]^* = [x,y]$. The algebra $(A,*)$ is third power-associative if and only if relation (2.35) holds for some linear form $\tau : A \to F$, some $\beta \in F$, and some symmetric bilinear form $\sigma : A \times A \to F$, where I is the identity element of A. The algebra $(A,*)$ is third and fourth power-associative if and only if it is power-associative if and only if one of the four cases (a) - (d) of Theorem 2.14 and Lemma 2.16 holds. If $(A,*)$ is power-associative, then it is Malcev- and Jordan-admissible, and $(A,\circ) \cong A^+$ in case (d), and otherwise $(A,*)$ is a quadratic algebra.

Proof. If A satisfies the hypotheses of Theorem 2.23, then A is either associative or an octonion algebra (Schafer [3,p.56]). The latter case has been treated in Corollary 2.22. Assume that A is associative and let $(A,*)$ be third power-associative. Then there exists a finite extension E of F such that the scalar extension $A_E = E \otimes_F A$ of A is isomorphic to M(n,E) for some integer n . The product "*" may be extended to A_E by linearity, and it satisfies the properties that $[x,y]^* = [x,y]$ and that "*" is third power-associative by (2.5). Therefore, by Theorem 2.17, there exist a linear form $\tau : A_E \to E$, a scalar $\beta \in E$, and a symmetric bilinear form $\sigma : A_E \times A_E \to E$ such that (2.35) holds. When "*" is restricted to A , relation (2.35) holds for A with $\tau : A \to E$, $\beta \in E$, and $\sigma : A \times A \to E$. We show that $\tau(A) \subseteq F$, $\beta \in F$, and $\sigma(A,A) \subseteq F$.

If $n \geq 3$, then A_E is not a quadratic algebra since $A_E \cong M(n,E)$. Consequently, A is not a quadratic algebra and hence there exists an $x \in A$ such that x, x^2, I are linearly independent in A . Since $x \circ x$ lies in A , from $x \circ x = 2\tau(x)x + 2\beta x^2 + \sigma(x,x)I$ we see that β is an element of F . But then for any $n \geq 2$, relation (2.35) shows that $\tau(x)$ and $\sigma(x,y)$ are in F for all $x,y \in A$.

Suppose that (A,*) is third and fourth power-associative. Then $(A,\circ)$ is fourth power-associative and by the above the product "*" is given by relation (2.35). If $\beta = 0$, then by Theorem 2.14 one of the cases (a) - (c) of Theorem 2.14 holds for A and hence $(A,\circ)$ is a quadratic Jordan algebra by Theorem 2.18. In these cases, (A,*) is power-associative also (see remarks following Theorem 2.18). If $\beta \neq 0$, then $A_E \cong M(n,E)$ for some $n \geq 3$. Since $(A_E,\circ)$ is also fourth power-associative, condition (d) of Theorem 2.17 holds for $(A_E,\circ)$ and hence for $(A,\circ)$. Conversely, if one of the four cases (a) - (d) is satisfied, then as noted in the proof of Theorem 2.18 (A,*) is power-associative. If case (d) holds, it follows from Lemma 2.16 that $(A,\circ)$ is isomorphic to A^+ . □

The product "*" given by (2.35) where $\beta = 0$ and σ is identically zero appeared in one of the earliest papers by Weiner [1] on Lie-admissible algebras. Weiner [1] investigated power-associative Lie-admissible products on the 3-dimensional simple cross-product Lie algebra, and showed that each such multiplication has the form given by (2.35) for a linear form τ on A, where $\beta = 0$ and σ is identically zero. Benkart [2] generalized Weiner's result to arbitrary finite-dimensional split simple Lie algebra of characteristic 0. Benkart's results will be discussed in the next section.

We next determine power-associative products "$*$" defined on a pseudo-octonion algebra such that $[x,y]^* = [x,y]$. We determine these products separately for characteristic $\neq 2,3$ and characteristic 3. For the former case, the following general result is useful.

Lemma 2.24. Let $A = s\ell(n+1,F)$ be the Lie algebra of $(n+1) \times (n+1)$ trace zero matrices over a field F of characteristic $p \neq 2,3$ such that $n+1$ is not divisible by p. Let "$*$" be a product on A such that $[x,y]^* = [x,y]$. The product "$*$" is third power-associative if and only if there exist a linear form τ on A and a scalar $\beta \in F$ such that

$$x * y = \frac{1}{2}[x,y] + \tau(x)y + \tau(y)x + \beta x \# y \tag{2.67}$$

where $x \# y$ is defined by (2.38). The product "$*$" is power-associative if and only if it is third and fourth power-associative if and only if relation (2.67) holds and $\beta = 0$.

Proof. Let $x \circ y = \frac{1}{2}(x * y + y * x)$. Under the assumption, we show that $x \circ y$ is given by

$$x \circ y = \tau(x)y + \tau(y)x + \beta x \# y .$$

Assume that "$*$" is third power-associative. We note that the assumption is equivalent to relation (2.5) and that fourth power-associativity of "$*$" is equivalent to that of "$\circ$" (see Lemma 1.10). Since $M(n+1,F) = A \oplus FI$, the product "$\circ$" can be extended to $M(n+1,F)$ by defining $I \circ x = I \circ x = 0$ for all matrices x, and relation (2.5) holds for $M(n+1,F)$. Thus, by Theorem 2.18, $x \circ y$ is specified by

$$x \circ y = \tau(x)y + \tau(y)x + \beta(xy + yx) + \sigma(x,y)I ,$$

where β is a scalar, τ a linear form, and σ a symmetric bilinear form, and where $\beta = 0$ if $n = 1$. When x and y are in A, $x \circ y \in A$ and hence the trace of $x \circ y$ is zero. This implies that $2\beta(\operatorname{tr} xy) + (n+1)\sigma(x,y)I = 0$ and $\sigma(x,y) = \frac{-2\beta}{n+1}(\operatorname{tr} xy)$. Therefore,

$$x \circ y = \tau(x)y + \tau(y)x + \beta(xy + yx - \frac{2}{n+1}(\operatorname{tr} xy)I) = \tau(x)y + \tau(y)x + \beta x \# y ,$$

as desired. Conversely, it is clear that the product "$*$" given by (2.67) is third power-associative.

Suppose now that "$*$" is third and fourth power-associative. Then the product "$\circ$" extended to $M(n+1,F)$ satisfies the fourth power identity (2.6) if and only if (2.6) holds for A. Thus, by Theorem 2.17, relation (2.6) on $M(n+1,F)$ implies that either $\beta = 0$ or $\tau(I) + 2\beta \neq 0$ (case (d) of Lemma 2.16). Since

$$0 = I \circ y = \tau(I)y + \tau(y)I + 2\beta y + \sigma(I,y)I$$

for all y, it must be that $\tau(I) + 2\beta = 0$, and hence only the case $\beta = 0$ can occur when (2.6) is satisfied. In this case, "$*$" is power-associative and the proof is complete. □

<u>Corollary 2.25</u>. Let A be the same as in Lemma 2.24. If "$*$" is flexible Lie-admissible product on A such that $(A,*)^-$ is isomorphic to

the Lie algebra $A = s\ell(n+1,F)$, then "$*$" is given by (2.67), where the linear form τ on A is identically zero. If $(A,*)$ is flexible power-associative such that $(A,*)^- \cong s\ell(n+1,F)$, then $x * y = \frac{1}{2}[x,y]$ and $(A,*)$ is a Lie algebra.

Proof. Under the hypotheses, flexibility of "$*$" is equivalent to relation (2.36), $[x \circ y,x] = x \circ [y,x]$. If "$*$" is flexible, then in particular it is third power-associative, so that $x \circ y = \tau(x)y + \tau(y)x + \beta x \# y$, and that may be substituted into (2.36) to give

$$\tau([x,y])x + \beta[x \# y,x] + \beta[x,y] \# x = 0 .$$

The last two terms cancel, since the linear mapping ad_x is a derivation on the associative algebra $M(n+1,F)$, and thus a derivation on A with product "#" by (2.38), implying $[x \# y,x] = x \# [y,x]$. Hence if $(A,*)$ is flexible, then $\tau([x,y]) = 0$ for all $x,y \in A$. Since $[A,A] = A$, it must be that $\tau = 0$, as claimed. If, in addition, "$*$" is power-associative, then by Lemma 2.24 $\beta = 0$ and $x * y = \frac{1}{2}[x,y]$. □

Let A be a pseudo-octonion algebra of Definition 2.3. Since A^- is isomorphic to $s\ell(3,F)$, third power-associative and power-associative products on A are determined by Lemma 2.24 and Corollary 2.25, as special cases. If $x * y$ is the product given by (2.21), then setting $\alpha = 1/(2\mu - 1)$ gives $\alpha x * y = \frac{1}{2}[x,y] + \frac{1}{2}\alpha x \# y$, and hence $(A,*)$ is isomorphic to an algebra described in Corollary 2.25, where $n = 2$. For the characteristic 3 case, we recall the multiplication table (2.27) of the pseudo-octonion algebra A of characteristic 3 . Let x_{ij}, i, j = 0, 1, 2, be the basis of A given by Definition 2.4, where $x_{00} = 0$. It is routine to check from (2.27) that the products $[x,y] = xy - yx$ and $x \cdot y = \frac{1}{2}(xy + yx)$ are described by

(2.68)
$$[x_{ij}, x_{k\ell}] = (i\ell - kj)x_{i+k,j+\ell} ,$$
$$x_{ij} \cdot x_{k\ell} = - x_{i+k,j+\ell} .$$

Theorem 2.26. Let A be the pseudo-octonion algebra over a field F of characteristic 3 whose multiplication is given by (2.27). Let "$*$" be a product defined on A such that $[x,y]^* = [x,y]$. Then $(A,*)$ is third power-associative if and only if the product "$*$" is given by

(2.69) $$x * y = \frac{1}{2}[x,y] + \tau(y)x + \tau(x)y + \beta x \cdot y ,$$

where τ is a linear form on A, and β is a scalar in F. The product "$*$" is flexible if and only if (2.69) holds and τ is identically zero.

Proof. Suppose that "$*$" is third power-associative. Recall that this hypothesis is equivalent to the identity $[x, x \circ x] = 0$, where $x \circ y = \frac{1}{2}(x * y + y * x)$. To facilitate computation, we label the basis x_{ij} of A as

(2.70)
$$e_1 = x_{01},\ e_2 = x_{10},\ e_3 = x_{11},\ e_4 = x_{21} ,$$
$$e_{-1} = x_{02},\ e_{-2} = x_{20},\ e_{-3} = x_{22},\ e_{-4} = x_{12} .$$

It is clear that if $e_t = x_{ij}$, then $e_{-t} = x_{(-i)(-j)}$. We determine the product "$\circ$" such that $[x, x \circ x] = 0$ for all $x \in A$. Let

$$e_i \circ e_j = \sum_k c_{ij}^k e_k,\ c_{ij}^k \in F .$$

It follows from (2.68) and (2.70) that $[e_i, e_j] = 0$ if and only if $j = \pm i$. Using this and $[e_i, e_i \circ e_i] = 0$, we have $c_{ii}^k = 0$ for all i, k with $i \neq \pm k$, and hence

(2.71) $$e_i \circ e_i = c_{ii}^i e_i + c_{ii}^{-i} e_{-i} .$$

Substituting $x = y = e_i$ and $z = e_{-i}$ in relation (2.5) gives $2[e_i, e_i \circ e_{-i}] = [e_i \circ e_i, e_{-i}] = 0$ by (2.71), and so $c^k_{i(-i)} = 0$ for all $i \neq \pm k$, which implies

(2.72) $$e_i \circ e_{-i} = c^i_{i(-i)} e_i + c^{-i}_{i(-i)} e_{-i} .$$

Assume now that $i \neq \pm j$. It is readily seen from (2.68) that for any $i \neq \pm j$ there is a unique integer $v(i,j) \neq 0$, $-4 \leq v(i,j) \leq 4$, such that

(2.73) $$v(i,j) = v(j,i) \neq \pm i, \neq \pm j ,$$
$$[e_i, e_{v(i,j)}] = -[e_{-i}, e_j] .$$

In fact, if $e_i = x_{pq}$ and $e_j = x_{st}$, then equating (2.73) gives

(2.74) $$e_{v(i,j)} = x_{p+s,q+t} .$$

Relation (2.5) implies that $2[e_i, e_i \circ e_j] = [e_i \circ e_i, e_j]$, and hence $2 \sum c^k_{ij}[e_i, e_k] = c^i_{ii}[e_i, e_j] + c^{-i}_{ii}[e_{-i}, e_j]$ by (2.71). When (2.68) and (2.73) are applied to this, we have

(2.75) $$e_i \circ e_j = c^i_{ij} e_i + c^{-i}_{ij} e_{-i} + \frac{1}{2} c^i_{ii} e_j - \frac{1}{2} c^{-i}_{ii} e_{v(i,j)} .$$

Since $e_i \circ e_j = e_j \circ e_i$ and $v(i,j) = v(j,i)$, by symmetry (2.75) gives

$$c^i_{ij} = \frac{1}{2} c^j_{jj}, \quad c^{-i}_{ij} = 0 , \quad i \neq \pm j ,$$

and $c^{-j}_{jj} = c^{-i}_{ii}$ for all i, j. Letting $\beta = -c^{-j}_{jj}$ and $c_i = c^i_{ii}$ for each i, (2.71) and (2.75) reduce to

(2.76) $$e_i \circ e_i = c_i e_i - \beta e_{-i},$$
$$e_i \circ e_j = \frac{1}{2} c_j e_i + \frac{1}{2} c_i e_j - \beta e_{v(i,j)}, \quad i \neq \pm j .$$

We substitute (2.72) and (2.76) in $[e_i, e_{-j} \circ e_j] = [e_i \circ e_{-j}, e_j]$ $+ [e_i \circ e_j, e_{-j}]$ to obtain

$$c^{j}_{j(-j)}[e_i,e_j] + c^{-j}_{j(-j)}[e_i,e_{-j}] = \frac{1}{2}c_{-j}[e_i,e_j] + \frac{1}{2}c_j[e_i,e_{-j}]$$

$$- \beta[e_{v(i,j)},e_j] - \beta[e_{v(i,j)},e_{-j}] .$$

Since the last two terms of the right side cancel each other, this implies $\frac{1}{2}c_j = c^{-j}_{j(-j)}$ for all j. Thus, (2.72) reduces to

$$e_i \circ e_{-i} = \frac{1}{2}c_{-i}e_i + \frac{1}{2}c_i e_{-i} .$$

Hence we have only 9 free parameters and all other parameters are zero, when third power-associativity is imposed on $(A,*)$. Converting (2.76) and the last relation in terms of the basis x_{ij} and using (2.68) and (2.74), one has

$$x_{ij} \circ x_{k\ell} = \frac{1}{2}c_{ij}x_{k\ell} + \frac{1}{2}c_{k\ell}x_{ij} + \beta x_{ij} \cdot x_{k\ell}$$

for all (i,j), (k,ℓ), since $-x_{(-i)(-j)} = x_{ij} \cdot x_{ij}$. Here we have set $c_i = c_{pq}$ if $e_i = x_{pq}$. This gives relation (2.69), as claimed, where τ is a linear form on A defined by $\tau(x_{ij}) = \frac{1}{2}c_{ij}$.

Conversely, let $(A,*)$ have a multiplication defined by (2.69). Then $[x,x*x]^* = [x,x*x] = [x,2\tau(x)x + \beta x \cdot x] = \beta[x,x \cdot x] =$ $\beta[x,x^2] = 0$, since A is third power-associative (in fact, A is flexible). Thus $(A,*)$ is third power-associative. Note that flexibility of $(A,*)$ is equivalent to the identity $[x \circ y,x] = x \circ [y,x]$, which converts (2.69) into the identity

$$\tau([x,y])x + \beta[x \cdot y,x] + \beta x \cdot [x,y] = 0 .$$

Since A is flexible, the adjoint map ad_x is a derivation on A^+, and hence the last two terms vanish. This gives $\tau([A,A]) = \tau(A) = 0$. □

We see from (2.68) that A^- is one of the algebras described in Theorem 11 of Albert and Frank [1], and hence A^- is a central simple Lie algebra (this is also easy to check directly). It is shown that the Killing form of A^- vanishes identically on A^- and A^- is not restricted (Okubo and Osborn [2] and Block [1]).

Corollary 2.27. Let A be the pseudo-octonion algebra over a field F of characteristic 3 where F is not the prime field, and let "$*$" be a third power-associative product on A such that $[x,y]^* = [x,y]$. Then $(A,*)$ is power-associative if and only if $(A,*)^+ = (A,\circ)$ is fourth power-associative.

Proof. By Theorem 2.26, "$*$" is given by (2.69), so that

(2.77) $$x \circ x = 2\tau(x)x + \beta x^2, \quad x \circ x \circ x = (\tau(x)^2 + \beta\tau(x^2) + \beta^2(x,x))x ,$$

where $(\, , \,)$ is a symmetric bilinear form on A satisfying relation (2.13). We first show that "$\circ$" is power-associative. In view of a result of Leadley and Ritchie [1], it suffices to verify that the fifth power identity $(x \circ x \circ x) \circ (x \circ x) = ((x \circ x \circ x) \circ x) \circ x$ holds. But, using (2.13), (2.69) and (2.77), it is easy to see that this identity is satisfied, and hence $(A,\circ)$ is power-associative. We next linearize (2.13) to the relation $x(yz) + z(yx) = (xy)z + (zy)x = 2(x,z)y$, which implies the identity

(2.78) $$x^2x^2 = 2(x,x^2)x - (x,x)x^2 ,$$

using (2.13). It follows from (2.78), (2.13) and (2.77) that the nth power of each x in $(A,\circ)$ is spanned by x and x^2 . This implies by

(2.69) that the nth powers of each x in $(A,\circ)$ and $(A,*)$ coincide, and hence $(A,*)$ is power-associative. □

Using (2.77) and (2.78), we find that the fourth power identity of "∘" is equivalent to the relation

$$(2.79)\quad \beta[\tau(x)\tau(x^2) + \beta\tau(x)(x,x) + \beta^2(x,x^2)]x = \beta[\tau(x)^2 + \beta\tau(x^2) + \beta^2(x,x)]x^2 \ .$$

Recall that the bilinear form $(\ ,\ )$ on A defined by $(x_{ij},x_{k\ell}) = -\delta_{i(-k)}\delta_{j(-\ell)}$ satisfies relation (2.13) (this can be verified directly from (2.27)). Using this and substituting $x = x_{ij}$ into (2.79),

$$(2.80)\qquad \beta^3 = \beta\tau(x_{ij})\tau(x_{(-i)(-j)}),\ \beta\tau(x_{ij})^2 = \beta^2\tau(x_{(-i)(-j)})$$

for all (i,j). Assume that (2.79) holds. If $\tau(x_{ij}) = 0$ for some x_{ij}, then $\beta = 0$ by (2.80). If $\tau(x_{ij}) \neq 0$ for all (i,j), then we see from (2.80) that either $\beta = 0$, or $\tau(x_{ij})^3 - \tau(x_{(-i)(-j)})^3 = [\tau(x_{ij}) - \tau(x_{(-i)(-j)})]^3 = 0$. In the latter case, $\beta = \tau(x_{ij})$ for all (i,j). Finally, since $(x_{ij} \cdot x_{ij}) \cdot (x_{ij} \cdot x_{ij}) = -x_{ij}$ and $(x_{ij} \cdot x_{ij}) \cdot x_{ij} = 0$, A^+ is not fourth power-associative, and A is not Jordan-admissible.

The results in this section are based on work of Myung [9].

2.5. Power-associative products on simple Lie and Malcev algebras

In this section we determine all finite-dimensional power-associative Malcev-admissible algebras A over a field F of characteristic 0 such that A^- is a split simple Malcev algebra over F. Since the characteristic is zero, it suffices to determine third and fourth power-associative products on A satisfying these properties (Lemma 1.11). Our principal result is that all third power-associative products on these algebras are given by relation (2.67) where $\beta x \# y = 0$ for all x, y, when A^- is not a Lie algebra of type $A_n (n \geq 2)$. In light of Lemma 2.24, our major effort is devoted to determining third power-associative Malcev-admissible algebras A such that A^- is a split simple Malcev algebra which is not a Lie algebra of type A_n.

There exists only one split simple, non-Lie, Malcev algebra over a field of characteristic $\neq 2,3$. This algebra is a 7-dimensional, anti-commutative algebra with basis e_0, e_{-i}, e_i $(i = 1,2,3)$ whose multiplication is given by

$$(2.81) \qquad \begin{aligned} &[e_0, e_{\pm i}] = \pm 2e_{\pm i}, \ [e_i, e_{-i}] = e_0, \ i = 1, 2, 3 , \\ &[e_{\pm i}, e_{\pm j}] = \pm 2e_{\mp k}, \ (ijk) = (123), (231), (312) , \end{aligned}$$

where all other products are zero (Sagle [1,2] and Carlsson [1]). This algebra is isomorphic to A^-/Fe for some split octonion algebra A over F (Kuzmin [1]). We denote by C_0^- the algebra defined by (2.81).

Let A be a Malcev-admissible algebra with multiplication denoted by "$*$". Since $x * y = \frac{1}{2}[x,y] + x \circ y$ for some commutative product "$\circ$" on A and third power-associativity of "$*$" is equivalent to the identity (2.3) or to (2.5), we focus on determining all commutative products "$\circ$" on A satisfying relation (2.3). When A^- is a finite-dimensional split simple Lie algebra over a field F of characteristic 0, the main tool to be employed is the Dynkin diagram or the Cartan matrix relative to a system of simple roots for A^-, and the expression of highest root in terms of simple roots. If $\alpha_1, \cdots, \alpha_n$ form a system of simple roots for A^- in the ordering specified in Humphreys [1,p.66] or

Myung [6,p.193], then the highest (long) root relative to that system is described by Table 2.3.

TABLE 2.3

Lie algebra	Highest root
A_n	$\alpha_1 + \alpha_2 + \cdots + \alpha_n$
B_n	$\alpha_1 + 2\alpha_2 + \cdots + 2\alpha_n$
C_n	$2\alpha_1 + 2\alpha_2 + \cdots + 2\alpha_{n-1} + \alpha_n$
D_n	$\alpha_1 + 2\alpha_2 + \cdots + 2\alpha_{n-2} + \alpha_{n-1} + \alpha_n$
E_6	$\alpha_1 + 2\alpha_2 + 2\alpha_3 + 3\alpha_4 + 2\alpha_5 + \alpha_6$
E_7	$\alpha_1 + 2\alpha_2 + 3\alpha_3 + 4\alpha_4 + 3\alpha_5 + 2\alpha_6 + \alpha_7$
E_8	$2\alpha_1 + 3\alpha_2 + 4\alpha_3 + 6\alpha_4 + 5\alpha_5 + 4\alpha_6 + 3\alpha_7 + 2\alpha_8$
F_4	$2\alpha_1 + 3\alpha_2 + 4\alpha_3 + 2\alpha_4$
G_2	$3\alpha_1 + 2\alpha_2$

Lemma 2.28. Let A be a Malcev-admissible algebra over a field F of characteristic $\neq 2$, and let H be an abelian Cartan subalgebra of A^- which permits a Cartan decomposition $A^- = H + \sum_{\alpha \neq 0} A_\alpha$ such that each $ad_h (h \in H)$ acts diagonally on the root space A_α for all roots α. Let "$\circ$" be a commutative product on A satisfying relation (2.3). Then

(1) H is a subalgebra of $(A,\circ)$.

(2) For all roots α, β, and $h, h' \in H$, $e_\alpha \in A_\alpha$, there exist scalars $\tau_\alpha(h)$, $\tau_\alpha(h')$ and an element $u_\alpha(h,e_\alpha) \in H$, depending on h, e_α, α, such that

(2.82) $$h \circ e_\alpha = \tau_\alpha(h)e_\alpha + u_\alpha(h,e_\alpha) ,$$

(2.83) $$\tau_\alpha(h')\alpha(h) + \tau_\alpha(h)\alpha(h') = \alpha(h \circ h') ,$$

(2.84) $$[h,e_\alpha \circ e_\beta] = \alpha(u_\beta(h,e_\beta))e_\alpha + \beta(u_\alpha(h,e_\alpha))e_\beta + (\tau_\alpha(h)-\tau_\beta(h))[e_\alpha,e_\beta] .$$

(3) For each root α and $e_\alpha \in A_\alpha$, $Fe_\alpha + H$ is a subalgebra of $(A,\circ)$.

Proof. (1) For each $h \in H$, the condition $[h,h \circ h] = 0$ implies that $h \circ h = h_1 + \sum_{\beta(h)=0} a_\beta e_\beta$ where $h_1 \in H$ and $a_\beta \in F$. Let $h' \in H$ and let $h \circ h' = h_1' + \sum_\gamma b_\gamma e_\gamma$. Then by relation (2.5)

$$0 = 2[h,h \circ h'] + [h',h \circ h] = 2 \sum_{\gamma(h)\neq 0} \gamma(h)b_\gamma e_\gamma + \sum_{\beta(h)=0} \beta(h')a_\beta e_\beta .$$

Since the set of e's appearing in the first sum is disjoint from that in the second, $\beta(h')a_\beta = 0$ for all $h' \in H$. Thus $a_\beta = 0$ for all such β and so, $h \circ h \in H$ for each $h \in H$. Since $h \circ h' = \frac{1}{2}[(h + h') \circ (h + h') - h \circ h - h' \circ h']$, we have that H is a subalgebra of $(A,\circ)$.

(2) Consider the relation $2[h,h \circ e_\alpha] + [e_\alpha,h \circ h] = 0$ by (2.5). Since the second term of this is a multiple of e_α by (1), we get $h \circ e_\alpha = \tilde{h} + \sum_{\beta(h)=0} a_\beta e_\beta + be_\alpha$, $b \in F$, and similarly, $h' \circ e_\alpha = \tilde{h}' + \sum_{\gamma(h')=0} c_\gamma e_\gamma + de_\alpha$. Substituting these into (2.5) gives

$$\sum_{\substack{\beta \neq \alpha \\ \beta(h)=0}} \beta(h')a_\beta e_\beta + \sum_{\substack{\gamma \neq \alpha \\ \gamma(h')=0}} c_\gamma \gamma(h)e_\gamma = 0 .$$

For a given $\beta \neq \alpha$ arising from $h \circ e_\alpha$, the element $h' \in H$ can be chosen so that $\beta(h') \neq 0$. Then e_β just appears in the first term, so that its coefficient a_β must be zero. It follows from this that $h \circ e_\alpha \in H + Fe_\alpha$, implying (2.82). Noting that $[h \circ h',e_\alpha] =$

$[h',h \circ e_\alpha] + [h,h' \circ e_\alpha]$ by (2.5), relation (2.83) follows from (2.82). Relations (2.5) and (2.83) can be combined to compute

$$[h,e_\alpha \circ e_\beta] = [h \circ e_\beta, e_\alpha] + [h \circ e_\alpha, e_\beta]$$

$$= \tau_\beta(h)[e_\beta,e_\alpha] + \alpha(u_\beta(h,e_\beta))e_\alpha + \tau_\alpha(h)[e_\alpha,e_\beta] + \beta(u_\alpha(h,e_\alpha))e_\beta ,$$

which gives relation (2.84).

(3) When $e_\alpha = e_\beta$, (2.84) implies that $e_\alpha \circ e_\alpha \in Fe_\alpha + H$, and hence the result follows from (2.82) and part (1). □

Corollary 2.29. Let A be a Malcev-admissible algebra over a field F of characteristic $\neq 2,3$ such that A^- is isomorphic to the 7-dimensional split simple Malcev algebra over F. Then any commutative product "$\circ$" on A satisfying relation (2.3) is given by

$$\text{(2.85)} \qquad x \circ y = \tau(y)x + \tau(x)y$$

for some linear form τ on A.

Proof. Note that A^- is given by the multiplication (2.81), and that $A^- = Fe_0 + A_\alpha + A_{-\alpha}$ is the Cartan decomposition relative to the Cartan subalgebra $H = Fe_0$, where $A_\alpha = Fe_1 + Fe_2 + Fe_3$ and $A_{-\alpha} = Fe_{-1} + Fe_{-2} + Fe_{-3}$ are root spaces for the roots α, $-\alpha$ defined by $\alpha(e_0) = 2$. Let $e_0 \circ e_0 = ce_0$ by Lemma 2.28 (1). When $h = h' = e_0$, (2.83) gives $\tau_\alpha(e_0) = \tau_{-\alpha}(e_0) = \frac{1}{2}c$. Thus, by (2.82) and (2.84), we can let $e_0 \circ e_i = \frac{1}{2}ce_0 + c_ie_0$ for $i \neq 0$ and $e_i \circ e_j = b_{ij}e_0 + c_je_i + c_ie_j$ for $i \neq 0$ and $j \neq 0$, where we have used the relations $c_{\pm i}e_0 = u_{\pm\alpha}(e_0,e_{\pm i})$ and $\alpha(u_{\pm\alpha}(e_0,e_{\pm i})) = 2c_{\pm i}$ for $i \neq 0$. But $[e_i \circ e_j, e_i] = [e_i, e_i \circ e_j] + [e_j, e_i \circ e_i]$ implies $b_{ij} = 0$ for all $i, j \neq 0$. Thus, defining a linear form τ by $\tau(e_0) = \frac{1}{2}c$ and $\tau(e_i) = c_i$ for $i \neq 0$ gives relation (2.85), as desired. □

If each root space A_α is one-dimensional, then we may assume that the element $u_\alpha(h, e_\alpha) \in H$ in (2.82) depends only on h and α, and hence we set

$$u_\alpha(h, e_\alpha) = u_\alpha(h) .$$

Lemma 2.30. Let A be a finite-dimensional Lie-admissible algebra over a field F of characteristic 0 such that A^- is a split simple Lie algebra over F which is not of type A_n. Let H be a split Cartan subalgebra of A^-. Suppose that "∘" is a commutative product on A satisfying relation (2.3). Then there is a linear form τ on H such that relation (2.85) holds for H.

Proof. Let $A^- = H \oplus \sum_\alpha Fe_\alpha$ be the Cartan decomposition of A^-. Suppose that $\{\alpha_1, \cdots, \alpha_n\}$ is any basis of the dual space H^* of H consisting of roots, and let $\{h_1, \cdots, h_n\}$ be the basis of H dual to the α's. By substituting $h = h_i$, $h' = h_j$ and $\alpha = \alpha_k$ for $k = 1, \cdots, n$ into (2.83), we find that

$$(2.86) \qquad h_i \circ h_j = c_{ij} h_i + c_{ji} h_j, \quad 1 \le i, j \le n ,$$

where $c_{ij} = \tau_{\alpha_i}(h_j)$. Relation (2.86) holds for any choice of basis of H^* comprised of roots together with its dual basis. In particular, let $\{\alpha_1, \cdots, \alpha_n\}$ denote a fixed ordered basis of simple roots with $\{h_1, \cdots, h_n\}$ as its dual and with $h_i \circ h_j$ given by (2.86). Let $C = (c_{ij})$ denote the $n \times n$ matrix determined by the coefficients c_{ij} of (2.86). For the proof of the Lemma, it suffices to verify that each column of C is a constant.

Suppose first that A^- is of type B_n or C_n for $n \ge 2$, F_4 or G_2. Then the Dynkin diagram consists of a chain with the i th node joined to the $(i + 1)$ st node, and hence $\alpha_i + \alpha_{i+1}$ is a root for all $i = 1,$

$\cdots, n-1$. Then roots $\alpha_1, \cdots, \alpha_{i-1}, \alpha_i + \alpha_{i+1}, \alpha_{i+1}, \cdots, \alpha_n$ form a basis of H , and it is easy to see that $\{h_1,\cdots,h_i,h_{i+1} - h_i,h_{i+2},\cdots,h_n\}$ is its dual basis. Therefore, by (2.86)

$$(h_{i+1} - h_i) \circ h_j = \mu(h_{i+1} - h_i) + \nu h_j$$

must hold for some $\mu, \nu \in F$ and for each $j \neq i+1$. Equating this relation with (2.86), the value of μ gives

(2.87) $$c_{i+1,j} = c_{ij}$$

for all $j \neq i$, $i+1$. Similarly, the relation $(h_{i+1} - h_i) \circ (h_{i+1} - h_i) = \lambda(h_{i+1} - h_i)$ for some $\lambda \in F$ combines with (2.86) to give the relation

(2.88) $$c_{i+1,i+1} - c_{i+1,i} = c_{i,i+1} - c_{ii}$$

for all $i = 1, \cdots, n-1$. Relation (2.87) shows that, in a given column of matrix $C = (c_{ij})$, all the entries above the diagonal entry are equal, and also all the ones below the diagonal entry. Relation (2.88) gives an additional relationship among the entries of C .

Let $\gamma = m_1\alpha_1 + \cdots + m_n\alpha_n$ be the highest root relative to the basis $\{\alpha_1,\cdots,\alpha_n\}$ of simple roots. In the case that $\bar{A}$ is of type B_n, C_n, F_4 or G_2 , Table 2.3 shows that there is an index k such that $m_i \neq m_k$ for any $i \neq k$. If $i \neq k$, then let $q_i = m_i m_k^{-1}$ and consider the basis $\{\alpha_1,\cdots,\alpha_{k-1},\gamma,\alpha_{k+1},\cdots,\alpha_n\}$ consisting of roots which has as its dual

$$\{h_1 - q_1h_k,\cdots,h_{k-1} - q_{k-1}h_k,m_k^{-1}h_k,\cdots,h_n - q_nh_k\} .$$

Then for each $i \neq k$, the element $h_i - q_ih_k$ squares to a multiple of itself, which combines with (2.86) to yield the condition

(2.89) $$c_{ii} - q_ic_{ik} = c_{ki} - q_ic_{kk}$$

for all $i \neq k$. Since $q_i \neq 1$ for any $i \neq k$, equation (2.88) together with (2.89) for $i = k - 1$, $k + 1$ implies $c_{kk} = c_{k-1,k}$ and $c_{kk} = c_{k+1,k}$, which shows by (2.87) that $c_{1k} = c_{2k} = \cdots = c_{nk}$. This in turn implies via (2.89) that $c_{ii} = c_{ki}$ for all i . Thus when $i < k$, in light of (2.87) all entries below and on the main diagonal in the i th column are equal, and for $k < i$ the same result holds for the entries above and on the main diagonal. If $i < k$, then from $c_{ii} - c_{i,i-1} = c_{i-1,i} - c_{i-1,i-1}$, it follows that $c_{ii} = c_{i-1,i}$, so that for each $i < k$, all the entries in i th column are equal. Similarly, for the columns to right of the k th column, we can argue with the subscripts i and $i + 1$. Therefore, we have established that when A^- is of type B_n or C_n where $n \geq 2$, F_4, or G_2 , then $c_{1i} = c_{2i} = \cdots = c_{ni}$ for each $i = 1, 2, \cdots, n$. Denoting that common value by c_i , (2.86) is written as

$$h_i \circ h_j = c_j h_i + c_i h_j$$

for all i, j . Defining a linear form τ on H by $\tau(h_i) = c_i$ for $i = 1, \cdots, n$ and extending it linearly to H , we have relation (2.85) holding for H .

Suppose next that A^- is one of the residual types $D_n (n \geq 4)$, E_6 E_7, or E_8 . In each of these cases, we may assume that the simple roots are ordered in such a way that $\alpha_1, \cdots, \alpha_{n-1}$ form a simple chain in each Dynkin diagram, and a branch node is labeled by α_n which is joined to the simple root α_k for some $k \neq 1, n$. In the case that A^- is of type E_6, E_7 or E_8 , this ordering is slightly different from one specified in Table 2.3 (Humphreys [1,p.58]). However, the important fact that the unique maximal coefficient of the highest root γ relative to the present ordering takes place at the simple root α_k remains unchanged. The same argument as above shows that for $1 \leq i \leq n - 2$, equations (2.87) and (2.88) hold. Using the basis $\{\alpha_1,\cdots,\alpha_{k-1},\alpha_k + \alpha_n,\alpha_{k+1},\cdots,\alpha_n\}$ with $\{h_1,\cdots,h_{k-1},h_k,h_{k+1},\cdots,h_n - h_k\}$ as its dual, we see that $(h_n - h_k) \circ h_j$ is a linear combination of $h_n - h_k$, h_j, and $h_n - h_k$ squares to a multiple of itself. It follows from this and (2.86) that

$$c_{nj} = c_{kj} \quad \text{for all} \quad j \neq k,\ n\ ,$$

(2.90)

$$c_{nn} - c_{nk} = c_{kn} - c_{kk}\ .$$

In the cases E_6, E_7, and E_8, the highest root $\gamma = m_1\alpha_1 + \cdots + m_n\alpha_n$ has its maximal coefficient m_k (Table 2.3 and Humphreys [1,p.58]). Thus, arguing just as in the preceding cases gives relation (2.89):

$$c_{ii} - q_i c_{ik} = c_{ki} - q_i c_{kk}$$

for $i \neq k$ where $q_i = m_i m_k^{-1} \neq 1$. Therefore, since relations (2.87) and (2.88) hold for $1 \leq i \leq n - 2$, the same argument as in the preceding cases shows that the value in each column of the $(n - 1) \times (n - 1)$ submatrix of C obtained by deleting the nth row and nth column of C is constant. When $j \neq k, n$, by the first part of (2.90) we have that the entries in the j th column are all equal for each $j \neq k, n$. The second part of (2.90) combines with (2.89) for $i = n$ to show $c_{nk} = c_{kk}$, and hence the values in the k th column are constant also. Thus, the second part of (2.90) reduces to $c_{nn} = c_{kn}$, which together with (2.87) for $j = n$ implies that the values in the n th column are equal, as well. Defining $\tau(h_i)$ to be the common value in the i th column of C and extending τ linearly to H, we have relation (2.85) holding for the E_6, E_7 and E_8 cases.

Assume that A^- is of type D_n for $n \geq 4$. The value of k in this case is $n - 2$, and the highest root γ is given by Table 2.3.

Consider the basis $\{\gamma, \alpha_2, \cdots, \alpha_{n-2}, \alpha_{n-1}, \alpha_n\}$ and its dual

$$\{h_1, h_2 - 2h_1, \cdots, h_{n-2} - 2h_1, h_{n-1} - h_1, h_n - h_1\} .$$

Using these bases and the fact that $h_i - 2h_1$, $h_{n-1} - h_1$, and $h_n - h_1$ square to themselves, respectively, we derive that

$$c_{ii} - 2c_{i1} = c_{1i} - 2c_{11}$$

must hold for all i with $2 \le i \le n - 2$ and that

$$c_{n-1,n-1} - c_{n-1,1} = c_{1,n-1} - c_{11} ,$$

$$c_{nn} - c_{n1} = c_{1n} - c_{11} .$$

Let C_0 be the $(n - 2) \times (n - 2)$ submatrix of C obtained by deleting the last two rows and columns from C. Relation (2.88) for $i = 1$ combines with the first equation above for $i = 2$ to give $c_{11} = c_{21}$, which in turn shows by (2.87) that the first column of C_0 is constant. Thus, the first relation above reduces to $c_{ii} = c_{1i}$ for $1 \le i \le n - 2$, which implies by (2.87) that $c_{1i} = \cdots = c_{i-1,i} = c_{ii}$ for $1 \le i \le n - 2$, and hence by (2.88) $c_{i+1,i} = c_{ii}$. Thus, we conclude from (2.87) that the values in each column of C_0 are equal. Since (2.87) gives $c_{n-1,j} = c_{n-2,j}$ and (2.90) gives $c_{nj} = c_{n-2,j}$ for each $j = 1, 2, \cdots, n - 3$, the values in the first $n - 3$ columns of the $n \times n$ matrix C are equal. The second relation above together with (2.87) and (2.90) implies that the values in the $(n - 1)$th column are constant, and the third equation and (2.87) show that the result holds for the nth column. Finally, for $i = n - 3$, relation (2.88) and the second equation of (2.90) give $c_{n,n-2} = c_{n-2,n-2}$, which together with (2.90) and (2.87) shows that the values in the $(n - 2)$th column of C are constant also. Therefore, we conclude that relation (2.85) holds for the case of type D_n $(n \geq 4)$. □

To extend relation (2.85) to the entire algebra A , we need

Lemma 2.31. Let A , the product "∘" , and τ be the same as in Lemma 2.30. Then, for each root α , there exists a scalar $c_\alpha \in F$ such that

$$h \circ e_\alpha = \tau(h)e_\alpha + c_\alpha h \quad \text{for all} \quad h \in H ,$$

$$e_\alpha \circ e_\alpha = 2c_\alpha e_\alpha .$$

Proof. In view of (2.83) and Lemma 2.30, we see that

$$\tau_\alpha(h')\alpha(h) + \tau_\alpha(h)\alpha(h') = \alpha(h \circ h') = \tau(h')\alpha(h) + \tau(h)\alpha(h')$$

must hold for all $h, h' \in H$ and all α . If $h' = h$ and h is not in the kernel of α , then the last equation implies that $\tau_\alpha(h) = \tau(h)$, while if h belongs to the kernel of α , but h' does not, then again $\tau_\alpha(h) = \tau(h)$. Thus by linearity $\tau_\alpha(h) = \tau(h)$ for all $h \in H$ and all α . Since $Fe_\alpha + H$ is a subalgebra of $(A,\circ)$ by Lemma 2.28(3), we can let

$$e_\alpha \circ e_\alpha = 2c_\alpha e_\alpha + h_\alpha$$

where $c_\alpha \in F$ and $h_\alpha \in H$. Using this, we find from $[e_\alpha \circ e_\alpha, e_\gamma] = 2[e_\alpha, e_\gamma \circ e_\alpha]$ that $\gamma(h_\alpha) = 0$ for every root γ , and hence $h_\alpha = 0$. Consequently, we get the second relation of Lemma 2.31.

Corresponding to each root α , there exist roots $\alpha_2, \cdots, \alpha_n$ such that $\{\alpha_1 = \alpha, \alpha_2, \cdots, \alpha_n\}$ forms a basis of H^* (Humphreys [1,p.51, Theorem (c)] or Myung [6,p.184, Theorem 5.3.6(3)]). Suppose that $\{h_1, \cdots, h_n\}$ is the associated dual basis of H . Since the derivation ad_{h_α} is nilpotent, $\theta = \exp ad_{e_\alpha}$ is an automorphism of the Lie algebra

A^- taking H to a Cartan subalgebra H', and mapping h_1 to $h_1 - e_\alpha$ and h_j to h_j for all $j \neq 1$. By Lemma 2.30 there exists a linear form $\tau' : H' \to F$ such that

$$h' \circ h'' = \tau'(h'')h' + \tau'(h')h''$$

for all $h', h'' \in H'$. Then $\tau'(h_j) = \tau(h_j)$ for all $j \neq 1$, and by (2.82) and the forgoing results,

$$\begin{aligned} 2\tau'(h_1 - e_\alpha)(h_1 - e_\alpha) &= (h_1 - e_\alpha) \circ (h_1 - e_\alpha) \\ &= 2\tau(h_1)h_1 - 2\tau(h_1)e_\alpha - 2u_\alpha(h_1) + 2c_\alpha e_\alpha . \end{aligned}$$

Thus $\tau'(h_1 - e_\alpha) = \tau(h_1) - c_\alpha$ and $u_\alpha(h_1) = c_\alpha h_1$. But then, for $j \neq 1$,

$$\begin{aligned} (\tau(h_1) - c_\alpha)h_j + \tau(h_j)(h_1 - e_\alpha) &= (h_1 - e_\alpha) \circ h_j \\ &= \tau(h_1)h_j + \tau(h_j)h_1 - \tau(h_j)e_\alpha - u_\alpha(h_j) \end{aligned}$$

which implies $u_\alpha(h_j) = c_\alpha h_j$. Since $h_1, \cdots, h_n$ form a basis of H, $u_\alpha(h) = c_\alpha h$ for all $h \in H$, and $h \circ e_\alpha = \tau_\alpha(h)e_\alpha + u_\alpha(h) = \tau(h)e_\alpha + c_\alpha h$, as asserted. □

Corollary 2.32. For all roots α, β, there exists $h_{\alpha,\beta} \in H$ such that $e_\alpha \circ e_\beta = c_\beta e_\alpha + c_\alpha e_\beta + h_{\alpha,\beta}$, where $h_{\alpha,\beta} = 0$ if $\beta = \alpha$.

Proof. Since $\tau_\alpha = \tau_\beta = \tau$ on H, (2.84) reduces to the equation $[h, e_\alpha \circ e_\beta] = \alpha(u_\beta(h))e_\alpha + \beta(u_\alpha(h))e_\beta$ for all $h \in H$. The result follows from the fact that $u_\alpha(h) = c_\alpha h$ for all $h \in H$. □

We are now ready to prove our principal result.

Theorem 2.33. Let A be a finite-dimensional Malcev-admissible algebra with product "*" over a field F of characteristic 0 such that

A^- is a split simple Malcev algebra over F. If A satisfies the third power identity, then there exist a linear form τ on A and a scalar $\beta \in F$ such that relation (2.67) holds, where $x \# y = 0$ for all $x,y \in A$ unless A^- is a Lie algebra of type A_n for $n \geq 2$. When A^- is a Lie algebra of type A_n, A^- is identified with the Lie algebra $\mathfrak{sl}(n+1)$ of $(n+1) \times (n+1)$ matrices of trace zero, and the product "#" is defined by

$$x \# y = xy + yx - \frac{2}{n+1}(\mathrm{tr}\ xy)I \tag{2.91}$$

where xy denotes matrix multiplication, tr is the trace, and I is the identity matrix. If the "*" product is power-associative, then $\beta x \# y = 0$ for all $x,y \in A$. Conversely, any product "*" defined by relation (2.67) using the multiplication [,] on a split simple Malcev algebra gives a third power-associative Malcev-admissible algebra, which is power-associative whenever $\beta x \# y = 0$ for all $x,y \in A$.

<u>Proof</u>. Since $x * y = \frac{1}{2}[x,y] + x \circ y$ for some commutative product "$\circ$" on A, as noted earlier, it suffices to show that all commutative products "$\circ$" satisfying relation (2.3) are given by

$$x \circ y = \tau(x)y + \tau(y)x + \beta x \# y \tag{2.92}$$

for some linear form τ on A and a scalar $\beta \in F$, where $\beta x \# y = 0$ for all $x,y \in A$ unless A^- is a Lie algebra of type A_n $(n \geq 2)$. When A^- is a Lie algebra of type A_n $(n \geq 2)$, the result has been proved in Lemma 2.24. We note that when $\beta x \# y = 0$ for all $x,y \in A$, the product "*" given by (2.67) is power-associative. If A^- is a split simple, non-Lie Malcev algebra, then A^- is the Malcev algebra defined by (2.81), and hence the result follows from Corollary 2.29. Suppose then that A^- is a split simple Lie algebra which is not of type A_n. We show that the elements $h_{\alpha,\beta}$ in Corollary 2.32 are zero for all roots α, β. But then relation (2.5) with $e_\alpha, e_\beta, e_\gamma$ implies by Corollary 2.32 that

$$\alpha(h_{\beta,\gamma})e_\alpha + \beta(h_{\gamma,\alpha})e_\beta + \gamma(h_{\alpha,\beta})e_\gamma = 0 ,$$

and from that it is easy to see that $h_{\alpha,\beta}$ lies in the kernel of γ for each root γ and so must be 0. Thus $e_\alpha \circ e_\beta = c_\beta e_\alpha + c_\alpha e_\beta$ for all roots α, β. The linear form τ on H in Lemma 2.30 can now be extended to a linear form on A by specifying $\tau(e_\alpha) = c_\alpha$ for each root α. Then Lemmas 2.30 and 2.31 and Corollary 2.32 imply relation (2.92), where $\beta x \# y = 0$ for all $x,y \in A$. □

The following result is immediate from Lemma 1.19, Corollary 2.25, and Theorem 2.33.

Corollary 2.34. If A is a finite-dimensional flexible Malcev-admissible algebra with product "*" over a field F of characteristic 0 such that A^- is split simple over F, then

$$x * y = \frac{1}{2}[x,y] + \beta x \# y$$

where $x \# y$ is the same as in (2.91). If A is flexible and power-associative, then $x * y = \frac{1}{2}[x,y]$ and A is a Malcev algebra. □

This corollary was originally proved in the algebraically closed case by Okubo and Myung [3] using adjoint operators and was part of a more general theorem proved by Benkart and Osborn [1] using representation theory. These approaches will be discussed in Chapters 3 and 4 in conjunction with investigation of the structure of A when A^- is central simple or has nonzero radical. Corollary 2.34 when A is flexible and power-associative preceded both those papers, appearing first in work by Laufer and Tomber [1]. A more direct proof of that case has been given in Corollary 1.16 also. The results in Theorem 2.33 and Corollary 2.34 can be carried over to the non-split case, namely, the case where A^- is central simple over F, by using a field extension argument. This case will be investigated in Chapter 3.

Corollary 2.35. Let A be a finite-dimensional, third power-associative Malcev-admissible algebra of characteristic 0 with product "$*$" such that A^- is split simple over F. Then A is Jordan-asmissible if and only if A is power-associative.

Proof. It follows from Theorem 2.33 that $A^+ = (A,\circ)$, where "$\circ$" is defined by (2.92). If A is power-associative, then $\beta x \# y = 0$ for all $x,y \in A$ by Theorem 2.33, and it is easy to see that $x \circ y = \tau(y)x + \tau(x)y$ satisfies the Jordan identity $(x \circ y) \circ x^2 = x \circ (y \circ x^2)$, where $x^2 = x \circ x = x * x$. Assume then that A is Jordan-admissible. Then "$\circ$" clearly satisfies the fourth power identity (2.6), and it follows from Lemma 2.24 and Theorem 2.33 that A is power-associative. □

2.6. THE SEMISIMPLE CASE

In this section we prove the analogue of Theorem 2.33 when A^- is semisimple. Note that a finite-dimensional semisimple Malcev algebra over a field of characteristic 0 is a direct sum of simple ideals. We begin with

Lemma 2.36. Let A be an algebra over a field F of characteristic $\neq 2$ with product "$*$" such that A^- is a direct sum $A^- = S_1^- \oplus \cdots \oplus S_n^-$ of simple ideals S_i^- of A^-. Suppose that "$\circ$" is a commutative product on A satisfying relation (2.3). Then, for each subset $\{i_1,\cdots,i_r\}$ of

$\{1,\cdots,n\}$, $S_{i_1} + \cdots + S_{i_r}$ is a subalgebra of $(A,\circ)$. If A is third power-associative, then $S_{i_1} + \cdots + S_{i_r}$ is a subalgebra of A for any $i_1, \cdots, i_r \in \{1,\cdots,n\}$.

Proof. It suffices to verify that each S_j is a subalgebra of $(A,\circ)$ and $S_i \circ S_j \subseteq S_i + S_j$ for all $i \neq j$. For $x \in S_i$ and $y,z \in S_j$ with $j \neq i$, by the linearized form (2.5) of (2.3), we have $[x,y \circ z] + [y,z \circ x] + [z,x \circ y] = 0$. The first term of this lies in S_i, while the remaining two belong to S_j. Thus $[S_i,y \circ z] = 0$ for all $i \neq j$, which implies that S_j is a subalgebra of $(A,\circ)$, since each S_i^- is a simple ideal of A^-. Let $x \in S_i$, $y \in S_j$ and $z \in S_k$ for distinct i, j, k. Then (2.5) shows that $[S_k,x \circ y] = 0$ for all $k \neq i, j$, which implies $S_i \circ S_j \subseteq S_i + S_j$. The second part follows from the fact that third power-associativity of A is equivalent to (2.3), where $x \circ y = \frac{1}{2}(x * y + y * x)$. □

The following result is useful for our principal result in the semi-simple case and for later investigation in Chapter 4.

Lemma 2.37. Let M be a finite-dimensional split simple Malcev algebra over a field F of characteristic 0 and let θ be a linear transformation on M such that $[x,\theta(y)] = [\theta(x),y]$ for all $x,y \in M$. Then θ is a scalar multiple of the identity map I.

Proof. Suppose first that M is a non-Lie, Malcev algebra, so that M is isomorphic to the algebra C_0^- defined by (2.81). Let e_0, $e_{\pm i}$, $i = 1, 2, 3$, be the basis given by (2.81) and let $P = \sum_{i=1}^{3} (Fe_i + Fe_{-i})$. Since $[e_0,P] = P$, the left side of $[e_0,\theta(e_{-i})] = [\theta(e_0),e_{-i}]$ lies in P and hence $\theta(e_0) = c_0e_0$ for some $c_0 \in F$. By a similar argument, we have

$$\theta(e_i) = c_i e_0 + b_i e_i, \ i = \pm 1, \pm 2, \pm 3$$

where the c_i and b_i are scalars in F. These relations combine with the equation $[e_{-i},\theta(e_i)] = [\theta(e_{-i}),e_i]$ to give $c_i = c_{-i} = 0$ and $b_i = b_{-i}$ for $i = 1, 2, 3$, which implies by $[e_i,\theta(e_j)] = [\theta(e_i),e_j]$ for $i \neq j$ that $b_i = b_j$. If we let $a = b_i = b_j$, then $\theta(e_i) = ae_i$ for $i = \pm 1, \pm 2, \pm 3$, and it follows from $[\theta(e_0),e_i] = [e_0,\theta(e_i)]$ that $a = c_0$, and hence θ is a scalar map on M.

Suppose next that M is a Lie algebra L, and let $L = H + \sum_\alpha L_\alpha$ be a Cartan decomposition of L relative to a Cartan subalgebra H. For each root β, let e_β be a nonzero element in L_β. Then for $h \in H$, the left side of $[h,\theta(e_{-\beta})] = [\theta(h),e_{-\beta}]$ lies in $\sum_\alpha L_\alpha$, and this implies that the component of $\theta(h)$ in L_β is zero. Since this holds for each root β, $\theta(H) \subseteq H$ and hence $[h,\theta(e_\beta)] = [\theta(h),e_\beta] \in L_\beta$ for each β, so that $\theta(e_\beta) \in H + L_\beta$. Using this in equation $[e_\alpha,\theta(e_\beta)] = [\theta(e_\alpha),e_\beta]$, we find that $\theta(e_\beta) \in L_\beta$ for each β, and thus $\theta(e_\beta)$ is a scalar multiple of e_β. Let $a_\beta \in F$ be such that $\theta(e_\beta) = a_\beta e_\beta$. Then the relation $[e_\alpha,\theta(e_\beta)] = [\theta(e_\alpha),e_\beta]$ gives $a_\alpha = a_\beta$, whenever $[e_\alpha,e_\beta] \neq 0$. This in particular shows that $a_\alpha = a_{-\alpha}$ and $a_\alpha = a_\beta$ if $\alpha + \beta$ is a root. Using the connectedness of the Dynkin diagram, we also find that $a_{\alpha_i} = a_{\alpha_j}$ for all simple roots α_i and α_j. Since for any non-simple root β there exists a simple root α_i such that $\beta - \alpha_i$ or $\beta + \alpha_i$ is a root (Myung [6,p.114]), we see that $a_\alpha = a_\beta$ for all roots α, β. We denote this common value by a. Since for each root β

$$\beta(\theta(h))e_\beta = [\theta(h),e_\beta] = [h,\theta(e_\beta)] = a\beta(h)e_\beta = \beta(ah)e_\beta ,$$

it follows that $\theta(h) = ah$ and hence $\theta = aI$, as asserted. $\square$

Theorem 2.38. Let A be a finite-dimensional, third power-associative Malcev-admissible algebra with product "$*$" over a field F of characteristic zero such that A^- is split semisimple, where each S_i^- denotes a simple summand of A^- for $i = 1, 2, \cdots, m$. Then for each j there exists a scalar $\beta_j \in F$, and for each ordered pair (i,j) there exists a linear form $\tau_{ij} : S_j \to F$ such that when $x \in S_i$ and $y \in S_j$,

$$(2.93) \qquad x * y = \frac{1}{2}[x,y] + \tau_{ji}(x)y + \tau_{ij}(y)x + \beta_j\, x \# y ,$$

where $x \# y$ is defined by $x \# y = x \#_i y$ if $i = j$, and $x \# y = 0$ if $i \neq j$, and where $x \#_i y$ is defined on the simple algebra S_i^- as in Theorem 2.33. Conversely, any multiplication "$*$" defined on a semisimple Malcev algebra in this manner gives a third power-associative Malcev-admissible algebra.

Proof. Assume first that $A = S_1 \oplus \cdots \oplus S_m$ is third power-associative and Malcev-admissible. Let $x \circ y = \frac{1}{2}(x * y + y * x)$, and suppose that $i \neq j$. It follows from Lemma 2.36 that $S_i \circ S_j \subseteq S_i + S_j$. For each $x \in S_i$, define the mapping $\pi_x : S_j \to S_j$ as the mapping $y \to x \circ y\ (y \in S_j)$ followed by the projection of $x \circ y$ onto S_j. Since S_j is a subalgebra of $(A,\circ)$ and hence $[y,z \circ x] + [z,x \circ y] = 0$ by (2.5) for all $y,z \in S_j$, the mapping π_x satisfies the relation $[y,\pi_x(z)] = [\pi_x(y),z]$ for all $y,z \in S_j$. Thus, by Lemma 2.37, π_x must be a scalar multiple of the identity map on S_j. Therefore, there exists a linear form $\tau_{ji} : S_i \to F$ such that $\pi_x(y) = \tau_{ji}(x)y$ for all $y \in S_j$. Reversing the role of i and j, and noting that $S_i \circ S_j \subseteq S_i + S_j$, we have $x \circ y = \tau_{ji}(x)y + \tau_{ij}(y)x$ when $i \neq j$, to complete the proof of relation (2.93). When $i = j$, (2.93) is a consequence of Theorem 2.33. Conversely, it is easy to see that any product defined by (2.93) determines a third power-associative Malcev-admissible algebra. □

The investigation of conditions for the product "$\circ$" given by (2.93) to be power-associative is much more complicated than the simple algebra case. However, when the multiplication in (2.93) is power-associative, then by Theorem 2.33, none of the β terms are present. Also, when F has characteristic zero, by Lemmas 1.10(ii) and 1.11, for products in (2.93), power-associativity is equivalent to (2.6) holding for the commutative product "$\circ$", so that the Malcev structure of the S_i^-'s can be ignored. Therefore, to investigate power-associativity of such products it is sufficient to study conditions to satisfy the fourth power identity (2.6) for commutative products "$\circ$" on $A = S_1 \oplus \cdots \oplus S_m$ defined by

$$(2.94) \qquad x \circ y = \tau_{ij}(y)x + \tau_{ji}(x)y$$

where $x \in S_i$, $y \in S_j$, and τ_{ij} is a linear form on the subspace S_j. If $A = S_1 \oplus \cdots \oplus S_m$ is the direct sum of nonzero vector spaces S_i with product "$\circ$" defined by (2.94), then it is easy to see that $S_{i_1} + \cdots + S_{i_r}$ is a subalgebra of A for each subset $\{i_1,\cdots,i_r\} \subseteq \{1,\cdots,m\}$. Recall that if F has characteristic $\neq 2,3,5$, then, for any commutative product, the fourth power identity is equivalent to power-associativity (Lemma 1.11(ii)). Thus, in what follows, we assume that the base field F has characteristic $\neq 2,3,5$. Then the fourth power identity is equivalent to the following multilinear identity in four variables:

$$(2.95) \qquad \sum_{\pi} (x_{\pi(1)} \circ x_{\pi(2)}, x_{\pi(3)}, x_{\pi(4)}) = 0 ,$$

where $(x,y,z) = (x \circ y) \circ z - x \circ (y \circ z)$, and π runs over all permutations of the set $\{1,2,3,4\}$ (Albert [1]). In the next section, we give a necessary and sufficient condition for products in (2.94) to satisfy (2.6) or (2.95).

Lemma 2.39. Let $A = S_1 \oplus \cdots \oplus S_m$ be a commutative algebra which is a direct sum of subalgebras S_i such that $S_i + S_j$ is a subalgebra of A for all $i \neq j$. Then A is power-associative if and only if each subalgebra consisting of four or fewer summands is power-associative.

Proof. Under the assumption, $S_{i_1} + \cdots + S_{i_r}$ is a subalgebra for each subset $\{i_1, \cdots, i_r\} \subseteq \{1, \cdots, m\}$. Assume that each subalgebra of A consisting of four or fewer summands is power-associative. When arbitrary four elements of A are substituted into the left side of (2.95), by multilinearity the left side is expanded to a sum of terms which are all zero by our assumption. Thus, (2.95) holds for A. □

Note that products in (2.94) satisfy the hypothesis of Lemma 2.39. Therefore, Lemma 2.39 shows that, to investigate conditions for such products to be power-associative, it suffices to work with algebras consisting of four or fewer summands S_i. As we shall see in the next section, those conditions are described by relations among the linear forms τ_{ij}. We here illustrate some of those relations as examples. In the following, assume that $A = S_1 \oplus \cdots \oplus S_m$ is a direct sum of vector spaces S_i over the field F with product defined by (2.94).

Example 2.1. (i) Consider the special case where $\tau_{ij} = 0$ for all $i \neq j$. Then, A is simply the direct sum of the power-associative ideals S_i, and A is power-associative as well.

(ii) Consider next the case where, for each i, $\tau_{ii} = \tau_{ji}$ for all j. Defining a linear form τ on A by

$$\tau(x_1 + \cdots + x_m) = \tau_{11}(x_1) + \cdots + \tau_{mm}(x_m)$$

for $x_i \in S_i$, the product (2.94) is combined to a single product given by $x \circ y = \tau(y)x + \tau(x)y$ for $x,y \in A$, and A is power-associative in this case also. □

Example 2.2. Let $A = S_1 \oplus S_2$ be the direct sum of two summands, and assume that $\tau_{21} = 2\tau_{11}$, $\tau_{12} = 0$, and τ_{22} is arbitrary. For $x \in S_1$ and $y \in S_2$, we compute

$$(x + y)^2 = 2\tau_{11}(x)x + 2[2\tau_{11}(x) + \tau_{22}(y)]y ,$$

$$\begin{aligned}(x + y)^4 &= (x + y)^2 \circ (x + y)^2 \\ &= 8\tau_{11}(x)^3x + 4(2\tau_{11}(x) + \tau_{22}(y))[(2\tau_{11}(x) + \tau_{22}(y))^2 + \tau_{22}(y)^2]y .\end{aligned}$$

Thus, A is power-associative. □

For the case of three summands, we give two special relations in

Example 2.3. (i) Let $A = S_1 \oplus S_2 \oplus S_3$, and assume that $\tau_{12} = \tau_{21} = \tau_{13} = \tau_{23} = \tau_{33} = 0$, and $\tau_{31} = \tau_{11}$ and $\tau_{32} = \tau_{22}$ are arbitrary. For $x \in S_1$, $y \in S_2$ and $z \in S_3$, by direct computation, we find that both $(x + y + z)^4$ and $(x + y + z)^2 \circ (x + y + z)^2$ are equal to

$$2^3[\tau_{11}(x)^3x + \tau_{22}(y)^3y + (\tau_{11}(x) + \tau_{22}(y))(\tau_{11}(x)^2 + \tau_{22}(y)^2)z] ,$$

which implies that A is power-associative.

(ii) Suppose that $\tau_{12} = \tau_{23} = \tau_{13} = \tau_{33} = 0$, $\tau_{21} = 2\tau_{11} = 2\tau_{31}$, $\tau_{32} = \tau_{22}$ where τ_{11} and τ_{22} are arbitrary. If $x \in S_1$, $y \in S_2$ and $z \in S_3$, then it follows that both $(x + y + z)^4$ and $[(x + y + z)^2]^2$ are equal to

$$\begin{aligned}2^3[\tau_{11}(x)^3x + (2\tau_{11}(x) + \tau_{22}(y))(2\tau_{11}(x)^2 + 2\tau_{11}(x)\tau_{22}(y) + \tau_{22}(y)^2)y \\ + (\tau_{11}(x) + \tau_{22}(y))^3z] ,\end{aligned}$$

and hence A is power-associative. □

For the following two cases of four summands, A fails to be power-associative.

<u>Example 2.4.</u> (i) Let $A = S_1 \oplus S_2 \oplus S_3 \oplus S_4$ be the direct sum of 4 nonzero summands, and assume that $\tau_{12} = \tau_{21} = \tau_{13} = \tau_{31} = \tau_{23} = \tau_{32} = 0$, $\tau_{14} = \tau_{44}$, $\tau_{41} = \tau_{11}$, $\tau_{24} = \tau_{44}$, $\tau_{42} = \tau_{22}$, $\tau_{43} = \tau_{33}$, and $\tau_{34} = \tau_{44}$ where τ_{ii} for $i = 1, 2, 3, 4$ are arbitrary. For $w \in S_1$, $x \in S_2$, $y \in S_3$ and $z \in S_4$, the left side of identity (2.95) is computed as

$$12\tau_{11}(w)\tau_{22}(x)\tau_{33}(y)z + 4f(w,x,y) , \tag{2.96}$$

where $f(w,x,y)$ is an element of $S_1 \oplus S_2 \oplus S_3$ given by

$$\tau_{22}(x)\tau_{33}(y)\tau_{44}(z)w + \tau_{11}(w)\tau_{33}(y)\tau_{44}(z)x + \tau_{11}(w)\tau_{22}(x)\tau_{44}(z)y .$$

In particular, if τ_{11}, τ_{22} and τ_{33} are all nonzero, and w, x, y, and z are chosen so that $\tau_{11}(w)$, $\tau_{22}(x)$, $\tau_{33}(y)$, and z are all nonzero, then (2.96) can not be zero. Thus, A is not power-associative in this case.

(ii) Assume that $\tau_{21} = \tau_{13} = \tau_{31} = \tau_{23} = \tau_{44} = \tau_{14} = \tau_{24} = \tau_{34} = 0$, $\tau_{41} = \tau_{11}$, $\tau_{12} = 2\tau_{22}$, $\tau_{32} = 2\tau_{22} = 2\tau_{42}$, and $\tau_{43} = \tau_{33}$, where τ_{ii} for $i = 1, 2, 3$ are arbitrary. For $w \in S_1$, $x \in S_2$, $y \in S_3$, and $z \in S_4$, direct computation shows that the left side of (2.95) equals

$$- 8\tau_{11}(w)\tau_{22}(x)\tau_{33}(y)z .$$

Thus, if τ_{ii} for $i = 1, 2, 3$ are all nonzero, and w, x, y, z are chosen so that $\tau_{11}(w)$, $\tau_{22}(x)$, $\tau_{33}(y)$, and z are all nonzero, then identity (2.95) is not satisfied, and hence A can not be power-associative. □

2.7. Power-associative products defined by linear forms

Throughout this section, we assume that $A = S_1 \oplus \cdots \oplus S_m$ is a finite-dimensional vector space which is the direct sum of nonzero subspaces S_i over a field F of characteristic $\neq 2,3,5$, and that there is defined a multiplication "$\circ$" on A given by relation (2.94), where τ_{ij} is a linear form on S_j. For simplicity, denote

$$\tau_{ii} = \tau_i, \; i = 1, \cdots, m .$$

We give a condition for the algebra A with product (2.94) to be power-associative in terms of the linear forms τ_{ij}. We note that if $\tau_i \neq 0$, then any element x of S_i with $\tau_i(x) \neq 0$ produces an idempotent $e = [2\tau_i(x)]^{-1}x$ in S_i. Idempotent decompositions along with relation (2.11) will play a major role for the investigation in this section.

Assume in what follows, unless otherwise stated, that the algebra A with product as in (2.94) is power-associative, or equivalently satisfies (2.6) or (2.95). Thus, relation (2.11) holds for A relative to an idempotent. The peirce spaces of A relative to an idempotent e will be denoted by $A_\lambda(e)$ for $\lambda = 0, \frac{1}{2}, 1$.

<u>Lemma 2.40</u>. If $\tau_i = \tau_j = 0$, then $\tau_{ij} = \tau_{ji} = 0$.

<u>Proof</u>. Assume that $\tau_i = \tau_j = 0$. Then $(x + y)^2 = 2\tau_{ji}(x)y + 2\tau_{ij}(y)x = 2\, x \circ y$ for all $x \in S_i$ and $y \in S_j$, and $(x \circ y)^2 = 2\tau_{ji}(x)\tau_{ij}(y)x \circ y$. Thus, if $\tau_{ij} \neq 0$ and $\tau_{ji} \neq 0$, then x and y can be chosen so that $e = x + y$ is an idempotent. The element $x - y$ belongs to $A_0(e)$, but $(x - y)^2 = -e \in A_1(e)$ by our assumption. Since $A_0(e)$ is a subalgebra of A, this is impossible. Hence one of τ_{ij} or τ_{ji} must be zero, say $\tau_{ij} = 0$. When $x + y$ is substituted

into the fourth power identity (2.6):

$$(x \circ x) \circ (x \circ x) = x \circ (x \circ (x \circ x)) \ ,$$

the left side is zero, while the right side is computed as $2^3\tau_{ji}(x)^3 y$. Thus τ_{ji} is zero also. □

Lemma 2.41. Let e be an idempotent of S_i , and assume that $i \neq j$. Then

(1) $\lambda \equiv \tau_{ji}(e) = 0, \frac{1}{2}$, or 1 .

(2) For each element $y \in S_j$, the element $y - \mu\tau_{ij}(y)e$ belongs to $A_\lambda(e)$, where $\lambda = 0, \frac{1}{2}$, or 1 , and $\mu = 1, 2$, or 0 , respectively. If $\lambda = 1$, then $\tau_{ij} = 0$.

(3) $\tau_{ji} = 2\lambda\tau_i$ for $\lambda = 0, \frac{1}{2}$, or 1 .

Proof. (1) Since $e = e \circ e = 2\tau_i(e)e$, $\tau_i(e) = \frac{1}{2}$. For each $y \in S_j$, $e \circ y = \tau_{ji}(e)y + \tau_{ij}(y)e = \lambda y + \tau_{ij}(y)e$, and hence e and y span a 2-dimensional subalgebra of A , in which λ takes the value $0, \frac{1}{2}$, or 1 by Lemma 2.13 (relation (2.43)).

(2) Assume first that $\lambda = 1$. Then $e \circ y = y + \tau_{ij}(y)e$ for each $y \in S_j$, which implies that, for the subalgebra $B = Fe + S_j$, $B_0(e) = B_{\frac{1}{2}}(e) = 0$. Thus, it must be that $\tau_{ij}(y) = 0$ for all $y \in S_j$, and $\tau_{ij} = 0$. The remaining part follows from the relation

$$[y - \mu\tau_{ij}(y)e] \circ e = \lambda y - (\mu - 1)\tau_{ij}(y)e$$

for each $y \in S_j$, where $\lambda = 0, \frac{1}{2}$, or 1 and $\mu = 1, 2$, or 0 , respectively.

(3) Note that if $x \in \ker \tau_i$, then $x \circ e = \tau_i(x)e + \tau_i(e)x = \frac{1}{2}x$, and hence $\ker \tau_i \subseteq A_{\frac{1}{2}}(e)$. Since multiplication by $x \in A_{\frac{1}{2}}(e)$ maps

$A_\lambda(e)$ into the other two Peirce spaces,

$$x \circ (y - \mu\tau_{ij}(y)e) = \tau_{ji}(x)y + \tau_{ij}(y)x - \frac{\mu}{2}\tau_{ij}(y)x$$

has no component in $A_\lambda(e)$ for $\lambda = 0, \frac{1}{2}$, or 1, and $\mu = 1, 2$, or 0, respectively. When $\mu = 0$ and $\lambda = 1$, this implies $\tau_{ji}(x) = 0$ for all $x \in \ker \tau_i$. Thus, $\tau_{ji} = c\tau_i$ for some scalar $c \in F$, which must be 2λ, since $\lambda = \tau_{ji}(e) = c\tau_i(e) = \frac{1}{2}c$. □

<u>Corollary 2.42</u>. (1) $\tau_i = 0$ implies $\tau_{ji} = 0$ for all j.

(2) If J denotes the sum of all S_j with $\tau_j = 0$, then J is an ideal of A such that $J^2 = 0$.

<u>Proof</u>. (1) If $\tau_j = 0$ as well, then by Lemma 2.40 $\tau_{ji} = 0$, so the result trivially holds in this case. Assume that $\tau_j \neq 0$. Then S_j has an idempotent, and by Lemma 2.41(3) $\tau_{ij}(y) = 0$ for all $y \in \ker \tau_j$. It is easy to see from (2.6) that $2^3\tau_{ji}(x)^3 y = (x + y)^4 = (x + y)^2 \circ (x + y)^2 = 0$ for all $x \in S_i$. Hence $\tau_{ji} = 0$.

(2) The result is immediate from (2.94), and part (1). □

<u>Lemma 2.43</u>. Assume that $\tau_i \neq 0$ and $i \neq j$. Then

(1) $\tau_{ji} = 0$ implies $\tau_{ij} = 0$ or $\tau_j \neq 0$ and $\tau_{ij} = 2\tau_j$.

(2) $\tau_{ji} = \tau_i$ implies $\tau_{ij} = \tau_j$.

(3) $\tau_{ji} = 2\tau_i$ implies $\tau_{ij} = 0$.

<u>Proof</u>. We first verify the second assertion. Assume $\tau_{ji} = \tau_i$. If $\tau_j = 0$, then by Corollary 2.42(1) $\tau_{ij} = 0$ and the result holds trivially. Suppose then that $\tau_j \neq 0$, and let e' be an idempotent of S_j, so that $\tau_j(e') = \frac{1}{2}$. In light of Lemma 2.41(1), $\tau_{ij}(e') = 0, \frac{1}{2}$, or 1. If $\tau_{ij}(e') = 1$, then by Lemma 2.41(2) $\tau_{ji} = 0$ to give a

contradiction. Assume $\tau_{ij}(e') = 0$. Since τ_{ij} vanishes on $\ker \tau_j$, we have $\tau_{ij} = 0$. Let e be an idempotent of S_i . Then since $\tau_{ji} = \tau_i$, $\tau_{ji}(e) = \frac{1}{2}$, and so $e \circ y = \frac{1}{2}y$ for all $y \in S_j$, which implies that $S_j \subseteq A_{\frac{1}{2}}(e)$. Since S_j is a subalgebra of A , we must conclude from (2.11) that $S_j^2 = 0$, to give a contradiction again. Thus, $\tau_{ij}(e') = \frac{1}{2}$ and by Lemma 2.41(3) $\tau_{ij} = \tau_j$. For part (1), assume $\tau_{ji} = 0$, and let $\tau_{ij} \neq 0$. Then $\tau_j \neq 0$ by Corollary 2.42(1), which, together with Lemma 2.41(3) and part (2), implies $\tau_{ij} = 2\tau_j$, as desired. The last assertion follows from Lemma 2.41(3) and part (2). □

When A is power-associative, Lemma 2.43 describes all relations between τ_{ij} for each pair (i,j) . Thus, if A consists of two summands, then the conditions (1) - (3) of Lemma 2.43 are necessary and sufficient for A with product (2.94) to be power-associative. In fact, the first part of (1), and (2) in Lemma 2.43 are the cases of Example 2.1, when $m = 2$, while the remaining cases are the situation in Example 2.2. Lemma 2.43 will play a basic role to determine the relations between the τ_{ij} for A consisting of three or four summands. To aid in describing such relationships, we associate A with a graph having m vertices with the vertex v_i corresponding to τ_i or S_i . The vertex v_i will be often depicted by a node "o" or "•" with label "i" . If $\tau_{ij} = 0 = \tau_{ji}$, then there is no edge connecting v_i and v_j . Otherwise, by Lemma 2.40 $\tau_i \neq 0$ or $\tau_j \neq 0$, say that $\tau_i \neq 0$. Then we depict the situations described by Lemma 2.43 graphically by

(2.97) (1)  or 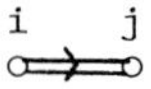 , (2)  , (3) 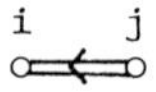 ,

respectively. When both τ_i and τ_j are nonzero, since the role of i and j in Lemma 2.43 can be reversed, the graph is drawn consistently no matter which of the two indices is chosen.

Definition 2.6. Let $\{i_1,\cdots,i_r\}$ be a nonempty subset of $\{1,\cdots,m\}$. A *graph* of the subalgebra $B = S_{i_1} \oplus \cdots \oplus S_{i_r}$ is defined as the set of vertices $v_{i_1}, \cdots, v_{i_r}$ together with each pair (i_k,i_ℓ) of those vertices connected by one of edges described in (2.97). This graph is denoted by $\Gamma = [v_{i_1},\cdots,v_{i_r}]$. A *subgraph* of a graph Γ is a graph obtained from Γ by discarding some vertices from Γ and by omitting all edges connected to those discarded vertices. A graph $\Gamma = [v_{i_1},\cdots,v_{i_r}]$ is said to be *allowable* if the corresponding subalgebra $S_{i_1} \oplus \cdots \oplus S_{i_r}$ is power-associative. We say that two vertices v_i, v_j in a graph Γ lie in the same component if there are vertices $v_{i_1}, \cdots, v_{i_k}$ in Γ such that $i_1 = i$, $j = i_k$, and vertices v_{i_ℓ}, $v_{i_{\ell+1}}$ are connected by at least one edge for all $\ell = 1, \cdots, k-1$. This defines an equivalence relation on Γ, and the equivalence classes are called the *components* of Γ. If Γ consists of a single component, then Γ is said to be *connected*. □

As observed above, for graphs of two nodes, we have

Lemma 2.44. A graph Γ of two nodes is allowable if and only if Γ is one of the cases described in (2.97). □

Therefore, to investigate conditions for A to be power-associative, it suffices to determine all allowable graphs of three and four nodes. The graphs of the two algebras in Example 2.3 are connected, allowable graphs of three nodes and are depicted respectively by

(2.98) (i) 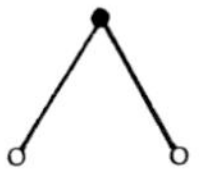 (ii) 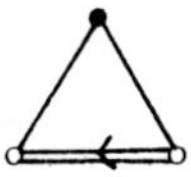

where the solid node "●" corresponds to the vertex v_i with $\tau_i = 0$ and "○" to vertices v_j with arbitrary τ_j. In the present cases, "●" corresponds to $\tau_3 = 0$. The two algebras in Example 2.4 give connected graphs of four nodes which are not allowable and depicted respectively by

(2.99) (i) 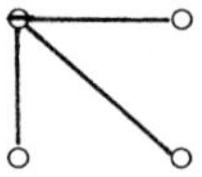 (ii) 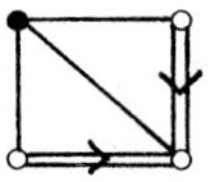

where each node corresponding to $\tau_i \neq 0$ is denoted by "○", arbitrary τ_j by "⊖", and "●" is the same as in (2.98). In Example 2.4, both "⊖" and "●" corresponds to τ_4. In discussing conditions for more general graphs to be allowable, the following result is often useful.

<u>Lemma 2.45</u>. Let B be a commutative algebra over the field F with product "$\circ$", and let e be an idempotent of B. Assume that there are subspaces B_0, $B_{1/2}$, B_1 such that $B = Fe + B_0 + B_{1/2} + B_1$ and such that $e \circ x_j = jx$ for $x_j \in B_j$. If B_0 and B_1 are power-associative subalgebras, if $B_{1/2}^2 = 0$, and if $B_i \circ B_j = 0$ for $i \neq j$, then B is power-associative.

<u>Proof</u>. For $a \in F$, $x \in B_0$, $y \in B_{1/2}$, and $z \in B_1$, we have $(ae + x + y + z)^2 = a^2e + x^2 + ay + (z^2 + 2az)$. It is easy to see that both $[(ae + x + y + z)^2]^2$ and $(ae + x + y + z)^4$ are equal to

$$a^4e + x^4 + a^3y + (z^4 + 4az^3 + 6a^2z^2 + 4a^3z) . \qquad \square$$

<u>Lemma 2.46</u>. A connected graph $\Gamma = [v_i, v_j, v_k]$ of three nodes where τ_i, τ_j, and τ_k are nonzero is allowable if and only if, after suitable relabeling, Γ is given by one of the following.

(i) 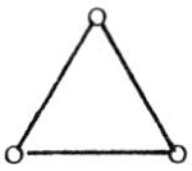 (ii) 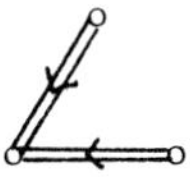

(2.100)

(iii) 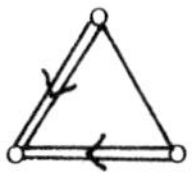 (iv) 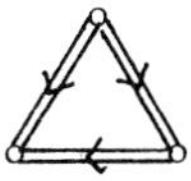 (v) 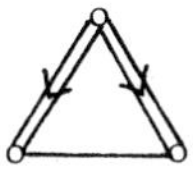

Proof. Assume that Γ is given by one of the graphs in (2.100), and let $B = S_i + S_j + S_k$ be the subalgebra associated with Γ. If Γ is the case (i), then it is clearly allowable, as shown in Example 2.1(ii). For the cases of (iii) and (v), we assume that vertices v_j and v_k are connected by a single edge, and let $S_p = S_j + S_k$. Define a linear form τ_p on S_p by $\tau_p(y + z) = \tau_j(y) + \tau_k(z)$ for $y \in S_j$ and $z \in S_k$. Defining $\tau_{ip} = 0$, $\tau_{pi} : S_i \to F$ by $\tau_{pi} = 2\tau_i$ for case (iii), and $\tau_{pi} = 0$, $\tau_{ip} : S_p \to F$ by $\tau_{ip} = 2\tau_p$ for case (v), these two cases reduce respectively to the graphs 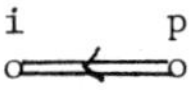

and 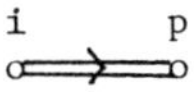

of two nodes, which are allowable by Lemma 2.44. For cases (ii) and (iv), denote by v_i the top vertex and by v_k the bottom right vertex, so that $\tau_{ij} = 2\tau_j$, $\tau_{kj} = 2\tau_j$, and $\tau_{ji} = \tau_{jk} = 0$ in both cases. Letting $B_{\frac{1}{2}}(e) = \ker \tau_j$, $B_1(e) = S_i + S_k$, and $B_0(e) = 0$ for some idempotent e of S_j, $B = Fe + B_0(e) + B_{\frac{1}{2}}(e) + B_1(e)$ satisfies the conditions of Lemma 2.45, and hence is power-associative. Thus Γ is allowable in these cases also.

Suppose then that $\Gamma = [v_i, v_j, v_k]$ is a connected allowable graph of three nodes, where the corresponding forms τ_i, τ_j, τ_k are nonzero. Assume first that Γ contains no double edge. Since Γ is connected, it must be the one depicted in (2.100)(i) or that graph with the bottom edge deleted. In the latter case, let the top node be v_k and the bottom nodes v_i and v_j. For an idempotent $e \in S_i$, each $y \in S_j$ belongs to

$B_0(e)$, while it follows from Lemma 2.41(2) that for each $z \in S_k$, $z - 2\tau_{ik}(z)e$ belongs to $B_{\frac{1}{2}}(e)$. By (2.11), $y \circ (z - 2\tau_{ik}(z)e) = \tau_{jk}(z)y + \tau_{kj}(y)z = \tau_k(z)y + \tau_j(y)z$ must be contained in $B_{\frac{1}{2}}(e) + B_1(e)$. This is impossible since $\tau_k \neq 0$, showing that Γ can not be allowable.

We next assume that Γ contains at least one double edge and that it is labeled 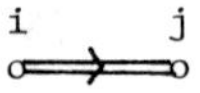 . We exclude all but the four graphs pictured in (2.100), arguing with an idempotent f belonging to S_j . Each $x \in S_i$ belongs to $B_1(f)$, and if $\lambda = \tau_{kj}(f) = 0, \frac{1}{2}$, or 1 , then for each $z \in S_k$ the element $z - \mu\tau_{jk}(z)f$ lies in $B_\lambda(f)$ where $\mu = 1, 2$, or 0 respectively (Lemma 2.41). Consider the case where $\lambda = 0$ and $\mu = 1$. Since the product

$$(z - \mu\tau_{jk}(z)f) \circ x = \tau_{ki}(x)z + \tau_{ik}(z)x - \mu\tau_{jk}(z)x$$

is zero by (2.11), we have $\tau_{ki} = 0$ and $\tau_{ik} = \tau_{jk}$. By (1) of Lemma 2.43 and connectedness of Γ , it follows that $2\tau_k = \tau_{ik} = \tau_{jk}$ to give the configuration in (iv) of (2.100). When $\lambda = \frac{1}{2}$ and $\mu = 2$, the above product has no component in $B_1(f)$, so that $\tau_{ik} = 2\tau_{jk}$. Since it follows from Lemma 2.43 that $\tau_{jk} = \tau_k$, Γ in this case is the one given by (v) of (2.100). Finally when $\lambda = 1$ and $\mu = 0$, $\tau_{kj} = 2\tau_j$ by Lemma 2.41(3), and hence, according to the types of edge connecting vertices v_i and v_k , there are four possibilities

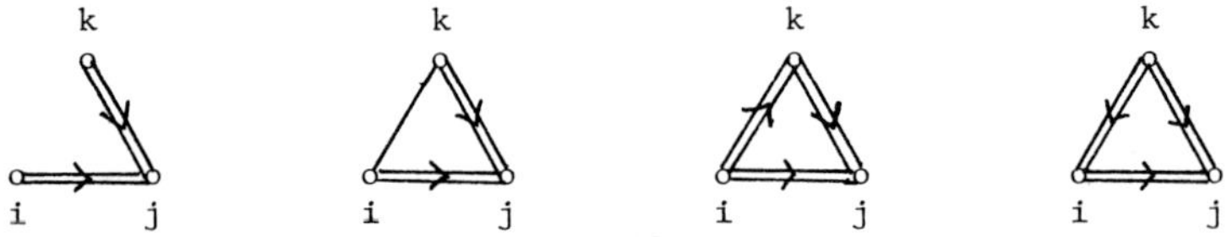

each of which appears in (2.100). □

The following result determines all allowable graphs of three nodes.

Lemma 2.47. Let i, j, k be distinct indices. The graph $\Gamma = [v_i, v_j, v_k]$ of the subalgebra $B = S_i + S_j + S_k$ is allowable if and only if it is one of the following

(2.101) 

or it is one of the graphs in (2.100), or it is one of the following

(2.102) (i)  (ii)  (iii) 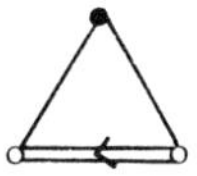

or it is one of the graphs in (2.100) with the top node darkened, where in the last two cases one of τ_i, τ_j, τ_k is zero and that zero one corresponds to the solid node in every case.

Proof. Assume first that Γ is an allowable graph which is not connected. Then Γ has a connected component consisting of a single node, and it is easy to see that the remaining components of Γ correspond to the configurations depicted in (2.97), which are allowable. Thus, in this case Γ is one of the graphs in (2.101).

For the rest of proof, suppose that Γ is connected. Thus, by Lemma 2.40, at least one of τ_i, τ_j, τ_k is nonzero. If all of these forms are nonzero, then Γ is in the situation of Lemma 2.46, and hence it is given by one of the graphs in (2.100). Assume then that one of τ_i, τ_j, τ_k is zero, and that $\tau_k = 0$ but $\tau_i \neq 0$. Let the solid node correspond to τ_k, and let e be an idempotent of S_i. Suppose that Γ is allowable. Then, since $\tau_{ik} = 0$, $S_k \subseteq B_\theta(e)$ for some $\theta = \tau_{ki}(e) = 0, \frac{1}{2}$, or 1, and by Lemma 2.41, for each element $y \in S_j$, $y - \mu\tau_{ij}(y)e$ lies in $B_\lambda(e)$, where $\lambda = \tau_{ji}(e) = 0, \frac{1}{2}$, or 1 and $\mu = 1, 2$, or 0 respectively. If z denotes an element of S_k, then

$$(2.103) \qquad (y - \mu\tau_{ij}(y)e) \circ z = \tau_{kj}(y)z - \mu\theta\tau_{ij}(y)z ,$$

since $\tau_{jk} = 0$. We proceed the proof following the relations between τ_i and τ_j given by Lemma 2.43.

Assume $\tau_{ij} = \tau_{ji} = 0$. Since $\tau_{ik} = 0$ and Γ is connected, $\tau_{ki}(e) = \theta \neq 0$ and hence $\theta = \frac{1}{2}$ or 1, which implies $S_k \subseteq B_{\frac{1}{2}}(e)$ or $S_k \subseteq B_1(e)$. It must be then that $S_k \subseteq B_{\frac{1}{2}}(e)$, since if $S_k \subseteq B_1(e)$, then $S_k \circ S_j \subseteq B_1(e) \circ B_0(e) = 0$ and $\tau_{kj} = \tau_{jk} = 0$ to give a contradiction to the connectedness of Γ. This shows by Lemma 2.41 that $\tau_{ki} = \tau_i$. Thus, vertices v_k and v_i must be connected by a single edge. Since $\tau_{jk} = 0$, arguing with S_j in place of S_i produces the same result for v_k and v_j. Thus, Γ is given by case (i) in (2.102), and this is also the first configuration depicted in (2.98) which has been shown to be allowable in Example 2.3.

Suppose that $\tau_{ji} = \tau_i$, being case (2) of Lemma 2.43. Thus, $\tau_{ij} = \tau_j$, implying $\lambda = \frac{1}{2}$ and $\mu = 2$. It follows from this that product (2.103) has no component in $B_\theta(e)$. Therefore, $\tau_{kj} = 2\theta\tau_{ij} = 2\theta\tau_j$, while $\tau_{ki} = 2\theta\tau_i$ for $\theta = \frac{1}{2}$ or 1, since if $\theta = 0$, then $\tau_{ki} = \tau_{ik} = \tau_{jk} = \tau_{kj} = 0$ to give a contradiction to the connectedness of Γ. If $\theta = 1$, then Γ is the one in (v) of (2.100), while Γ is given by the one in (i) of (2.100) if $\theta = \frac{1}{2}$. The same argument as in the proof of Lemma 2.46 shows that all graphs in (2.100) with the top node darkened are allowable.

Assume now that $\tau_{ji} = 2\tau_i$, being case (3) of Lemma 2.43, so that $\lambda = 1$ and $\mu = 0$. Consider the case of $\theta = 0$. Then, product (2.103) lies in $B_1(e) \circ B_0(e) = 0$, and hence $\tau_{kj} = \tau_{ki} = 0$, which implies that Γ is disconnected. Therefore, we may assume $\theta = \frac{1}{2}$ or 1. Suppose $\theta = \frac{1}{2}$, so that $\tau_{ki} = \tau_i$. If $\tau_{kj} = 0$ in this case, then Γ is the one in (ii) of (2.102). Since, for $B_{\frac{1}{2}}(e) = \ker \tau_i + S_k$, $B_1(e) = S_j$ and $B_0(e) = 0$, B satisfies the conditions of Lemma 2.45, it follows that the graph in (ii) of (2.102) is allowable. If $\tau_{kj} \neq 0$, then $\tau_j \neq 0$ and hence $\tau_{kj} = \tau_j$ or $\tau_{kj} = 2\tau_j$. The former case gives the graph in (iii) of (2.102), which is also the configuration in (ii) of (2.98) and is allowable as shown in Example 2.3. In the latter case, let f be an idempotent of S_j. Then, since $\tau_{kj}(f) = 1$ and $\tau_{ij} = 0$, $S_k \subseteq B_1(f)$ and $x - \tau_{ji}(x)f \in B_0(f)$ for each $x \in S_i$. Thus $z \circ (x - \tau_{ji}(x)f) = \tau_{ki}(x)z - \tau_{ji}(x)z = 0$ for all $z \in S_k$ and $x \in S_i$, to give a contradiction; $\tau_{ki} = \tau_{ji} = 2\tau_i$. Hence the case of $\tau_{kj} = 2\tau_j$ can not occur. Consider the situation where $\theta = 1$, so that $\tau_{ki} = 2\tau_i$. If $\tau_{kj} = 0$, then Γ is the graph in (ii) of (2.100). If $\tau_{kj} \neq 0$, and hence $\tau_j \neq 0$, then either $\tau_{kj} = \tau_j$ or $\tau_{kj} = 2\tau_j$, which respectively produces the graph in (iii) or in (iv) of (2.100).

It remains to consider the case $\tau_{ji} = 0$ and $\tau_{ij} = 2\tau_j \neq 0$. This case can be handled by reversing the roles of i and j in the preceding paragraph. This completes the proof. □

The following theorem is our principal result in this section which classifies all allowable graphs associated with the algebra A defined by product (2.94).

Theorem 2.48. Let $A = S_1 \oplus \cdots \oplus S_m$ be a finite-dimensional algebra over a field F of characteristic $\neq 2,3,5$ with multiplication defined by (2.94). Then A is power-associative if and only if each subgraph consisting of two nodes is one described in (2.97) and of three nodes is one of the graphs described in Lemma 2.47, and there are no subgraphs of four nodes given by (2.99), where each node in (2.99) denoted by "○" corresponds to an S_j with $\tau_j \neq 0$, by "●" corresponds to an S_j with $\tau_j = 0$, and by "⊖" corresponds to an arbitrary S_j.

Proof. Assume that A is power-associative. It has been shown in Example 2.4 that the subgraphs of four nodes given by (2.99) are not present. This together with Lemmas 2.44, 2.46, and 2.47 establishes the necessity of the conditions. To show the sufficiency of the conditions, in view of Lemma 2.39, it suffices to verify that each subalgebra consisting of four or fewer summands is power-associative. Therefore, we may assume that $A = S_1 + S_2 + S_3 + S_4$ where the sum of any three subspaces is power-associative and where the graph of A is not that described in (2.99). If the graph of A is disconnected, then our assumption implies that A is power-associative. Thus, in what follows, we further assume that the graph of A is connected.

Let us first assume that, after renumbering indices if necessary, $\tau_1, \cdots, \tau_k$ are nonzero, and that each of the vertices $v_2, \cdots, v_k$ is connected to v_1 by a single edge. It follows from Lemma 2.46 that each node v_i is connected to each v_j by a single edge. Thus, the multiplication in $S_p = S_1 \oplus \cdots \oplus S_k$ is defined by a single linear form τ_p given by $\tau_p(x_1 + \cdots + x_k) = \tau_1(x_1) + \cdots + \tau_k(x_k)$ for $x_i \in S_i$. If w is any vertex of the graph $\Gamma = [v_1,v_2,v_3,v_4]$ connected to some v_j, $1 \leq j \leq k$, then by Lemmas 2.46, and 2.47, whatever type of bond connects w to v_j, that same type must bond w to each v_i for $1 \leq i \leq k$. Hence, as in the proof of Lemma 2.46, we can replace $S_1, \cdots, S_k$ with the subspace S_p and collapse the vertices $v_1, \cdots, v_k$ to a single vertex. Therefore, if $k > 1$, then by our assumptions A is power-associative. In the remainder of the proof, it suffices to assume that if $\tau_i \neq 0$, then for each j with $\tau_j \neq 0$, the corresponding nodes are not connected by a single edge.

We focus on a vertex v_i of Γ corresponding to S_i with $\tau_i \neq 0$. If v_j is any vertex of Γ with $j \neq i$, then by the last assumption there are four possibilities: $\tau_{ji} = 2\tau_i$, $\tau_{ij} = 2\tau_j$, $\tau_{ji} = \tau_i$ with $\tau_j = 0$, and $\tau_{ij} = \tau_{ji} = 0$. Let e be an idempotent of S_i , for each j with $\tau_{ji} = 0$, let S_j' denote the subspace of elements of the form $y' = y - \tau_{ij}(y)e$, and let B_0 be the sum of all the subspaces S_j' . Suppose that B_1 is the sum of those S_k with $\tau_{ki} = 2\tau_i$ and that $B_{\frac{1}{2}}$ is the sum of $\ker \tau_i$ together with those S_ℓ having $\tau_\ell = 0$ and $\tau_{\ell i} = \tau_i$. Then $A = Fe + B_0 + B_{\frac{1}{2}} + B_1$ and it follows from power-associativity of the sum of three summands that B_0 and B_1 are subalgebras and

$$B_0 \circ B_1 = 0 = B_{\frac{1}{2}} \circ B_{\frac{1}{2}} . \tag{2.104}$$

Suppose first that for some index i with $\tau_i \neq 0$ the associated spaces B_0 and B_1 are nonzero. To show power-associativity holding for A , by (2.104) it suffices to verify that identity (2.95) holds for w, x, y, z where w, x, y lie in distinct subspaces of the form $S_j \subseteq B_{\frac{1}{2}}$ or B_1 , or $S_j' \subseteq B_0$, and where $z \in S_i$. If $z \in \ker \tau_i$, then since $B_0 \circ \ker \tau_i \subseteq \ker \tau_i$, $B_{\frac{1}{2}} \circ \ker \tau_i = 0 = B_1 \circ \ker \tau_i$, and $B_1 \neq 0$, all terms in (2.95) are zero. We may assume that $z = e$. Since $Fe + B_0 + B_1$ is power-associative by (2.104) and Lemma 2.45, we may suppose further that $w \in S_j \subseteq B_{\frac{1}{2}}$, $x \in S_k' \subseteq B_0$, and $y \in S_\ell \subseteq B_1$ where i, j, k, ℓ are distinct. Using the relations among τ_{pq}'s, (2.104), and the fact that $B_\lambda \subseteq A_\lambda(e)$ for $\lambda = 0, \frac{1}{2}$, or 1 , the left side of (2.95) in this case reduces to

$$5(w \circ y) \circ x - 5(w \circ x) \circ y . \tag{2.105}$$

By power-associativity of $S_i + S_j + S_k$ and $S_i + S_j + S_\ell$, we have $w \circ x = \mu w$ and $w \circ y = \nu w$ for some scalars μ, ν. Therefore, $(w \circ y) \circ x = \mu\nu w = (w \circ x) \circ y$ to show that (2.105) is zero. Hence whenever for some index i the corresponding spaces B_0 and B_1 are nonzero, A is power-associative.

It suffices to assume that for each index i with $\tau_i \neq 0$ either the associated space B_0 or B_1 is zero. Consider first the case where $B_0 = 0$. If $\tau_\ell = 0$ for every ℓ with $S_\ell \subseteq B_1$, then $B_{\frac{1}{2}} \circ B_1 = 0$ and by Lemma 2.45 A is power-associative. On the other hand, if $\tau_\ell \neq 0$ for some $S_\ell \subseteq B_1$, then our assumption implies that the space B_1' associated with the index ℓ is zero, since $\tau_{i\ell} = 0$ gives $B_0' \neq 0$. Therefore, we may reduce to considering the case that for some index i with $\tau_i \neq 0$, the space B_1 is zero and hence $A = Fe + B_0 + B_{\frac{1}{2}}$. In this situation there can exist at most one index k with $\tau_{ik} = 2\tau_k \neq 0$, for if two such indices j, k existed, then there would be a subgraph of the form

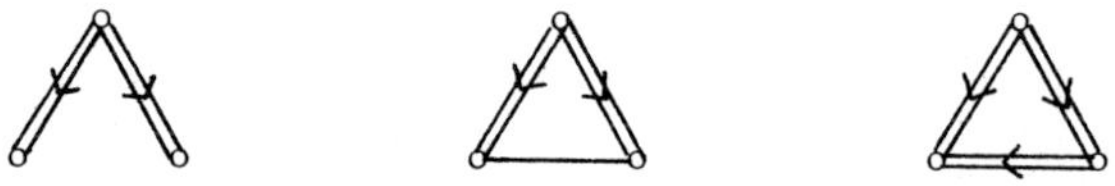

The first graph is not allowable by Lemma 2.46, the second graph is ruled out by the assumption that there are two nodes with $\tau_j \neq 0, \tau_k \neq 0$ connected by a single edge, and the last case can not occur, since there is an index k with $\tau_k \neq 0$ such that the associated space B_1' is nonzero. We let w, x, y lie in distinct subspaces of the form $S_j \subseteq B_{\frac{1}{2}}$ or $S_j' \subseteq B_0$ and let $z \in S_i$. If $z \in \ker \tau_i$ and $S_j \subseteq B_{\frac{1}{2}}$ for $j \neq i$, then it is easy to see that all terms in (2.95) are zero. We may assume that w, x, y lie in distinct subspaces of the form $S_j' \subseteq B_0$. Since there is at most one $k \neq i$ with $\tau_{ik} = 2\tau_k \neq 0$, it follows that there is at most one index k such that $S_k' \subseteq B_0$ with $\tau_{ik} \neq 0$. Thus we can assume $w \circ z = \mu z$, $x \circ z = 0 = y \circ z$, in which case it is easy to show that all terms in (2.95) are zero. Consider finally the case that $z = e$. If there are no $S_j \subseteq B_{\frac{1}{2}}$ with $j \neq i$ or if there is more than one, then all terms in (2.95) with w, x, y, e substituted are zero again. Thus we may assume $w \in S_j \subseteq B_{\frac{1}{2}}$, $x \in S_k$, and $y \in S_\ell$ where $\tau_{ki} = \tau_{\ell i} = 0$.

The remainder of the proof is devoted to showing power-associativity of A in this last case. Since there is at most one $k \neq i$ with $\tau_{ik} = 2\tau_k \neq 0$, we may further assume that $\tau_{i\ell} = 0$. Hence, we can let

$$x \circ w = \delta w,\ y \circ w = \mu w,\ x \circ y = \nu y + \pi x,\ e \circ x = \rho e ,$$

where $\delta = \tau_{jk}(x)$, $\mu = \tau_{i\ell}(y)$, $\nu = \tau_{\ell k}(x)$, $\pi = \tau_{k\ell}(y)$, and $\rho = \tau_{ik}(x)$. Then the left side of (2.95) with e, w, x, y substituted reduces to

$$(6\delta\mu - 3\nu\mu - 3\pi\delta - 3\mu\rho + \pi\rho)w . \tag{2.106}$$

Because of the assumptions on the spaces B_0' and B_1' associated with index k , and because no single edge can join v_k and v_ℓ when τ_k and τ_ℓ are nonzero, either τ_{ik} or $\tau_{k\ell}$ is zero and either $\tau_{\ell k}$ or $\tau_{k\ell}$ is zero. Consequently, $\pi\rho = 0$ in any case. We use these results together with the fact that the following subgraphs

(2.107) 

are not allowable by Lemma 2.47, to verify that except for the two graphs in (2.99), relation (2.106) is zero for all other possible graphs.

Consider first the case that $\tau_{ik} = 0$, so that $\rho = 0$. Since vertices v_k and v_ℓ will play identical roles, we may assume without loss of generality that $\tau_{\ell k} = 0$, and hence $\nu = 0$. Thus, (2.106) reduces to $3(2\delta\mu - \delta\pi)w$, and we show that $\pi = 2\mu$ in this case. If $\tau_{jk} = 0$, then $\delta = 0$ and (2.106) is zero. Suppose then that $\tau_{jk} \neq 0$, so that $\tau_k \neq 0$ and either $\tau_{jk} = \tau_k$ or $\tau_{jk} = 2\tau_k$. Since $\tau_j = 0$ and $\tau_{ji} = \tau_i$, the latter case produces a subgraph depicted in (2.107). Thus, the former case $\tau_{jk} = \tau_k$ is present to give a subgraph of the form

(2.108) 

If $\tau_\ell = 0$, then $\tau_{k\ell} = \tau_{\ell k} = 0$ and v_k, v_ℓ are not connected, which implies that either the graph Γ of A is disconnected or Γ contains a subgraph displayed in (2.107) or one in (i) of (2.99). Thus, $\tau_\ell \neq 0$, $\tau_{k\ell} \neq 0$, and $\tau_{k\ell} = 2\tau_\ell$, since v_k and v_ℓ can not be joined by a single edge. Using the fact that there is no subgraph given by (2.107), we also find that v_j and v_ℓ must be connected by a single edge. These results combine with (2.108) to give the graph of A as

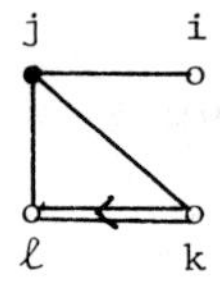

Thus $\tau_{k\ell} = 2\tau_\ell$ and $\tau_{j\ell} = \tau_\ell$ to show that $\pi = 2\mu$, which implies that (2.106) is zero.

Consider next the case where $\tau_{ik} \neq 0$, so that $\tau_{k\ell} = 0$ and $\pi = 0$. Since v_i and v_k in this case can not be joined by a single edge, $\tau_{ik} = 2\tau_k$. If $\tau_{j\ell} = 0$, then $\mu = 0$ and (2.106) is zero. Assume

$\tau_{j\ell} \neq 0$, so that $\tau_\ell \neq 0$ and $\tau_{j\ell} = \tau_\ell$ or $\tau_{j\ell} = 2\tau_\ell$. Since the first graph in (2.107) cannot occur, it must be that $\tau_{j\ell} = \tau_\ell$. Thus, the following types of edge are present in the graph Γ of A .

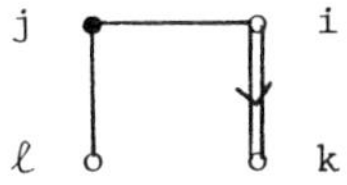

Since the second graph of (2.107) can not occur, v_j and v_k must be connected, while v_k and v_ℓ can not be connected by a single edge, since $\tau_k \neq 0$ and $\tau_\ell \neq 0$. It follows from this analysis that only the graphs

produce subgraphs described in Lemma 2.47, but do not give a subgraph of one of the types in (2.107) or a graph in (ii) of (2.99). In the first case, $\rho = 2\delta$ and $\nu = 0$, and (2.106) becomes $6\delta\mu - 3\mu\rho = 6\delta\mu - 6\delta\mu = 0$. In the second case, $\rho = \nu = \delta$ to imply that (2.106) is zero. This completes the proof to show that the conditions are sufficient for power-associativity of A . □

Theorem 2.48 is readily applied to determine all finite-dimensional power-associative Malcev-admissible algebras A over a field of characteristic 0 such that A^- is split semisimple.

<u>Corollary 2.49</u>. Let A be a finite-dimensional Malcev-admissible algebra with product "*" over a field F of characteristic 0 such that A^- is split semisimple over F . Let $A^- = S_1 \oplus \cdots \oplus S_m$ be the decomposition of A^- into simple ideals of A^- . Then A is power-associative

if and only if the product "$*$" is given by

$$x * y = \frac{1}{2}[x,y] + \tau_{ij}(y)x + \tau_{ji}(x)y \tag{2.109}$$

for $x \in S_i$, $y \in S_j$, where for each ordered pair (i,j) τ_{ij} is a linear form on S_j and the linear forms τ_{ij} satisfy the conditions described in Theorem 2.48.

Proof. If A is power-associative, then in view of Theorems 2.33 and 2.38 the product "$*$" is described by (2.109). But then since the commutative algebra A^+ is power-associative and has product defined by $x \circ y = \tau_{ij}(y)x + \tau_{ji}(x)y$ for $x \in S_i$ and $y \in S_j$, the necessity of the conditions follows from Theorem 2.48. Assume that the product "$*$" in A is given by (2.109), and that the linear forms τ_{ij} satisfy the conditions described in Theorem 2.48. Then that theorem implies that A^+ is power-associative, which together with Lemmas 1.10(ii) and 1.11 gives power-associativity of A , since F has characteristic zero. □

The investigation in this section is based on work of Benkart [2].

3

INVARIANT OPERATORS IN SIMPLE LIE ALGEBRAS AND FLEXIBLE MALCEV-ADMISSIBLE ALGEBRAS WITH A^- SIMPLE

3.1. Introduction

We have determined in Chapter 2 (Corollary 2.34 and Theorem 2.38) the structure of flexible Malcev-admissible algebras A over a field F of characteristic 0, when A^- is split semisimple over F. In this chapter we determine the structure of such algebras A, when A^- is simple over F. The result for semisimple A^- immediately follows from the case for simple A^-. Since the derivation algebra Der M of a semisimple Malcev algebra M is semisimple, we regard A as a Der A^--module. Representation theory of simple Lie algebras plays a main role in our investigation. Specifically, we first determine the so-called adjoint operators of a split simple Lie algebra in its irreducible modules, which in turn characterize the Der A^--module actions imposed on the multiplication of A. An adjoint operator of a Lie algebra L into an L-module V is an L-module homomorphism of L into the tensor product module $V \otimes V^*$, where V^* is the dual module of V. The results in this chapter are based on work of Okubo and Myung [3], Benkart and Osborn [1], and Myung [5].

Throughout this chapter, all algebras, modules, and representations are, unless otherwise stated, finite-dimensional.

Recall that a representation ρ of a Lie algebra L over F in a vector space V over F is a Lie homomorphism of L into $(\mathrm{Hom}_F V)^-$ and the composition xv for $x \in L$ and $v \in V$ defined by $xv = \rho(x)v$ makes V an L-module over F, which we call the *L-module afforded* by the representation ρ. Conversely, any L-module V gives rise to a representation ρ of L into V by defining $\rho(x)v = xv$ for $x \in L$

and $v \in V$. This representation is called the *representation afforded by the L-module* V . If V and W are L-modules, then the tensor product $V \otimes W$ is regarded as an L-module under the composition $x(v \otimes w) = xv \otimes w + v \otimes xw$ for $x \in L$, $v \in V$ and $w \in W$. An L-module homomorphism ϕ of V into W is a linear mapping of V into W such that $\phi(xv) = x\phi(v)$ for all $x \in L$ and $v \in V$. We denote by $\mathrm{Hom}_L(V,W)$ the set of L-module homomorphisms of V into W .

We recall some known facts about Malcev algebras. Let M be a Malcev algebra with product $[x,y]$ over a field F . The Killing form $K(\ ,\ )$ defined by

$$K(x,y) = \mathrm{tr}\ \mathrm{ad}_x \mathrm{ad}_y$$

is a symmetric invariant bilinear form on M (Sagle [1, p.430]) . Let $J(x,y,z)$ denote the Jacobian,

$$J(x,y,z) = [[x,y],z] + [[y,z],x] + [[z,x],y] .$$

Then, the set $N = \{x \in M \mid J(x,M,M,) = 0\}$ is called the *J-nucleus* of M , and is an ideal of M (Sagle [1, p.440]).

Definition 3.1. Let A be an arbitrary algebra over a field F. The Lie subalgebra of $(\mathrm{Hom}_F\ A)^-$ generated by all left and right multiplications L_x and R_y in A for $x,y \in A$ is termed the *Lie multiplication algebra* of A . Each derivation of A which lies in the Lie multiplication algebra of A is called *inner*. The set $\mathrm{IDer}\ A$ of all inner derivations of A is shown to be an ideal of the derivation algebra $\mathrm{Der}\ A$ of A (Schafer [3, p.21]), and is called the *inner derivation algebra* of A . □

For a Malcev algebra M, let

$$d(x,y) = ad_{[x,y]} + [ad_x, ad_y] \tag{3.1}$$

for $x,y \in M$. Sagle [1] has shown that

$$\text{IDer } M = ad_N + d(M,M) \tag{3.2}$$

where N is the J-nucleus of M. If L is a Lie algebra, then since $N = L$, IDer L is simply the set ad_L of adjoint mappings on L. If M is simple non-Lie, Malcev, then since $N = 0$, IDer $M = d(M,M)$. Recall that a Lie algebra L over a field F of characteristic $\neq 2,3$ is said to be of *type G* if the scalar extension L_K by the algebraic closure K of F is isomorphic to Der C for some (split) octonion algebra C over K (see Schafer [3, p.81] and Table 2.1). Thus, by a theorem of Cartan and Jacobson, any Lie algebra of type G over F of characteristic $\neq 2,3$ is a 14-dimensional central simple Lie algebra over F (Theorem 5.2, Chapter 5, or Schafer [3, p.82]). If F is algebraically closed of characteristic zero, then, as is well known, a Lie algebra of type G is the unique simple Lie algebra of type G_2. We note that these remarks apply to a central simple, non-Lie, Malcev algebra $M(\alpha,\beta,\gamma)$ given by Table 2.2. Therefore, Der $M(\alpha,\beta,\gamma)$ is a Lie algebra of type G.

For an ideal B of an arbitrary algebra A, define $B^{(i+1)} = B^{(i)}B^{(i)}$ for $i = 0, 1, \ldots$ where $B^{(0)} = B$. If $B^{(i)} = 0$ for some $i \geq 0$, then B is called *solvable*. The (solvable) *radical* Rad A of A is defined to be the maximal solvable ideal of A. If Rad $A = 0$, then A is called *semisimple*. It is well known (Loos [1] and Ravisankar [1]) that if M is a Malcev algebra over F of characteristic 0, then, as for Lie algebras, the following three properties are equivalent:

(3.3) M is semisimple;

(3.4) the Killing form of M is nondegenerate;

(3.5) M is a direct sum of simple ideals,

and each one of (3.3) - (3.5) implies that

(3.6) $$\text{Der } M = \text{IDer } M ,$$

(3.7) $\text{Der } M$ is a semisimple Lie algebra .

Relation (3.6) was first proved by Sagle [3]. Assume that A is an arbitrary algebra with product xy over a field F of characteristic 0 and that L is a semisimple subalgebra of $\text{Der } A$. Since A is regarded as an L-module, by Weyl's theorem $A = V_1 \oplus \cdots \oplus V_n$, where the V_i are irreducible L-modules. Let p_i denote the projection of A onto V_i . For each triple (i,j,k) , the mapping $\phi : V_i \otimes V_j \to V_k$ defined by

(3.8) $$\phi(x \otimes y) = p_k(xy)$$

for $x \in V_i$ and $y \in V_j$ is an L-module homomorphism of $V_i \otimes V_j$ into V_k , since $L \subseteq \text{Der } A$. Conversely, for any Lie algebra L , by taking a sum of irreducible L-modules $A = V_1 \oplus \cdots \oplus V_n$ and prescribing an element ϕ_k in $\text{Hom}_L (V_i \otimes V_j, V_k)$, one can define a product xy on A by putting

(3.9) $$xy = \phi_1(x \otimes y) + \cdots + \phi_n(x \otimes y)$$

for $x \in V_i$, $y \in V_j$, and by extending (3.9) bilinearly. It is easy to see that L acts as derivations on A under the product given by (3.9).

Therefore, for an algebra A with a specified semisimple Lie algebra of derivations, the determination of multiplication on A reduces to that of $\mathrm{Hom}_L\ (V_i \otimes V_j, V_k)$ for all i, j, k. The following result is instrumental to determine $\mathrm{Hom}_L\ (V_i \otimes V_j, V_k)$.

Lemma 3.1. Let L be a semisimple Lie algebra over an algebraically closed field of characteristic 0. Assume that U is an L-module with decomposition $U = U_1 \oplus \cdots \oplus U_m$, where the U_i are irreducible L-modules, and let W be an irreducible L-module. If k denotes the number of U_i's isomorphic to W, then

$$k = \dim_F \mathrm{Hom}_L\ (U,W) = \dim_F \mathrm{Hom}_L\ (W,U) .$$

Proof. Assume first that U is an irreducible L-module, and let ϕ be a nonzero L-module homomorphism of U into W. Since U and W are irreducible, ϕ is an L-module isomorphism of U to W. It then follows from Schur's lemma that any element of $\mathrm{Hom}_L(U,W)$ is a multiple of ϕ. Thus,

$$(3.10) \qquad \dim_F \mathrm{Hom}_L\ (U,W) = \begin{cases} 1, & U \cong W \text{ as an L-module}, \\ 0, & U \not\cong W \text{ as an L-module}. \end{cases}$$

We may assume then that $U_1, \cdots, U_k$ are the irreducible summands of U isomorphic to W and that ϕ_i denotes an L-module isomorphism of U_i to W for $i = 1, \cdots, k$. Let $\sigma_i = \phi_i p_i$ for $i = 1, \cdots, k$, where p_i is the projection of U onto U_i. For $\rho \in \mathrm{Hom}_L\ (U,W)$, we see that whenever $\rho(U_j) \neq 0$, the restriction $\rho|U_j$ is an L-module isomorphism of U_j to W, and hence by (3.10) $\rho(U_j) = 0$ for all $j > k$ and $\rho|U_i = c_i\phi_i$ for some $c_i \in F$ and $i = 1, \cdots, k$. If $x \in U$, then $\rho(x) = \Sigma\, \rho p_i(x) = \Sigma_{i=1}^k c_i\phi_i p_i(x) = \Sigma_{i=1}^k c_i\sigma_i(x)$. Noting $\sigma_i\phi_j^{-1} = \delta_{ij} I_W$ where I_W is the identity map on W, it follows that $\sigma_1, \cdots, \sigma_k$ form a basis of $\mathrm{Hom}_L\,(U,W)$, giving the first part of the relation. For the second part, we use the same notations as above, so that ϕ_i^{-1} is an L-module isomorphism of W to U_i for $i = 1, \cdots, k$. If we let I_i denote the injection of U_i into U , then $I_i\phi_i^{-1}$ is an L-module isomorphism of W into U . For $0 \neq \rho \in \mathrm{Hom}_L\,(W,U)$, whenever $p_i\rho \neq 0$, it is an L-module isomorphism of W to U_i , and hence by (3.10) $p_i\rho = 0$ for $i > k$ and $p_i\rho$ is a multiple of ϕ_i^{-1} for $i = 1, \cdots, k$. Thus, $I_1\phi_1^{-1}, \cdots, I_k\phi_k^{-1}$ form a basis of $\mathrm{Hom}_L\,(W,U)$. □

For irreducible L-modules U, V, W , to enumerate dimension of $\mathrm{Hom}_L\,(U\otimes V,W)$ by means of Lemma 3.1 the decomposition of $U \otimes V$ into irreducible L-modules must be known. Such decompositions are, however, not known for the general case. When L is a simple Lie algebra, the decomposition of $L \otimes L$ into irreducible L-modules has appeared in a number of papers (Djoković [1], Krämer [1], and Benkart and Osborn [1]). A complete list of decompositions of $L \otimes L$ will be given in the next section. The decomposition of $U \otimes V$ is also known to play an important role in particle physics. It is possible to compute the dimension for the special case of $\mathrm{Hom}_L\,(V\otimes V^*,L)$ without decomposing $V \otimes V^*$, where V^* is the dual L-module of V . This case will be discussed in the next section.

Let L and F be the same as in Lemma 3.1, and let U and V be irreducible L-modules. If Fz denotes a one-dimensional trivial L-module, then relation (3.10) implies

$$\dim_F \mathrm{Hom}_L\,(Fz\otimes U,V) = \begin{cases} 1, & \text{if } U \cong V , \\ 0, & \text{if } U \ncong V . \end{cases} \tag{3.11}$$

3.2. INVARIANT OPERATORS

Let U, V and W be L-modules for an arbitrary Lie algebra L over a field F. Recall that the dual space W^* can be converted into an L-module under the composition xf defined by $(xf)(x) = -f(xw)$ for $w \in W$, $f \in W^*$, and $x \in L$. For $v \otimes f \in V \otimes W^*$, define a linear mapping $(v\otimes f)'$ of W into V by $(v\otimes f)'(w) = f(w)v$ for $w \in W$. It is easy to see that the mapping $(v\otimes f) \to (v\otimes f)'$ is extended to a linear isomorphism of $V \otimes W^*$ into $\mathrm{Hom}_F(W,V)$. We can define an L-module structure on $\mathrm{Hom}_F(W,V)$, so that this linear isomorphism becomes an L-module isomorphism. In fact, for $T \in \mathrm{Hom}(W,V)$ and $x \in L$, define xT by

$$(xT)w = x(Tw) - T(xw) \tag{3.12}$$

for $w \in W$. One can routinely check that $(x(v\otimes f))' = x(v\otimes f)'$ holds for all $x \in L$, $v \in V$, and $f \in W^*$, and hence $\mathrm{Hom}_F(W,V)$ is an L-module under the composition (3.12). Thus, we often identify $V \otimes W^*$ with $\mathrm{Hom}(W,V)$ as an L-module.

<u>Definition 3.2</u>. Let U, V, and W be L-modules over a field F. An L-module homomorphism of U into $\mathrm{Hom}(W,V)$ is called an L-*invariant operator* of U into the pair (W,V), or into the pair (ρ_W,ρ_V) of representations afforded by W and V. When L is regarded as an L-module under the adjoint action, an element of $\mathrm{Hom}_L(L,\mathrm{Hom}(W,V))$ or $\mathrm{Hom}_L(L,V\otimes W^*)$ is called an *adjoint operator* of L into (W,V). □

When $V = W$, $\dim_F \mathrm{Hom}_L(L,\mathrm{Hom}\ V)$ is called the *adjoint dimension* of L in V (Okubo and Myung [3]). Let ρ_U, ρ_V and ρ_W denote the representations afforded by L-modules U, V and W , respectively. Clearly, a linear mapping $\phi : U \to \mathrm{Hom}\,(W,V)$ is L-invariant if and only if

$$\phi(xu) = \rho_V(x)\phi(u) - \phi(u)\rho_W(x) \tag{3.13}$$

for all $x \in L$ and $u \in U$. In particular, when $U = L$ and $V = W$, (3.13) is expressed as

$$\phi([x,y]) = [\rho(x),\phi(y)] \tag{3.14}$$

for all $x,y \in L$, where $\rho = \rho_V = \rho_W$. The notion of adjoint operators by relation (3.14) was first introduced by Wigner [1] for the Lie algebra of the SU(2) group, and the general case has been studied by Okubo [2,13], who computed the adjoint dimension for an irreducible module V for a simple Lie algebra L over the complex number field F by using the transcendental method of the Wigner-Eckart theorem (O'Raifeartaigh [1,p.474]). When V is an irreducible module for a simple Lie algebra over an arbitrary algebraically closed field F of characteristic 0 , Okubo and Myung [3] have computed the adjoint dimension by using representation theory. A shorter proof for a more general case with L semisimple was given by Faulkner [1].

Lemma 3.2. Let U, V and W be modules for a Lie algebra L over an arbitrary field F . Then there exists a linear isomorphism

$$\mathrm{Hom}_L\,(W\otimes U,V) \cong \mathrm{Hom}_L\,(U,\mathrm{Hom}(W,V)) \ .$$

Proof. For $\phi \in \mathrm{Hom}_L\,(U,\mathrm{Hom}(W,V))$, define a linear mapping $\tilde{\phi} : W \otimes U \to V$ by $\tilde{\phi}(w\otimes u) = \phi(u)w$, where $u \in U$ and $w \in W$. If $x \in L$, then

$$\tilde{\phi}(x(w\otimes u)) = \tilde{\phi}(xw\otimes u + w\otimes xu)$$

$$= \phi(u)(xw) + \phi(xu)w$$

$$= \phi(u)\rho_W(x)w + \rho_V(x)\phi(u)w - \phi(u)\rho_W(x)w$$

$$= \rho_V(x)\phi(u)w = x(\tilde{\phi}(w\otimes u))$$

by (3.13), and hence $\tilde{\phi} \in \mathrm{Hom}_L(W\otimes U,V)$. If $\sigma \in \mathrm{Hom}_L(W\otimes U,V)$, then we define a linear mapping $\phi : U \to \mathrm{Hom}(W,V)$ by $\phi(u)w = \sigma(w\otimes u)$ for $u \in U$ and $w \in W$. Since $\phi(xu)w = \sigma(w\otimes xu) = \sigma(x(w\otimes u) - xw\otimes u) = x\sigma(w\otimes u) - \phi(u)(xw) = \rho_V(x)\phi(u)w - \phi(u)\rho_W(x)w$, we have $\phi \in \mathrm{Hom}_L(U,\mathrm{Hom}(W,V))$ and $\sigma = \tilde{\phi}$. Thus, the linear mapping $\phi \to \tilde{\phi}$ is surjective, and is clearly injective. □

Corollary 3.3. Let V be an irreducible module for a simple Lie algebra L over an algebraically closed field of characteristic 0 . Then, we have the linear isomorphisms:

$$\mathrm{Hom}_L(L, \mathrm{Hom}\, V) \cong \mathrm{Hom}_L(L, V\otimes V^*) \cong \mathrm{Hom}_L(V\otimes V^*,L) \cong \mathrm{Hom}_L(V\otimes L,V) .$$

Proof. This is immediate from Lemmas 3.1 and 3.2. □

Assume that F is the same as in Corollary 3.3 and L is a semi-simple Lie algebra over F . Let H be a Cartan subalgebra of L , and let $L = H \oplus \Sigma_\alpha L_\alpha$ be the Cartan decomposition of L relative to H . Denote by $\Pi = \{\alpha_1,\cdots,\alpha_n\}$ a system of simple roots of H , and let $\lambda_1, \cdots, \lambda_n$ be the fundamental weights relative to Π , so that

$$<\lambda_i,\alpha_j> \equiv 2(\lambda_i,\alpha_j) / (\alpha_j,\alpha_j) = \delta_{ij}$$

for $i,j = 1, \cdots, n$. Then, an irreducible L-module is given by a standard cyclic module $V(\lambda)$ of highest weight λ , where $V(\lambda)$ is generated as an L-module by a maximal vector v_λ of weight λ , and λ is a linear form on H which can be expressed as

$$(3.15) \qquad \lambda = m_1\lambda_1 + \cdots + m_n\lambda_n$$

for nonnegative integers m_i (Humphreys [1] and Myung [6]). Furthermore, $V(\lambda)$ is as an L-module isomorphic to the L-submodule of the tensor product $W_1 \otimes \cdots \otimes W_n$ generated by a weight vector of weight λ , where W_i is isomorphic to the tensor product of m_i copies of the so-called basic irreducible L-module $V(\lambda_i)$ of highest weight λ_i (Jacobson [2,p.225]). Note that if σ is a linear form on H , then by the nondegeneracy of (,), there is a unique element $t_\sigma \in H$ such that $(t_\sigma,h) = \sigma(h)$ for all $h \in H$.

Lemma 3.4. Let H, L, and λ be the same as above, and let H_λ be the set of elements $h \in H$ such that $\alpha(h) = 0$ for all roots α with $(\lambda,\alpha) = 0$. Then, dim H_λ equals the number of nonzero $m_i\lambda_i$'s in relation (3.15), and the set of t_{λ_i}'s with $m_i\lambda_i \neq 0$ in (3.15) forms a basis of H .

Proof. After a suitable reordering of simple roots if necessary, we may assume that $m_1, \cdots, m_p$ are all nonzero and $m_{p+1} = \cdots = m_n = 0$. Thus, $< \lambda,\alpha_i > = m_i > 0$ for $1 \leq i \leq p$ and $< \lambda_i,\alpha_i > = 0$ for $p < i \leq n$. Let α be a root of H with $(\lambda,\alpha) = 0$, and let $\alpha = \Sigma_{i=1}^n c_i\alpha_i$, so that the c_i are all nonnegative, or all nonpositive. Then, $0 = (\lambda,\alpha) = \Sigma_{j=1}^p c_j(\lambda,\alpha_j) = \frac{1}{2}\Sigma_{i,j=1}^{\ p} m_ic_j(\alpha_j,\alpha_j)\delta_{ij} = \frac{1}{2}\Sigma_{i=1}^p m_ic_i\cdot(\alpha_i,\alpha_i)$ by (3.15). Since $m_i > 0$ for $i = 1, \cdots, p$ and $(\alpha_i,\alpha_i) > 0$, it must be that $c_1 = \cdots = c_p = 0$, to show that dim $H_\lambda = p$. For

$i = p + 1, \cdots, n$ and $j = 1, \cdots, p$, we see that $\alpha_i(t_{\lambda_j}) = (t_{\alpha_i}, t_{\lambda_j}) = (\alpha_i, \lambda_j) = 0$, which implies that $t_{\lambda_1}, \cdots, t_{\lambda_p}$ form a basis of H_λ . □

When L is simple over F , using representation theory Okubo and Myung [3] have computed the adjoint dimension of L in V as $\dim H_\lambda$. Note that, in view of Corollary 3.3, the adjoint dimension equals $\dim \mathrm{Hom}_L (V \otimes V^*, L)$. More generally, Faulkner [1] has proven that $\dim H_\lambda = \dim \mathrm{Hom}_L (V \otimes V^*, L)$ for a semisimple Lie algebra L . Recall that the weights on the dual module V^* are the negatives of those on V with $-\lambda$ the lowest weight, and that if $v_\lambda \in V$ and $w_{-\lambda} \in V^*$ are the weight vectors corresponding to λ and $-\lambda$, then $v_\lambda \otimes w_{-\lambda}$ generates $V \otimes V^*$ as an L-module and has weight 0 (Jacobson [2]). The proof of the following result is given by Faulkner [1].

Theorem 3.5. Let V be an irreducible module for a semisimple Lie algebra L over an algebraically closed field F of characteristic 0 , and let λ be the highest weight in V for a Cartan subalgebra H of L . Then, the mapping $\psi : \phi \to \phi(v_\lambda \otimes w_{-\lambda})$ is a linear isomorphism of $\mathrm{Hom}_L (V \otimes V^*, L)$ to H_λ , where $v_\lambda \in V$ and $w_{-\lambda} \in V^*$ are weight vectors with weights λ and $-\lambda$.

Proof. Since $v_\lambda \otimes w_{-\lambda}$ generates $V \otimes V^*$ as an L-module and has weight 0 , each $\phi \in \mathrm{Hom}_L (V \otimes V^*, L)$ is uniquely determined by $\phi(v_\lambda \otimes w_{-\lambda})$, and $\phi(v_\lambda \otimes w_{-\lambda}) = h \in H$. Let α be a root such that $(\lambda, \alpha) = 0$, and let $0 \neq e_\alpha \in L_\alpha$ be a root vector corresponding to α . Since λ is highest, either $e_\alpha v_\lambda = 0$ or $e_\alpha w_{-\lambda} = 0$, which combines with $(\lambda, \alpha) = 0$ to show that $e_\alpha v_\lambda = e_\alpha w_{-\lambda} = 0$ (Myung [6,p.104]). Thus,

$$0 = \phi(e_\alpha v_\lambda \otimes w_{-\lambda} + v_\lambda \otimes e_\alpha w_{-\lambda}) = \phi(e_\alpha(v_\lambda \otimes w_{-\lambda})) = [e_\alpha, h] = -\alpha(h)e_\alpha ,$$

and $\alpha(h) = 0$, which implies that $\phi(v_\lambda \otimes w_{-\lambda}) \in H$. The mapping ψ is clearly linear and injective. To show ψ is surjective, we use relation (3.15), and by Lemma 3.4 it suffices to verify that each t_{λ_i} for $m_i\lambda_i \neq 0$ lies in the image of ψ . Consider first the tensor product $U = V_1 \otimes \cdots \otimes V_m$ of irreducible L-modules V_i , and identify the dual module U^* with $V_1^* \otimes \cdots \otimes V_m^*$. Let $U_0 = V_1 \otimes \cdots \otimes V_{m-1}$ and $U_0^* = V_1^* \otimes \cdots \otimes V_{m-1}^*$, and let $\langle\ ,\ \rangle$ be a pairing between U_0 and U_0^* , so that $\langle xv,w\rangle + \langle v,xw\rangle = 0$ for $v \in U_0$, $w \in U_0^*$ and $x \in L$. For $\phi_m \in \mathrm{Hom}_L\,(V_m \otimes V_m^*, L)$, define a mapping $\phi : U \otimes U^* \to L$ by

$$\phi((v_0 \otimes v) \otimes (w_0 \otimes w)) = \langle v_0, w_0\rangle\, \phi_m(v \otimes w)$$

for $v_0 \in U_0$, $w_0 \in U_0^*$, $v \in V_m$, and $w \in V_m^*$. It is easy to see that ϕ is an L-module homomorphism of $U \otimes U^*$ into L . Let η_i be the highest weight of V_i with weight vector v_{η_i} , and let S be the irreducible L-submodule of U with highest weight $\eta_1 + \cdots + \eta_m$. If $\bar{\phi}$ denotes the restriction of ϕ to $S \otimes S^*$, then clearly $\bar{\phi} \in \mathrm{Hom}_L\,(S \otimes S^*, L)$, and moreover

$$\bar{\phi}(v_{\eta_0} \otimes v_{\eta_m} \otimes w_{-\eta_0} \otimes w_{-\eta_m}) = \langle v_{\eta_0}, w_{-\eta_0}\rangle\, \phi_m(v_{\eta_m} \otimes w_{-\eta_m}) ,$$

where $v_{\eta_0} \in U_0$ is a weight vector of weight $\eta_0 = \eta_1 + \cdots + \eta_{m-1}$. Thus, $\psi(\bar{\phi})$ is a nonzero multiple of $\psi(\phi_m)$.

Since any irreducible L-module is isomorphic to the tensor product of irreducible modules W_i for the simple summands L_i of L (Goto and Grosshans [1,p.326]), and each W_i is isomorphic to the L_i-submodule of the tensor product of basic irreducible modules generated by a highest weight vector, by the reduction process above we may assume that L is simple and V is a basic irreducible L-module corresponding to a fundamental weight $\lambda = \lambda_i$ with $m_i\lambda_i \neq 0$ in (3.15). We identify $\mathrm{Hom}\,V = V \otimes V^*$, and since L acts faithfully on V, we regard L as a subalgebra of $(\mathrm{Hom}\,V)^-$, and hence the L-module action on $\mathrm{Hom}\,V$ is given by $x \cdot A = [x,A]$ for $x \in L$ and $A \in \mathrm{Hom}\,V$ (see (3.12)). Thus, if we define $f(A,B) = \mathrm{tr}_V(AB)$ for $A, B \in \mathrm{Hom}\,V$, then $f(\ ,\ )$ is a nondegenerate L-invariant bilinear form on $\mathrm{Hom}\,V$, and extends the trace form on L defined by the L-module V, which must be a nonzero multiple of the Killing form $(\ ,\ )$ on L (Humphreys [1,p.118]). Since $(\ ,\ )$ is nondegenerate, for each $u \in V \otimes V^*$ there is a unique element $\phi(u) \in L$ such that $f(x,u) = (x,\phi(u))$ for all $x \in L$. If x and y are elements of L, then $(x,[y,\phi(u)]) = ([y,x],\phi(u)) = f([y,x],u) = f(x,yu) = (x,\phi(yu))$ for all $x,y \in L$ and $u \in V \otimes V^*$, and hence $[y,\phi(u)] = \phi(yu)$, to show $\phi \in \mathrm{Hom}_L(V \otimes V^*,L)$. For $h \in H$ and $v \in V$, we have

$$h(v_\lambda \otimes w_{-\lambda})v = h(w_{-\lambda}(v)v_\lambda) = w_{-\lambda}(v)\lambda(h)v_\lambda ,$$

which implies that the trace of $h(v_\lambda \otimes w_{-\lambda})$ on V is $w_{-\lambda}(v_\lambda)\lambda(h) = \langle v_\lambda,w_{-\lambda}\rangle\lambda(h)$. Thus, $(h,\phi(v_\lambda \otimes w_{-\lambda})) = f(h,v_\lambda \otimes w_{-\lambda}) = \langle v_\lambda,w_{-\lambda}\rangle\lambda(h) = (h,\langle v_\lambda,w_{-\lambda}\rangle t_\lambda)$ for all $h \in H$. Therefore, $\phi(v_\lambda \otimes w_{-\lambda}) = \langle v_\lambda,w_{-\lambda}\rangle t_\lambda$, as desired. □

When λ is expressed as in (3.15), the dimension of $V(\lambda)$ can be computed from Weyl's formula (Humphreys [1,p.139]) using the integers m_i's. The weights of $V(\lambda) \otimes V(\mu)$ are just the sum of the weights of $V(\lambda)$ with those of $V(\mu)$. Thus, either using the weights and Weyl's formula or Steinberg's formula (Humphreys [1,p.141]), one can decompose $V(\lambda) \otimes V(\mu)$ into the sum of irreducible L-modules. Some special cases of this decomposition can be found in Seligman [2,p.323], for example. The decomposition of $V(\lambda) \otimes V(\mu)$ in general case involves complicated computation. However, once the expression (3.15) is known for λ, the dimension of $\mathrm{Hom}_L (V \otimes V^*, L)$ is easily computed from Theorem 3.5. For the special case of the adjoint module L where L is simple, both the explicit decomposition of $L \otimes L$ and the expression (3.15) for the highest root are well known and are useful for our investigations. For convenience, we here give these expressions.

We note that an explicit expression of (3.15) depends on the ordering of simple roots, and that the highest weight of the adjoint module is simply the highest root which is given by Table 2.3 (Chapter 2, Secion 2.5). Let L be a simple Lie algebra of rank n over the field F, given in Theorem 3.5. Assume that the system Π of simple roots $\alpha_1, \cdots, \alpha_n$ is ordered as labeled in Table 3.1 of the Dynkin diagram.

The ordering of Π in Table 3.1 is the same as one given by Humphreys [1,p.58] or Myung [6,p.133], except for $D_n (n \geq 5)$ where only α_1 and α_2 are interchanged. Let λ_0 denote the highest root relative to Π labeled in Table 3.1. The expression of λ_0 in terms of simple roots $\alpha_1, \cdots, \alpha_n$ is the same as in Table 2.1, except for $D_n (n \geq 5)$. It is easy to see that for the Lie algebra L of type $D_n (n \geq 5)$, λ_0 is given by

(3.16) $$\lambda_0 = 2\alpha_1 + \alpha_2 + 2\alpha_3 + \cdots + 2\alpha_{n-2} + \alpha_{n-1} + \alpha_n .$$

TABLE 3.1

A_n $(n \geq 1)$:

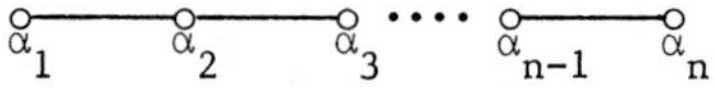

B_n $(n \geq 2)$:

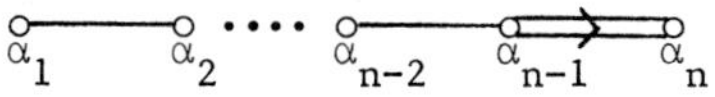

C_n $(n \geq 3)$:

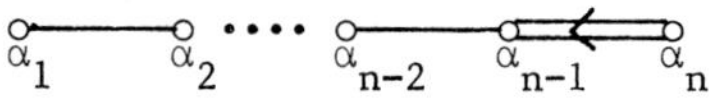

D_4:

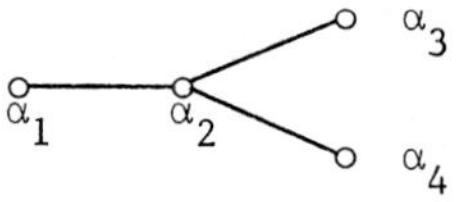

D_n $(n \geq 5)$:

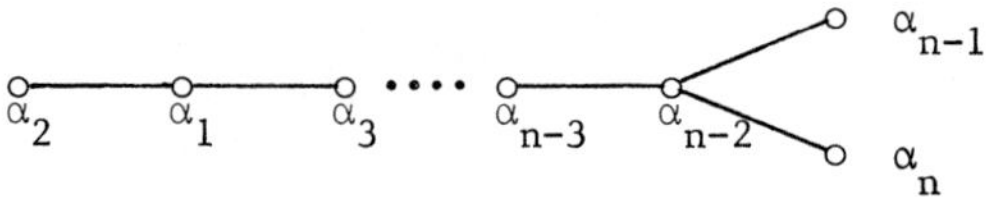

E_6:

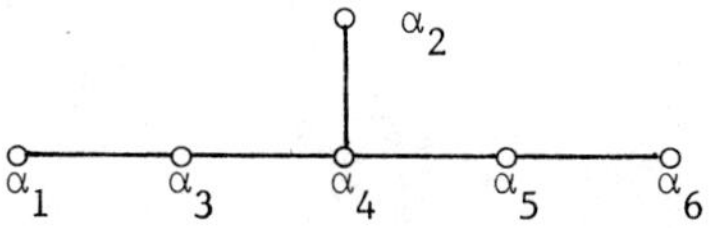

E_7:

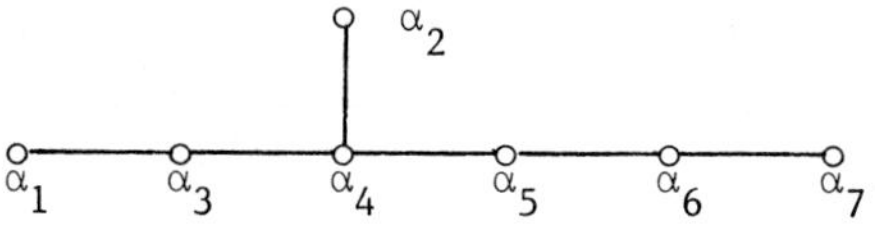

E_8:

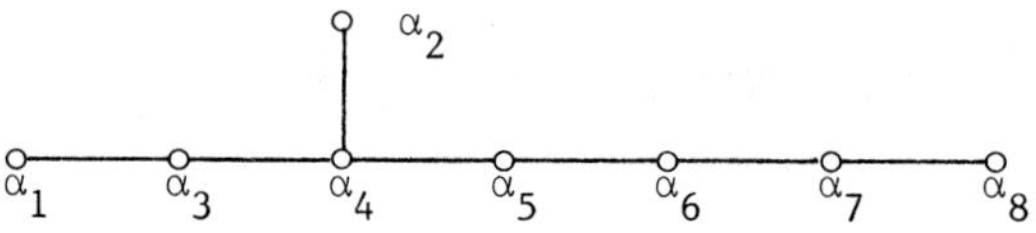

F_4:

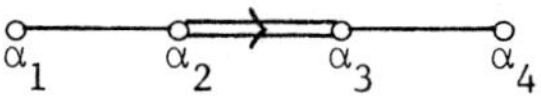

G_2:

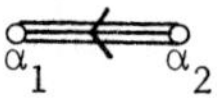

Let $\lambda_1, \cdots, \lambda_n$ be the fundamental weights corresponding to Π, so that each λ_i is expressed by

(3.17) $$\lambda_i = d_{i1}\alpha_1 + \cdots + d_{in}\alpha_n$$

for $d_{ij} \in F$ and $i = 1, \cdots, n$. Since $\delta_{ij} = \langle\lambda_i,\alpha_j\rangle = \sum_k d_{ik} \langle\alpha_k,\alpha_j\rangle$, the $n \times n$ matrix $d(\Pi) = (d_{ij})$ determined by (3.17) is the inverse matrix of the Cartan matrix $c(\Pi) = (\langle\alpha_i,\alpha_j\rangle)$ relative to Π. Note that $c(\Pi)$ and $\det c(\Pi)$ can be computed from the Dynkin diagram in Table 3.1 (Humphreys [1,p.58 and 68]). For convenience, we give the explicit expression (3.17) for each case in Table 3.2. Except for the case of A_n $(n \geq 1)$ - D_n $(n \geq 5)$, we use the inverse matrix notation $d(\Pi) = c(\Pi)^{-1}$.

TABLE 3.2

A_n $(n \geq 1)$: $$\lambda_i = \frac{1}{n+1}[(n-i+1)\alpha_1 + 2(n-i+1)\alpha_2 + \cdots + (i-1)(n-i+1)\alpha_{i-1} + i(n-i+1)\alpha_i + i(n-i)\alpha_{i+1} + \cdots + i\alpha_n]$$

B_n $(n \geq 2)$: $$\lambda_i = \alpha_1 + 2\alpha_2 + \cdots + (i-1)\alpha_{i-1} + i(\alpha_i + \alpha_{i+1} + \cdots + \alpha_n), \quad (1 \leq i < n)$$

$$\lambda_n = \frac{1}{2}(\alpha_1 + 2\alpha_2 + \cdots + n\alpha_n)$$

C_n $(n \geq 3)$: $$\lambda_i = \alpha_1 + 2\alpha_2 + \cdots + (i-1)\alpha_{i-1} + i(\alpha_i + \cdots + \alpha_{n-1} + \frac{1}{2}\alpha_n)$$

D_4:

$$d(\Pi) = \frac{1}{4}\begin{bmatrix} 4 & 4 & 2 & 2 \\ 4 & 8 & 4 & 4 \\ 4 & 8 & 6 & 6 \\ 4 & 8 & 8 & 8 \end{bmatrix}$$

TABLE 3.2 (continued)

D_n $(n \geq 5)$: $\lambda_1 = 2\alpha_1 + \alpha_2 + 2\alpha_3 + \cdots + 2\alpha_{n-2} + \alpha_{n-1} + \alpha_n$

$$\lambda_2 = \alpha_1 + \alpha_2 + \cdots + \alpha_{n-2} + \tfrac{1}{2}(\alpha_{n-1} + \alpha_n)$$

$$\lambda_3 = 2\alpha_1 + \alpha_2 + 3(\alpha_3 + \cdots + \alpha_{n-2}) + \tfrac{3}{2}(\alpha_{n-1} + \alpha_n)$$

$$\lambda_i = 2\alpha_1 + \alpha_2 + \cdots + (i-1)\alpha_{i-1} + i(\alpha_i + \cdots + \alpha_{n-2}) + \tfrac{1}{2}i(\alpha_{n-1} + \alpha_n), \quad (4 \leq i < n - 1)$$

$$\lambda_{n-1} = \tfrac{1}{2}[2\alpha_1 + \alpha_2 + \cdots + (n-2)\alpha_{n-2} + \tfrac{1}{2}n\alpha_{n-1} + \tfrac{1}{2}(n-2)\alpha_n]$$

$$\lambda_n = \tfrac{1}{2}[2\alpha_1 + \alpha_2 + \cdots + (n-2)\alpha_{n-2} + \tfrac{1}{2}(n-2)\alpha_{n-1} + \tfrac{1}{2}n\alpha_n]$$

E_6:

$$d(\Pi) = \frac{1}{3}\begin{bmatrix} 4 & 3 & 5 & 6 & 4 & 2 \\ 3 & 6 & 6 & 9 & 6 & 3 \\ 5 & 6 & 10 & 12 & 8 & 4 \\ 6 & 9 & 12 & 18 & 12 & 6 \\ 4 & 6 & 8 & 12 & 10 & 5 \\ 2 & 3 & 4 & 6 & 5 & 4 \end{bmatrix}$$

E_7:

$$d(\Pi) = \frac{1}{2}\begin{bmatrix} 4 & 4 & 6 & 8 & 6 & 4 & 2 \\ 4 & 7 & 8 & 12 & 9 & 6 & 3 \\ 6 & 8 & 12 & 16 & 12 & 8 & 4 \\ 8 & 12 & 16 & 24 & 18 & 12 & 6 \\ 6 & 9 & 12 & 18 & 15 & 10 & 5 \\ 4 & 6 & 8 & 12 & 10 & 8 & 4 \\ 2 & 3 & 4 & 6 & 5 & 4 & 3 \end{bmatrix}$$

TABLE 3.2 (continued)

E_8:

$$d(\Pi) = \begin{bmatrix} 4 & 5 & 7 & 10 & 8 & 6 & 4 & 2 \\ 5 & 8 & 10 & 15 & 12 & 9 & 6 & 3 \\ 7 & 10 & 14 & 20 & 16 & 12 & 8 & 4 \\ 10 & 15 & 20 & 30 & 24 & 18 & 12 & 6 \\ 8 & 12 & 16 & 24 & 20 & 15 & 10 & 5 \\ 6 & 9 & 12 & 18 & 15 & 12 & 8 & 4 \\ 4 & 6 & 8 & 12 & 10 & 8 & 6 & 3 \\ 2 & 3 & 4 & 6 & 5 & 4 & 3 & 2 \end{bmatrix}$$

F_4:

$$d(\Pi) = \begin{bmatrix} 2 & 3 & 4 & 2 \\ 3 & 6 & 8 & 4 \\ 2 & 4 & 6 & 3 \\ 1 & 2 & 3 & 2 \end{bmatrix}$$

G_2:

$$d(\Pi) = \begin{bmatrix} 2 & 1 \\ 3 & 2 \end{bmatrix}$$

<u>Theorem 3.6</u>. Let L be a simple Lie algebra over an algebraically closed field of characteristic 0, and let λ_0 denote the highest root relative to the ordering of Π specified in Table 3.1. Then

$$\lambda_0 = \lambda_1 + \lambda_n \quad \text{for } A_n (n \geq 1) ,$$

$$\lambda_0 = \lambda_1 \quad \text{for } D_n (n \geq 5), E_7, F_4 ,$$

$$\lambda_0 = \lambda_2 \quad \text{for } B_n (n \geq 3), D_4, E_6, G_2 ,$$

$$\lambda_0 = 2\lambda_1 \quad \text{for } C_n (n \geq 3) ,$$

$$\lambda_0 = \lambda_8 \quad \text{for } E_8 , \lambda_0 = 2\lambda_2 \quad \text{for } B_2 .$$

Proof. If L is not of type $D_n (n \geq 5)$, then these relations follow from Tables 2.1 and 3.2. If L is of type $D_n (n \geq 5)$, then the relation $\lambda_0 = \lambda_1$ is a consequence of (3.16) and Table 3.2. □

Theorem 3.6 has also been proved by Okubo [2,13] under a slightly different ordering of Π. To give a complete list of the decomposition of $L \otimes L$ for each case, we use the same ordering of Π as in Table 3.1 which coincides with one given by Benkart and Osborn [1]. Thus, L is isomorphic to the irreducible L-module $V(\lambda_0)$ for λ_0 given by Theorem 3.6. In Table 3.3, the first column denotes the type of Lie algebra, the second column lists the corresponding tensor product $V(\lambda_0) \otimes V(\lambda_0)$ and its irreducible summands, and the last column gives the dimension of $V(\lambda_0) \otimes V(\lambda_0)$ and of the corresponding summands, which can be computed from Weyl's formula using the integers m_i given in (3.15). Those dimensions and decompositions of $V(\lambda_0) \otimes V(\lambda_0)$ may be found in Benkart and Osborn [1], and in Seligman [2,pp.305-331] with a different labeling of Π.

TABLE 3.3

Lie algebra	Decompositions of $L \otimes L$	Dimensions
A_1	$V(2\lambda_1) \otimes V(2\lambda_1)$	3×3
	$V(4\lambda_1)$	5
	$V(2\lambda_1)$	3
	$V(0)$	1
A_2	$V(\lambda_1 + \lambda_2) \otimes V(\lambda_1 + \lambda_2)$	8×8
	$V(2\lambda_1 + 2\lambda_2)$	27
	$V(3\lambda_2)$	10
	$V(3\lambda_1)$	10
	$V(\lambda_1 + \lambda_2)$	8
	$V(\lambda_1 + \lambda_2)$	8
	$V(0)$	1
$A_n (n \geq 3)$	$V(\lambda_1 + \lambda_n) \otimes V(\lambda_1 + \lambda_n)$	$n(n + 2) \times n(n + 2)$
	$V(2\lambda_1 + 2\lambda_n)$	$\frac{(n + 1)^2(n + 4)n}{4}$
	$V(\lambda_2 + 2\lambda_n)$	$\frac{(n + 3)(n + 2)n(n - 1)}{4}$
	$V(2\lambda_1 + \lambda_{n-1})$	$\frac{(n + 3)(n + 2)n(n - 1)}{4}$
	$V(\lambda_1 + \lambda_n)$	$n(n + 2)$
	$V(\lambda_1 + \lambda_n)$	$n(n + 2)$
	$V(\lambda_2 + \lambda_{n-1})$	$\frac{(n + 1)^2(n + 2)(n - 2)}{2}$
	$V(0)$	1
B_2	$V(2\lambda_2) \otimes V(2\lambda_2)$	10×10
	$V(4\lambda_2)$	35
	$V(\lambda_1 + 2\lambda_2)$	35
	$V(\lambda_1)$	5
	$V(2\lambda_2)$	10
	$V(2\lambda_1)$	14
	$V(0)$	1

TABLE 3.3 (continued)

Lie algebra	Decompositions of $L \otimes L$	Dimensions
B_3	$V(\lambda_2) \otimes V(\lambda_2)$	21×21
	$V(2\lambda_2)$	168
	$V(\lambda_1 + 2\lambda_3)$	189
	$V(2\lambda_1)$	27
	$V(\lambda_2)$	21
	$V(2\lambda_3)$	35
	$V(0)$	1
$B_n (n \geq 4)$	$V(\lambda_2) \otimes V(\lambda_2)$	$n(2n + 1) \times n(2n + 1)$
	$V(2\lambda_2)$	$\frac{(2n + 3)(2n + 1)(n - 1)(n + 1)}{3}$
	$V(\lambda_1 + \lambda_3)$	$\frac{(2n + 3)(2n + 1)n(n - 1)}{2}$
	$V(2\lambda_1)$	$n(2n + 3)$
	$V(\lambda_4)$	$\frac{(2n + 1)(2n - 1)n(n - 1)}{6}$
	$V(\lambda_2)$	$n(2n + 1)$
	$V(0)$	1
$C_n (n \geq 3)$	$V(2\lambda_1) \otimes V(2\lambda_1)$	$n(2n + 1) \times n(2n + 1)$
	$V(4\lambda_1)$	$\frac{(2n + 3)(2n + 1)(n + 1)n}{6}$
	$V(2\lambda_1 + \lambda_2)$	$\frac{(2n + 3)(2n + 1)n(n - 1)}{2}$
	$V(\lambda_2)$	$(2n + 1)(n - 1)$
	$V(2\lambda_2)$	$\frac{(2n + 3)(2n - 1)n(n - 1)}{6}$
	$V(2\lambda_1)$	$n(2n + 1)$
	$V(0)$	1

TABLE 3.3 (continued)

Lie algebra	Decompositions of $L \otimes L$	Dimensions
D_4	$V(\lambda_2) \otimes V(\lambda_2)$	28×28
	$V(2\lambda_2)$	300
	$V(\lambda_1 + \lambda_3 + \lambda_4)$	350
	$V(2\lambda_4)$	35
	$V(2\lambda_3)$	35
	$V(2\lambda_1)$	35
	$V(\lambda_2)$	28
	$V(0)$	1
$D_n (n \geq 5)$	$V(\lambda_1) \otimes V(\lambda_1)$	$n(2n - 1) \times n(2n - 1)$
	$V(2\lambda_1)$	$(n + 1)(2n - 1)$
	$V(\lambda_1 + \lambda_3)$	$\frac{(2n - 1)(2n - 3)n(n + 1)}{2}$
	$V(2\lambda_2)$	$\frac{(2n - 3)(2n + 1)n(n + 1)}{3}$
	$V(\lambda_4)$	$\frac{(2n - 1)(2n - 3)n(n - 1)}{6}$
	$V(\lambda_1)$	$n(2n - 1)$
	$V(0)$	1
E_6	$V(\lambda_2) \otimes V(\lambda_2)$	78×78
	$V(2\lambda_2)$	2430
	$V(\lambda_4)$	2925
	$V(\lambda_1 + \lambda_6)$	650
	$V(\lambda_2)$	78
	$V(0)$	1

TABLE 3.3 (continued)

Lie algebra	Decompositions of L ⊗ L	Dimensions
E_7	$V(\lambda_1) \otimes V(\lambda_1)$	133 × 133
	$V(2\lambda_1)$	7371
	$V(\lambda_3)$	8645
	$V(\lambda_6)$	1539
	$V(\lambda_1)$	133
	$V(0)$	1
E_8	$V(\lambda_8) \otimes V(\lambda_8)$	248 × 248
	$V(2\lambda_8)$	27,000
	$V(\lambda_7)$	30,380
	$V(\lambda_1)$	3875
	$V(\lambda_8)$	248
	$V(0)$	1
F_4	$V(\lambda_1) \otimes V(\lambda_1)$	52 × 52
	$V(2\lambda_1)$	1053
	$V(\lambda_2)$	1274
	$V(2\lambda_4)$	324
	$V(\lambda_1)$	52
	$V(0)$	1
G_2	$V(\lambda_2) \otimes V(\lambda_2)$	14 × 14
	$V(2\lambda_2)$	77
	$V(3\lambda_1)$	77
	$V(2\lambda_1)$	27
	$V(\lambda_2)$	14
	$V(0)$	1

Lemma 3.7. Let L be a simple Lie algebra over an algebraically closed field F of characteristic 0. Then

$$\dim_F \operatorname{Hom}_L (L \otimes L, L) = \dim_F H_{\lambda_0} = \begin{cases} 2, & \text{if } L \text{ is of type } A_n (n \geq 2) , \\ 1, & \text{otherwise} . \end{cases}$$

If L is of type A_n, so that L is isomorphic to $\mathfrak{sl}(n + 1)$, then let $x \# y$ denote the product defined by relation (2.91), and otherwise let $x \# y = 0$. Then for any $\phi \in \operatorname{Hom}_L (L \otimes L, L)$ there exist fixed scalars $\alpha, \beta \in F$ such that

$$\phi(x \otimes y) = \alpha[x,y] + \beta x \# y \tag{3.18}$$

for all $x,y \in L$.

Proof. Table 3.3 shows that if L is not of type $A_n (n \geq 2)$, then an irreducible summand of $L \otimes L$ isomorphic to the adjoint module L appears only once, and that if L is of type A_n for $n \geq 2$, then the adjoint module L appears twice in the decomposition of $L \otimes L$. Thus, using either Lemma 3.1 or Theorem 3.6 we have the desired relation of dimension. The mapping $x \otimes y \to [x,y]$ is clearly a nonzero element of $\operatorname{Hom}_F (L \otimes L, L)$, which forms a basis of $\operatorname{Hom}_L (L \otimes L, L)$ for L not of type $A_n (n \geq 2)$. If L is of type A_n for $n \geq 2$, then it is easy to check that the mapping $x \otimes y \to x \# y$ with

$$x \# y = xy + yx - \frac{2}{n + 1}(\operatorname{tr} xy)I \tag{3.19}$$

is a nonzero L-homomorphism of $L \otimes L$ into L, which is independent of the Lie product. Here we identify L with $\mathfrak{sl}(n + 1)$, the Lie algebra of $(n + 1) \times (n + 1)$ trace zero matrices over F, and xy denotes the matrix product. Thus, in any case each element of $\mathrm{Hom}_L(L \otimes L, L)$ is described by relation (3.18) with $x \# y = 0$ for L not of type $A_n (n \geq 2)$. □

Lemma 3.8. Assume that L is a central simple Lie algebra over a field F of characteristic 0, and let Fz be a one-dimensional trivial module for L. Then, $\dim_F \mathrm{Hom}_L (L \otimes L, Fz) = 1$, and for the Killing form $K(\, , \,)$ of L, the mapping $x \otimes y \to K(x,y)z$ is a basis of $\mathrm{Hom}_L (L \otimes L, Fz)$.

Proof. The invariance of $K(\, , \,)$ shows that the mapping $x \otimes y \to K(x,y)z$ is a nonzero L-homomorphism of $L \otimes L$ into Fz. Thus, $\dim_F \mathrm{Hom}_L (L \otimes L, Fz) \geq 1$. Let K be the algebraic closure of L and L_K be the scalar extension of L to K. Since L is central simple over F, L_K is simple over K, and it follows from Table 3.3 that a one-dimensional trivial L_K-module appears only once in the decomposition of $L_K \otimes L_K$. Since any L-homomorphism of $L \otimes L$ into Fz can be lifted to an L_K-homomorphism of $L_K \otimes L_K$ into Kz, by Table 3.3 we have

$$1 \leq \dim_F \mathrm{Hom}_F (L \otimes L, Fz) \leq \dim_K \mathrm{Hom}_{L_K} (L_K \otimes L_K, Kz) = 1 ,$$

as desired. □

3.3. Modules for Malcev algebras

Let M be a Malcev algebra with product $[x,y]$ over a field F of characteristic $\neq 2$. It is well known that the Malcev identity (1.3) (Chapter 1) is equivalent to the identity

$$(3.20) \qquad [[x,z],[y,t]] = [x,y,z,t] + [y,z,t,x] + [z,t,x,y] + [t,x,y,z] ,$$

where $[x,y,z,t] = [[[x,y],z],t]$. For a vector space V over F , a linear mapping $\rho : M \to \mathrm{Hom}_F V$ is called a *representation* of M in V if the identity

$$(3.21) \quad \rho([[x,y],z]) = \rho([x,z])\rho(y) + \rho(x)\rho([y,z]) - \rho(y)\rho(x)\rho(z) + \rho(z)\rho(y)\rho(x)$$

holds. It is immediate from (3.20) that the adjoint mapping $\mathrm{ad} : M \to \mathrm{Hom}\, M$ given by $\mathrm{ad}_x (y) = [x,y]$ is a representation of M , which we call the *adjoint representation* of M . Let V be a vector space over F with two bilinear mappings : $M \times V \to V$ denoted by $(x,v) \to xv$ and $(x,v) \to vx$. Define a product on the vector space direct sum $B = M \oplus V$ by

$$(3.22) \qquad (x + u)(y + v) = [x,y] + xv + uy$$

for $x,y \in M$ and $u,v \in V$. The algebra B defined by (3.22) is termed *the split null extension* of M determined by the given bilinear mappings. A vector space V over F with two bilinear mappings : $M \times V \to V$ is a *Malcev module* for M or an M-*module* if the corresponding split null extension of M is a Malcev algebra. Thus, V is an M-module if and only if $xv = - vx$ for all $x \in M, v \in V$, and (3.20) holds for all $x,y,z \in M$ and all $t \in V$.

A Malcev module V for M affords a representation $\rho : M \to \mathrm{Hom}\, V$ via $\rho(x)v = xv$ for $x \in M, v \in V$, and conversely, each representation $\rho : M \to \mathrm{Hom}\, V$ converts V into an M-module by $xv = - vx = \rho(x)v$. For an M-module V , recall the Jacobian in $B = M \oplus V$:

(3.23) $J(x,y,v) = [x,y]v + (vx)y + (yv)x = [x,y]v + y(xv) - x(yv)$

for $x,y \in M$ and $v \in V$. The set $N(M) = \{v \in V \mid J(M,M,v) = 0\}$ is called the *module nucleus* of V for M. If $V = N(M)$, then V is said to be a *Lie module* for M. Clearly, $N(M)$ is the maximal Lie submodule of V for M. Malcev module homomorphism, irreducibility and complete reducibility are defined in the obvious manners. The following Malcev analog of Weyl's theorem for Lie algebras is well known (Kuzmin [1]).

Theorem 3.9. Any Malcev module for a semisimple Malcev algebra over a field F of characteristic 0 is completely reducible. □

Recall from (2.81) that the unique split simple, non-Lie, Malcev algebra, denoted by C_0^-, over a field F of characteristic $\neq 2,3$ is given by the 7-dimensional algebra with basis e_0, e_{-i}, e_i $(i = 1,2,3)$ and with multiplication

$$(3.24)\qquad \begin{aligned} &[e_0,e_{\pm i}] = \pm\, 2e_{\pm i},\ [e_i,e_{-i}] = e_0,\ i = 1,\ 2,\ 3\ , \\ &[e_{\pm i},e_{\pm j}] = \pm\, 2e_{\mp k},\ (ijk) = (123),\ (231),\ (312)\ , \end{aligned}$$

where all other products are zero. We note that C_0^- contains a 3-dimensional Lie subalgebra $Fe_0 + Fe_1 + Fe_{-1}$ isomorphic to $s\ell(2)$, and a 2-dimensional subspace $Fe_2 + Fe_{-3}$ which is stable under the adjoint action by the elements of $s\ell(2)$. Thus, if we let $u = e_2$ and $v = e_{-3}$, then it is easy to see from (3.24) that $Fu + Fv$ is an irreducible non-Lie, Malcev module for $s\ell(2)$ under the adjoint action and its module action is given by

(3.25) $[h,u] = 2u,\ [h,v] = -\,2v,\ [e,u] = 2v,\ [f,v] = 2u\ ,$

where all other module actions are zero, and $\{h,e,f\}$ is the standard basis of $s\ell(2)$ such that $[h,e] = 2e$, $[h,f] = -2f$, and $[e,f] = h$. The module defined by (3.25) will be denoted by $\tilde{M}_2$. Carlsson [1,2] has proven that if M is a split simple Malcev algebra over a field F of characteristic 0 and V is an irreducible non-Lie M-module, then either $M \cong C_0^-$ and V is isomorphic to the adjoint module C_0^-, or $M \cong s\ell(2)$ and V is isomorphic to the module $\tilde{M}_2$ defined by (3.25). For a semisimple Malcev algebra, we have the following result proved by Carlsson [2].

Lemma 3.10. Let M be a split semisimple Malcev algebra over a field F of characteristic 0 and let $M = \Sigma_{i=1}^{n} S_i$ be the direct sum of simple ideals S_i. Then any M-module V has the decomposition $V = V_1 + \cdots + V_r$ into irreducible M-modules V_i, which satisfies the conditions:

(1) V_1 is the maximal Lie submodule for M;

(2) for each i with $2 \le i \le r$, there is a unique k, $1 \le k \le n$, such that V_i is a non-Lie S_k-module isomorphic to $\tilde{M}_2$ or to the adjoint module C_0^-, and $S_j V_i = 0$ for all $j = 1, \cdots, n$ with $j \ne k$. In the former case, S_k is isomorphic to $s\ell(2)$, and in the latter case, S_k is isomorphic to C_0^-. □

Corollary 3.11. Let V be an irreducible M-module for a split semisimple Malcev algebra M over a field of characteristic 0. If V is a non-Lie M-module, then V is annihilated by all simple Lie summands of M, except possibly one $s\ell(2)$-summand. If V is a Lie module for M, then V is annihilated by all non-Lie summands of M.

Proof. The first part is immediate from Lemma 3.10. Assume then that V is a Lie module for M , and let S be any non-Lie simple summand of M . Denote by $J(v,x,y)$ the Jacobian in the split null extension $B = M + V$. Since S is non-Lie simple, we have $J(S,S,S) = S$ by Sagle [1,p.440]. From a known identity $2J(x,y,z)v = J([x,y],z,v) + J([y,z],x,v) + J([z,x],y,v)$ (Sagle [1,p.429]), it follows that

$$SV = J(S,S,S)V \subseteq J(S,S,V) = 0 ,$$

since V is a Lie module for S . □

Let V be an M-module with afforded representation ρ . For $x,y \in M$, the inner derivation $d(x,y) = ad_{[x,y]} + [ad_x, ad_y]$ in the split null extension $B = M + V$ stabilizes V , and the restriction of $d(x,y)$ to V is

$$d(x,y)|_V = \rho([x,y]) + [\rho(x),\rho(y)] .$$

Using the relation $[\sigma,d(x,y)] = d(\sigma x,y) + d(x,\sigma y)$ for any $\sigma \in \text{Der } M$ (Sagle [1,p.453]), we see that $d(M,M)$ is a subalgebra of Der M which stabilizes V . Thus, V can be regarded as a Lie module for $d(M,M)$. This allows us to convert a Malcev module into a Lie module from which we can gain more stringent information about the multiplication between the irreducible summands. Another advantage is that the tensor product $V_1 \otimes V_2$ of two non-Lie, Malcev modules V_1, V_2 is not in general a Malcev module under the composition $x(v_1 \otimes v_2) = xv_1 \otimes v_2 + v_1 \otimes xv_2$ (this can be easily seen from $\tilde{M}_2 \otimes \tilde{M}_2$ under the $\mathfrak{sl}(2)$ action), but $V_1 \otimes V_2$ can be regarded as a Lie module for $d(M,M)$. For most of the interesting cases, the structure of $V_1 \otimes V_2$ constrained by the Lie module for $d(M,M)$ is sufficient to determine the multiplication of certain classes of flexible Malcev-admissible algebras.

We give the description of $d(M,M)$ for the non-Lie, irreducible $\mathfrak{sl}(2)$-module $\tilde{M}_2$ and the C_0^--module. Consider first the $\mathfrak{sl}(2)$-module $\tilde{M}_2$ where the module operation is denoted by the adjoint mapping as in (3.25). Let $x = x_0 h + x_1 e + x_{-1} f$ and $y = y_0 h + y_1 e + y_{-1} f$ denote elements of $\mathfrak{sl}(2)$. The matrix of ad_x relative to the basis $\{u,v\}$ is given by

$$ad_x = 2\begin{bmatrix} x_0 & x_1 \\ x_{-1} & -x_0 \end{bmatrix}.$$

For $i,j = 0, \pm 1$, define $\langle i,j\rangle = x_i y_j - x_j y_i \in F$. It follows that

$$d(x,y) = -\,ad_{[x,y]} = -2\begin{bmatrix} \langle 1,-1\rangle & 2\langle 0,1\rangle \\ 2\langle -1,0\rangle & -\langle 1,-1\rangle \end{bmatrix},$$

which in particular implies that $d(M,M)$ is isomorphic to $\mathfrak{sl}(2)$. Thus, $\tilde{M}_2$ is an irreducible Lie module for $\mathfrak{sl}(2)$ under the $d(\ ,\ )$-action.

Consider next $d(C_0^-,C_0^-)$ for the adjoint module C_0^-. For elements $x = \Sigma_{i=-3}^{3} x_i e_i$ and $y = \Sigma_{i=-3}^{3} y_i e_i$ of C_0^-, the matrix of ad_x relative to the basis $\{e_0,e_1,e_2,e_3,e_{-1},e_{-2},e_{-3}\}$ can be computed by (3.24) as follows:

$$(3.26)\qquad ad_x = \begin{bmatrix} 0 & -2x_1 & -2x_2 & -2x_3 & 2x_{-1} & 2x_{-2} & 2x_{-3} \\ -x_{-1} & 2x_0 & 0 & 0 & 0 & 2x_3 & -2x_2 \\ -x_{-2} & 0 & 2x_0 & 0 & -2x_3 & 0 & 2x_1 \\ -x_{-3} & 0 & 0 & 2x_0 & 2x_2 & -2x_1 & 0 \\ x_1 & 0 & -2x_{-3} & 2x_{-2} & -2x_0 & 0 & 0 \\ x_2 & 2x_{-3} & 0 & -2x_{-1} & 0 & -2x_0 & 0 \\ x_3 & -2x_{-2} & 2x_{-1} & 0 & 0 & 0 & -2x_0 \end{bmatrix}$$

The entries of the matrix of $d(x,y)$ consist of 14 independent parameters $c_i (i=2,\cdots,15)$ in F and are given by

$$(3.27)\quad d(x,y) = \begin{bmatrix} 0 & -2c_2 & -2c_3 & -2c_4 & 2c_5 & 2c_6 & 2c_7 \\ -c_5 & -c_8 & -c_9 & -c_{10} & 0 & -c_4 & c_3 \\ -c_6 & -c_{11} & -c_{12} & -c_{13} & c_4 & 0 & -c_2 \\ -c_7 & -c_{14} & -c_{15} & c_8+c_{12} & -c_3 & c_2 & 0 \\ c_2 & 0 & c_7 & -c_6 & c_8 & c_{11} & c_{14} \\ c_3 & -c_7 & 0 & c_5 & c_9 & c_{12} & c_{15} \\ c_4 & c_6 & -c_5 & 0 & c_{10} & c_{13} & -(c_8+c_{12}) \end{bmatrix}$$

where each c_i is computed in terms of $\langle i,j\rangle = x_i y_j - x_j y_i$ as

$$c_2 = -4\langle 1,0\rangle - 2\langle -3,-2\rangle, \qquad c_9 = 6\langle -1,2\rangle,$$

$$c_3 = -4\langle 2,0\rangle - 2\langle -1,-3\rangle, \qquad c_{10} = 6\langle -1,3\rangle,$$

$$c_4 = -4\langle 3,0\rangle - 2\langle -2,-1\rangle, \qquad c_{11} = 6\langle -2,1\rangle,$$

$$c_5 = 4\langle -1,0\rangle + 2\langle 3,2\rangle, \qquad c_{12} = 2\langle 3,-3\rangle + 2\langle 1,-1\rangle + 4\langle -2,2\rangle,$$

$$c_6 = 4\langle -2,0\rangle + 2\langle 1,3\rangle, \qquad c_{13} = 6\langle -2,3\rangle,$$

$$c_7 = 4\langle -3,0\rangle + 2\langle 2,1\rangle, \qquad c_{14} = 6\langle -3,1\rangle,$$

$$c_8 = 2\langle 3,-3\rangle + 2\langle 2,-2\rangle + 4\langle -1,1\rangle, \qquad c_{15} = 6\langle -3,2\rangle.$$

As is well known, it follows from (3.27) that $d(C_0^-,C_0^-)$ is the 14-dimensional split simple Lie algebra G_2 and C_0^- is the irreducible G_2-module of highest weight λ_1 (see Tables 3.1 and 3.2). Assume now that V is an M-module for a split semisimple Malcev algebra M over F of characteristic 0. Thus, by Weyl's theorem for a Malcev module, V is decomposed into the direct sum

(3.28) $$V = V_1 \oplus \cdots \oplus V_m$$

of irreducible M-modules V_i . Let $M = S_1 \oplus \cdots \oplus S_n$ be the direct sum of simple ideals S_j of M . Since each derivation of M is inner by (3.6), it acts on V , and hence V is regarded as a Lie module for $\operatorname{Der} M$, where each V_i is a $\operatorname{Der} M$ -submodule of V . Since $\operatorname{Der} M$ is a semisimple Lie algebra by (3.7), $\operatorname{Der} M$ is decomposed to the direct sum of simple ideals $\operatorname{Der} S_i$ as

(3.29) $$\operatorname{Der} M = \operatorname{Der} S_i \oplus \cdots \oplus \operatorname{Der} S_n ,$$

where each $\operatorname{Der} S_i$ is split over F . Since $\operatorname{Der} S_i$ is inner, by (3.2) $\operatorname{Der} S_i = \operatorname{ad}_{S_i}$ if S_i is a Lie algebra, and $\operatorname{Der} S_i \cong G_2$ if S_i is non-Lie, so that $S_i = C_0^-$. We note that ad_x for $x \in M$ acts on V as $\operatorname{ad}_x(v) = xv$ for all $v \in V$.

<u>Lemma 3.12</u>. Let M be a split semisimple Malcev algebra over a field F of characteristic 0 and let $M = S_1 + \cdots + S_n$ be the decomposition into simple ideals of M . Assume that V is an M-module with decomposition given by (3.28). Then

(1) Decomposition (3.28) is the direct sum of irreducible Lie submodules for $\operatorname{Der} M$.

(2) $(\operatorname{Der} S_i)V_j \neq 0$ if and only if $(\operatorname{Der} S_i)V_j = V_j$ if and only if $S_iV_j = V_j$.

(3) $S_i \cong V_j$ as a Malcev module for M if and only if $S_i \cong V_j$ as a Lie module for $\operatorname{Der} M$.

(4) $(\operatorname{Der} S_i)V_j = 0$ if and only if $S_iV_j = 0$.

<u>Proof</u>. Let us denote $D = \mathrm{Der}\ M$ and $D_i = \mathrm{Der}\ S_i$.

(1) Since D is inner, each V_i is a D-submodule. It suffices to show that V_i is an irreducible D-submodule. Assume that V_i is a non-Lie M-module. Then, by Lemma 3.10, either $V_i \cong \tilde{M}_2$ or $V_i \cong C_0^-$ as an M-module, and there is an index k, $1 \le k \le n$, such that $S_j V_i = 0$ for all $j \ne k$, and either $S_k \cong s\ell(2)$ or $S_k \cong C_0$. As noted above, in either case, V_i is an irreducible D-submodule. If V_i is a Lie module for M , then by Corollary 3.11 V_i is annihilated by all non-Lie summands of M , and hence any D-submodule of V_i would be an M-submodule of V_i . This implies that V_i is an irreducible D-submodule also when V_i is a Lie module for M .

(2) If $D_i V_j \ne 0$, then since $D_i V_j$ is a D-submodule of V_j , by part (1) $D_i V_j = V_j$. If $D_i V_j = V_j$, then D_i is inner, $V_j = D_i V_j \subseteq S_i V_j \subseteq V_j$ and $S_i V_j = V_j$. Conversely, assume $S_i V_j = V_j$. If V_j is a Lie module for M , then by Corollary 3.11 S_i must be a Lie algebra and $D_i V_j = S_i V_j = V_j$. The same argument shows that $D_i V_j = V_j$ when V_j is a non-Lie M-module.

(3) Assume $S_i \cong V_j$ as a D-module. If V_j is a Lie module for M , then by Corollary 3.11 the D-action on V_j is the adjoint action and S_i is a Lie algebra, hence $S_i \cong V_j$ as an M-module. If V_j is a non-Lie module, then by the same Corollary it must be that $V_j \cong C_0 \cong S_i$ as an M-module. The converse is obvious, since D is inner.

(4) The proof is similar to that of part (2). □

<u>Definition 3.3</u>. Let M be the same as in Lemma 3.12 and let V be an M-module. For an irreducible M-submodule W of V , the *support* of W in the M-module V is defined as the set $\Gamma = \Gamma(W) = \{j \mid 1 \le j \le n$ and $S_j W \ne 0\}$, where n is the number of simple summands S_i of M . For each

subset Γ of $\{1,2,\cdots,n\}$, denote by V_Γ the sum of all irreducible M-submodules of V with support Γ. By convention, $V_\Gamma = 0$ if there is no irreducible M-submodule of V with support Γ. If Γ is the empty set, then we write V_0 for V_Γ, which is the sum of all trivial M-submodules. □

We note from Lemma 3.12 that W is an irreducible M-submodule if and only if W is an irreducible Der M-submodule, and hence the support of W in the M-module V coincides with that of W in the Der M-module V, which is the set of indices j such that $1 \leq j \leq n$ and $(\text{Der } S_j)W \neq 0$. The concept of support will play an important role in Chapter 4.

3.4. Adjoint operators in simple Lie algebras

In this section, we determine $\text{Hom}_D(M \otimes M, M)$ when M is a simple Malcev algebra over a field F of characteristic 0 but not necessarily central simple over F, where $D = \text{Der } M$. In fact, every element of $\text{Hom}_D(M \otimes M, M)$ is determined in terms of scalars in the centroid of M. We consider first the case that M is central simple over F.

Lemma 3.13. If M is a central simple, non-Lie Malcev algebra over a field F of characteristic 0, then

$$\dim_F \text{Hom}_D(M \otimes M, M) = 1 .$$

Proof. Let K be the algebraic closure of F, and let $M_K = K \otimes_F M$. Then, $D_K = K \otimes_F D$ is $\text{Der } M_K$, and hence is the simple Lie algebra of type G_2, since M_K is simple over K. Furthermore, M_K is an irreducible D_K- module of highest λ_1 . From the known decomposition

$$V(\lambda_1) \otimes V(\lambda_1) = V(2\lambda_1) \oplus V(\lambda_2) \oplus V(\lambda_1) \oplus V(0)$$

(Seligman [2,p.329]) and Lemma 3.1, $\dim_K \mathrm{Hom}_{D_K}(M_K \otimes M_K, M_K) = 1$. Since any D-module homomorphism of $M \otimes M$ into M can be lifted to a D_K-module homomorphism of $M_K \otimes M_K$ into M_K and the mapping $x \otimes y \to [x,y]$ is an element of $\mathrm{Hom}_D(M \otimes M, M)$, we have $\dim_F \mathrm{Hom}_D(M \otimes M, M) = 1$. □

Assume that M is a central simple Lie algebra L over F , so that $D = ad_L$. Let L be of type $A_n (n \geq 1)$. Then, there are two possibilities (Jacobson [2,p.310]). The first is that there is a central simple associative algebra G over F such that $L = [G,G]$ and $G = L \oplus F1$. For the remaining case, there is a simple associative algebra G over F with involution J of the second kind and with center $F(q)$, a quadratic extension of F , such that $q^J = -q$, and such that if $S = S(G,J)$ denotes the set of skew symmetric elements of G and H the set of symmetric elements of G relative to J , then $G = S \oplus H$, $L = [S,S]$ and $S = L \oplus Fq$. In the second case, note that L, Fq and H are L-submodules of G under the adjoint action. In both cases, if x and y are elements of L , then define $x \# y$ by

$$(3.30) \quad x \# y = \begin{cases} \text{projection of } xy + yx \text{ onto } L \text{ in the first case,} \\ \text{projection of } q(xy + yx) \text{ onto } L \text{ in the second case.} \end{cases}$$

It can be easily seen that "#" is a commutative product on L satisfying

$$(3.31) \qquad [x,y \# z] = y \# [x,z] + [x,y] \# z$$

for all $x,y,z \in L$.

Lemma 3.14. Assume that M is a central simple Malcev algebra over F of characteristic 0, and let $D = \mathrm{Der}\, M$. Then,

$$\dim_F \mathrm{Hom}_D(M \otimes M, M) = \begin{cases} 2, & \text{if } M \text{ is a Lie algebra of type } A_n (n \geq 2), \\ 1, & \text{otherwise} \end{cases}$$

Proof. If M_K denotes the scalar extension of M to the algebraic closure K of F, then $D_K = \mathrm{Der}\, M_K$, and as in the proof of Lemma 3.13, we have

$$\dim_F \mathrm{Hom}_D (M \otimes M, M) \leq \dim_K \mathrm{Hom}_{D_K} (M_K \otimes M_K, M_K) .$$

Since the mapping $x \otimes y \to [x,y]$ is in $\mathrm{Hom}_D (M \otimes M, M)$, $\dim_F \mathrm{Hom}_D (M \otimes M, M) \geq 1$. In light of Lemmas 3.7 and 3.13, it suffices to assume that M is a Lie algebra L of type $A_n (n \geq 1)$. Thus, L is one of the algebras described above, and L_K is isomorphic to $\mathfrak{sl}(n + 1, K)$ in both cases (Jacobson [2, p.301]). Let "#" be the commutative product on L defined by (3.30). It follows from (3.31) that the mapping $x \otimes y \to x \# y$ is an L-module homomorphism of $L \otimes L$ into L. When the product "#" is lifted to L_K, the resulting product is the one given by (3.19) defined on $\mathfrak{sl}(n + 1, K)$. Thus, $x \# y = 0$ for all $x,y \in L$ if $n = 1$, while $x \# y \neq 0$ for some $x,y \in L$ if $n \geq 2$. Since $\dim_K \mathrm{Hom}_{L_K} (L_K \otimes L_K, L_K) = 2$ for $n \geq 2$ by Lemma 3.7, we conclude that $\dim_F \mathrm{Hom}_L (L \otimes L, L)$ is 2 for $n \geq 2$ and is 1 for $n = 1$, and furthermore, any $\phi \in \mathrm{Hom}_L (L \otimes L, L)$ is given by

$$\phi(x \otimes y) = \alpha[x,y] + \beta x \# y \tag{3.32}$$

for all $x,y \in L$, where α, β are fixed scalars in F. □

To extend Lemma 3.14 to the case of a simple Malcev algebra over F, recall first that for an arbitrary algebra A over F, the *centroid* E

of A is the set of elements of $\mathrm{Hom}_F A$ which commute with left and right multiplications L_x and R_y in A for all $x,y \in A$, and that if A is simple over F, then E is an extension field of F and A is central simple over E under the scalar multiplication $\gamma x = \gamma(x)$ for $\gamma \in E$ and $x \in A$ (Jacobson [2,p.291]).

Assume that A is an algebra over a field K of arbitrary characteristic, and let F be a subfield of K. Then, A can be regarded as an F-algebra by restricting the scalar multiplication in A to F but with the same multiplication as in A. The F-algebra obtained by this manner is called the F-*descent* of A. If F is the field of real numbers and A is a complex algebra, then the F-descent of A is called the *realification* of A. If $\{x_j\}$ is a K-basis of A and $\{\omega_i\}$ is an F-basis of K, then clearly, $\{\omega_i x_j\}$ is an F-basis of A. An F-subspace A_0 of A is termed an F-*form* of A if A_0 is an F-algebra under the same multiplication as in A and if $(A_0)_K = K \otimes_F A_0 \cong A$ as a K-algebra. When F is the field of real numbers, an F-form of a complex algebra A is called a *real form* of A. For an F-automorphism t of K, define a scalar multiplication $\alpha \cdot x$ on A by $\alpha \cdot x = t(\alpha)x$ for $\alpha \in K$ and $x \in A$. This together with the multiplication in A converts A into a K-algebra, which is called the t-*conjugate* of A and is denoted by $\tilde{A}_t$. It is clear that $\tilde{A}_t$ is isomorphic to A as an F-algebra, and $\tilde{A}_t$ is simple or central simple over K if and only if A is. When K is a finite Galois extension of F, we have

<u>Lemma 3.15</u>. Assume that A is an algebra over a field K and F is a subfield of K such that K is a finite Galois extension of F. Let $G = \{t_1 = e, t_2, \cdots, t_m\}$ be the Galois group of K over F, where $m = [K : F]$. Then, the scalar extension $K \otimes_F A$ is isomorphic to the

direct sum

$$\tilde{A} = \tilde{A}_1 \oplus \cdots \oplus \tilde{A}_m$$

as a K-algebra, where $\tilde{A}_i$ denotes the t_i-conjugate of A for $i = 1, \cdots m$. Furthermore, if A has an F-form, then the $\tilde{A}_i$ are isomorphic to A as a K-algebra.

<u>Proof</u>. For a basis $\{\omega_1, \cdots, \omega_m\}$ of K over F, each element $u \in K \otimes_F A$ is uniquely expressed as

$$(3.33) \qquad u = \sum_{i=1}^{m} \omega_i(1 \otimes y_i), \quad y_i \in A .$$

In fact, assume that $\sum_i \omega_i(1 \otimes y_i) = 0$, and let $y_i = \sum_j \alpha_{ij} z_j$ for $\alpha_{ij} \in F$ and for an F-basis $\{z_j\}$ of A. Then, $\sum_{i,j} \alpha_{ij}\omega_i(1 \otimes z_j) = 0$, and hence $\sum_i \alpha_{ij}\omega_i = 0$ since $\{1 \otimes z_j\}$ is a K-basis of $K \otimes_F A$. Thus, $\alpha_{ij} = 0$ for all i, j, to give $y_i = 0$ for $i = 1, \cdots, m$. For each $1 \le j \le m$, define a mapping $\sigma_j : K \otimes_F A \to \tilde{A}_j$ by

$$(3.34) \qquad \sigma_j(u) = \sum_i t_j(\omega_i) y_i ,$$

where u is given by (3.33). It is easy to see that σ_j is independent of the basis $\{\omega_i\}$. We show that each σ_j is a K-algebra homomorphism. If $\alpha \in K$ and $\alpha \neq 0$, then $\{\alpha\omega_i\}$ is an F-basis of K and hence by (3.33) $\sigma_j(\alpha u) = \sum_i t_j(\alpha\omega_i)y_i = \sum_i t_j(\alpha)t_j(\omega_i)y_i = \alpha \cdot (\sum_i t_j(\omega_i)y_i)$ $= \alpha \cdot \sigma_j(u)$, to show that σ_j is K-linear. Let $v = \sum_i \omega_i(1 \otimes y_i')$ and $\omega_i\omega_\ell = \sum_k \alpha_{i\ell}^k \omega_k$ for $\alpha_{i\ell}^k \in F$. Then, $t_j(\omega_i\omega_\ell) = t_j(\omega_i)t_j(\omega_\ell)$ $= \sum_k \alpha_{i\ell}^k t_j(\omega_k)$ and

$$\sigma_j(uv) = \sigma_j(\sum_{i,\ell} \omega_i\omega_\ell(1 \otimes y_i y_\ell')) = \sum_{i,k,\ell} \alpha_{i\ell}^k t_j(\omega_k)(y_i y_\ell')$$

$$= (\sum_i t_j(\omega_i)y_i)(\sum_\ell t_j(\omega_\ell)y'_\ell) = \sigma_j(u)\sigma_j(v) ,$$

which shows that σ_j is a K-algebra homomorphism.

Define next a mapping $\sigma : K \otimes_F A \to \tilde{A}$ by

$$\sigma(u) = (\sigma_1(u), \cdots, \sigma_m(u)) ,$$

where u is an element of $K \otimes_F A$ given by (3.33). Clearly, σ is a K-algebra homomorphism. To show that σ is a K-algebra isomorphism of $K \otimes_F A$ to $\tilde{A}$, since $\dim_K (K \otimes_F A) = \dim_F A = m \cdot \dim_K A = m \cdot \dim_K \tilde{A}_i = \dim_K \tilde{A}$, it suffices to verify that σ is injective. Assume then that $\sigma_1(u) = \cdots = \sigma_m(u) = 0$. Thus, $\sum_i t_j(\omega_i)y_i = 0$ for all $j = 1, \cdots, m$. Since K is finite Galois over F of degree m, the matrix

$$\begin{bmatrix} t_1(\omega_1) & t_1(\omega_2) & \cdots & t_1(\omega_m) \\ t_2(\omega_1) & t_2(\omega_2) & \cdots & t_2(\omega_m) \\ \cdots & \cdots & \cdots & \cdots \\ t_m(\omega_1) & t_m(\omega_2) & \cdots & t_m(\omega_m) \end{bmatrix}$$

is nonsingular (Herstein [1,p.196]). Hence, we have $y_1 = \cdots = y_m = 0$, and σ is injective.

Suppose that A has an F-form A_0, so that $K \otimes_F A_0 = A$. Then, each element u of A is uniquely expressed by relation (3.33) with all y_i in A_0. For each $1 \le j \le m$, define $\sigma_j : A \to \tilde{A}_j$ by relation (3.34). Since A_0 is an F-algebra, the same argument as above shows that σ_j is an injective K-algebra homomorphism of A into $\tilde{A}_i$. Since $\dim_K A = \dim_K \tilde{A}_i$, σ_j is surjective. □

Lemma 3.15 is well known for complex Lie algebras (Sagle and Walde [1,p.187]). If A is simple over K , then since each conjugate $\tilde{A}_i$ is simple over K , $K \otimes_F A$ is semisimple over K (the solvable radical = 0) . Thus, we have

Corollary 3.16. Let A, F, and K be the same as in Lemma 3.15. If A is simple over K , then $K \otimes_F A$ is the direct sum of simple ideals and hence semisimple over K . □

Corollary 3.17. Let E be a finite Galois extension field of a field F , and let $[E : F] = m$. Then, the tensor product $E \otimes_F E$ is the direct sum of the conjugates $\tilde{E}_1, \cdots, \tilde{E}_m$ of E , where each $\tilde{E}_i$ is isomorphic to E as an E-algebra.

Proof. When E is viewed as an F-algebra, it follows from Lemma 3.15 that $E \otimes_F E \cong \tilde{E}_1 \oplus \cdots \oplus \tilde{E}_m$. Each F-automorphism t of E is clearly an E-algebra isomorphism of E to the t-conjugate $\tilde{E}_t$. □

Corollary 3.18. Assume that A is a simple algebra over a field K , and let F be any subfield of K such that K is finite Galois over F . Then, the F-descent of A is a simple algebra over F .

Proof. Let A' denote the F-descent of A , and let B be a nonzero ideal of A' as an F-algebra. Then, $B_K = K \otimes_F B$ is a nonzero ideal of $K \otimes_F A$, and hence by Corollary 3.16 is a direct sum of simple ideals, which in particular implies $B_K = B_K B_K = (BB)_K$. It follows from this that $B = BB$. Since $A = A'$ as a set, for $\alpha \in K$ we have $\alpha B = \alpha(BB)$ $= (\alpha B)B \subseteq A'B \subseteq B$, and so B is an ideal of A as a K-algebra, to give $B = A$. Thus, A' is simple over F . □

Corollary 3.18 has been proved in Sagle and Walde [1,p.262] for complex Lie algebras. We now prove our principal result in this section.

Theorem 3.19. Assume that M is a simple Malcev algebra over a field F of characteristic 0, and let E denote the centroid of M and let $D = \text{Der } M$. Denote by "#" the product on M defined by (3.30) when M is a Lie algebra of type $A_n (n \geq 2)$ over E, and let "#" be the zero product in any other cases. Then, any $\phi \in \text{Hom}_D(M \otimes M, M)$ is given by relation (3.32) with scalars α, β in E.

Proof. Since F has characteristic 0, E is a separable extension of F and hence is contained in a finite Galois extension K of F. The scalar extension of the E-algebra M to K is central simple over K, since M is central simple over its centroid. Thus, without loss of generality, we may assume $K = E$. Consider then the scalar extension $M_E = E \otimes_F M$. By Lemma 3.15, M_E is the direct sum of simple ideals M_i, $i = 1, 2, \cdots, m$, where each M_i is the t-conjugate of M over E for an F-automorphism t of E. Hence, each M_i is central simple over E and is isomorphic to M as an F-algebra. We note that $D_E = \text{Der } M_E = D_1 \oplus \cdots \oplus D_m$, where $D_i = \text{Der } M_i$ is a central simple Lie algebra over E.

Any element ϕ of $\text{Hom}_D(M \otimes M, M)$ can be lifted to a unique element ϕ' of $\text{Hom}_{D_E}(M_E \otimes M_E, M_E)$, and

$$\text{Hom}_{D_E}(M_E \otimes M_E, M_E) = \text{Hom}_{D_E}((\sum_i M_i) \otimes (\sum_j M_j), \sum_k M_k)$$

$$= \sum_{i,j,k} \oplus \text{Hom}_{D_E}(M_i \otimes M_j, M_k) .$$

For $i \neq j$, $M_i \otimes M_j$ is a trivial D_E-module, since $D_\ell M_p = 0$ for $\ell \neq p$. Thus, by Lemma 3.1 $\mathrm{Hom}_{D_E}(M_i \otimes M_j, M_k) = 0$ for $i \neq j$. If $\psi \in \mathrm{Hom}_{D_E}(M_i \otimes M_i, M_k)$ for $i \neq k$, then for $x,y \in M_i$ and $d \in D_k$ we have

$$d(\psi(x \otimes y)) = \psi(d(x \otimes y))$$

$$= \psi((dx) \otimes y + x \otimes (dy)) = 0 .$$

Thus, $D_k(\psi(x \otimes y)) = 0$, and since $\psi(x \otimes y) \in M_k$ and M_k is a non-trivial irreducible D_K-module, we have $\psi(x \otimes y) = 0$ for all $x,y \in M_i$, and so $\mathrm{Hom}_{D_E}(M_i \otimes M_i, M_k) = 0$ for $i \neq k$. Hence

$$\begin{aligned}\mathrm{Hom}_{D_E}(M_E \otimes M_E, M_E) &= \sum_i \oplus \mathrm{Hom}_{D_E}(M_i \otimes M_i, M_i) \\ &= \sum_i \oplus \mathrm{Hom}_{D_i}(M_i \otimes M_i, M_i) .\end{aligned} \tag{3.35}$$

Let ϕ_i' denote the restriction of ϕ' to $M_i \otimes M_i$, and x, y denote elements of M. By (3.35), we can write

$$\phi' = \phi_1' + \cdots + \phi_m' ,$$

$$[x,y] = \sum_{i=1}^{m} [x_i,y_i] , \quad x \# y = \sum_{i=1}^{m} x_i \# y_i ,$$

where x_i and y_i are the M_i-components of x and y, respectively. Since M_i is central simple over E, by relation (3.32) of Lemma 3.14 there exist some fixed scalars α_i, $\beta_i \in E$ such that

$$\phi_i'(x_i \otimes y_i) = \alpha_i [x_i,y_i] + \beta_i x_i \# y_i$$

for all $x, y \in M$. Hence, by (3.35)

$$\phi(x \otimes y) = \phi'(x \otimes y) = \sum_i \phi_i' (x_i \otimes y_i)$$

$$= \sum_i (\alpha_i[x_i, y_i] + \beta_i\, x_i \# y_i)$$

$$= \alpha[x,y] + \beta\, x \# y$$

where $\alpha = \sum_i \alpha_i$ and $\beta = \sum_i \beta_i$ are scalars in E. More precisely, α and β are scalars in the extension field K of E. But, since

$$\phi(x \otimes y) - \phi(y \otimes x) = 2\alpha[x,y] \in M ,$$

$$\phi(x \otimes y) + \phi(y \otimes x) = 2\beta\, x \# y \in M$$

for all $x, y \in M$, and since $[M,M] = M$ and $M \# M = M$ by (3.31) when "#" is a nonzero product, we have that α, β lie in $K \cap \mathrm{Hom}_F M = E$, as desired. □

3.5. Flexible Malcev-admissible algebras with A^- simple

In Chapter 3 we have determined third power-associative or flexible Malcev-admissible algebras A over F when A^- is split simple over F (Theorem 2.33 and Corollary 2.34). In this section we determine the structure of these algebras when A^- is simple but not necessarily split over F. Assume that L is a central simple Lie algebra of type $A_n (n \geq 1)$ over a field F of characteristic 0. Thus, L is one of two types of algebras described in Section 3.4. In the first case, L is called a Lie algebra of *type* A_I, while L is called *type* A_{II} in the second case. It is well known that there exists a finite Galois extension field K of F such that $L_K = K \otimes_F L$ is isomorphic to $s\ell(n+1,K)$ for $n \geq 1$ (Jacobson [2,p.298]). If L is of type A_I, then, as before, there is a central simple associative algebra over F such that $L = [G,G]$ and G is an F-form of $M(n+1,K)$. For L of type A_{II}, there exists a quadratic subfield $F(q)$ of K with involution J such that $q^J = -q$, and there exists a central simple associative algebra G over $F(q)$ such that $L_{F(q)} = [G,G]$ and G is an $F(q)$-form of $M(n+1,K)$ (Jacobson [2,p.309]). In the latter case, J is extended to an involution of G and G is simple as an F-algebra such that $G = L \oplus Fq \oplus H$, as described in Section 3.4. In both cases, the trace of each element in L is well defined.

Theorem 3.20. Assume that A is a flexible Malcev-admissible algebra over a field F of characteristic 0 such that A^- is simple over F. Then, either

(1) A^- is of type A_n over its centroid E for $n \geq 2$, and there is a central simple associative algebra G over E such that $[G,G] = A^-$, or there is a simple associative algebra over E with involution of the second kind such that $A^- = [S,S]$ where S is the set of skew elements. If "#" denotes the commutative product on A^- defined by relation (3.30), then multiplication "*" in A is given by

$$x * y = \frac{1}{2}[x,y] + \beta\, x \# y \tag{3.36}$$

for $x,y \in A^-$ and for some fixed $0 \neq \beta \in E$, or

(2) A is a Malcev algebra.

Proof. Since $D = \operatorname{Der} A^-$ is inner and ad_x is a derivation of A

for all $x \in A$, $D = \text{Der } A$, and hence the mapping $x \otimes y \to x * y$ is an element of $\text{Hom}_D (A^- \otimes A^-, A^-)$. Thus, in view of Theorem 3.19, it remains only to show that the coefficient α in (3.32) becomes $\frac{1}{2}$. This follows from

$$[x,y] = x * y - y * x = \alpha[x,y] + \beta x\#y - \alpha[y,x] - \beta y\#x = 2\alpha[x,y] . \qquad \square$$

Corollary 3.21. Let A, F, E, and "#" be the same as in Theorem 3.20. Then the centroid of A coincides with the centroid E of A^- .

Proof. Let E' denote the centroid of A . If A^- is not a Lie algebra of type A_n over E , then $A \cong A^-$ and so $E = E'$. Assume then that A^- is a Lie algebra of type A_n for $n \geq 2$. The definition of "#" by means of relation (3.30) implies that "#" is a commutative algebra product on A^- over E . Since the left multiplication in A^+ by each element x is β times the left multiplication of "#" by x , each element of E commutes with the left multiplications in A^+ , and hence commutes with the left and right multiplications in A . This shows $E = E'$, since $E' \subseteq E$. $\square$

Corollary 3.22. Let A, F, and E be the same as in Theorem 3.20. Then:

(1) If A^- is non-Lie, Malcev, then A is a Malcev algebra isomorphic to A^- .

(2) If A^- is a Lie algebra not of type $A_n (n \geq 2)$ over E , then A is a Lie algebra isomorphic to A^- .

(3) If A^- is of type $A_n (n \geq 2)$ over E and of type A_I , then A is either a Lie algebra, or isomorphic to an algebra with multiplication given by

$$x * y = \mu\, xy + (1 - \mu)yx - \frac{1}{n+1}(\mathrm{tr}\ xy)\ I\ , \tag{3.37}$$

$x, y \in [G,G]$, defined on $[G,G]$ over E , where G is a central simple associative algebra over E described above, $\mu \neq \frac{1}{2}$ is a fixed scalar in E , xy the product in G , and I the identity element of G .

(4) If A^- is of type $A_n (n \geq 2)$ over E and of type A_{II} , then A is either a Lie algebra, or isomorphic to an algebra with multiplication given by

$$x * y = \mu\, xy + (q - \mu)yx - \frac{q}{n+1}(\mathrm{tr}\ xy)\ I \tag{3.38}$$

defined on $[S,S]$ over E , where S is the set of skew elements in a simple associative algebra over E with involution J of the second kind, described above, $q^J = -q$, and μ is a fixed element of $E(q)$, a quadratic extension of E , such that $2\mu - q$ is a nonzero element of E .

Conversely, an algebra described by (1) - (4) is flexible Malcev-admissible with A^- simple, and no algebras from different lists are isomorphic.

<u>Proof</u>. (1) and (2) follow from Theorem 3.20.

(3) We identify $A^- = [G,G]$. Since $G = [G,G] \oplus EI$ and for $x,y \in A^-$, $x \# y$ is the projection of $xy + yx$ onto $[G,G]$, we have $x \# y = xy + yx + \alpha I$ for some $\alpha \in E$, and hence $\alpha = -(2/(n+1))(\mathrm{tr}\ xy)$, since $\mathrm{tr}\ x \# y = 0$ for $x,y \in [G,G]$. Thus, we have $x \# y = xy + yx - \frac{2}{n+1}(\mathrm{tr}\ xy)I$, which together with relation (3.36) gives

$$x * y = (\tfrac{1}{2} + \beta)xy + (\beta - \tfrac{1}{2})yx - \frac{2\beta}{n+1}(\mathrm{tr}\ xy)\ I \tag{3.39}$$

for some $\beta \in E$. If $\beta = 0$, then A is a Lie algebra. If $\beta \neq 0$, then multiplying both sides of (3.39) by $(2\beta)^{-1}$ gives the right side of

(3.37) with $\mu = \frac{1}{4}(\beta^{-1} + 2)$. Since E is the centroid of A by Corollary 3.21, the mapping $x \to (2\beta)x$ gives an isomorphism of an algebra with product $x * y$ to an algebra with product $(2\beta)^{-1} x * y$.

(4) We identify $A^- = [S,S]$ over E . Then, $G = [S,S] \oplus Eq \oplus H$, and for $x,y \in [S,S]$, $x \# y$ is the projection of $q(xy + yx)$ onto $[S,S]$. Thus, $x \# y = q(xy + yx) + \alpha q$ for some $\alpha \in E$. Since $\operatorname{tr} x \# y = 0$ for $x,y \in [S,S]$, $x \# y = q(xy + yx - \frac{2}{n+1}(\operatorname{tr} xy)I)$ which can combine with (3.36) to give

$$x * y = (\frac{1}{2} + \beta q)xy + (\beta q - \frac{1}{2})yx - \frac{2\beta q}{n+1}(\operatorname{tr} xy)I$$

for $x,y \in A^-$. If $\beta = 0$, then A is a Lie algebra. In case of $\beta \neq 0$, we have $(2\beta)^{-1} x * y = \mu\, xy + (q - \mu)yx - \frac{q}{n+1}(\operatorname{tr} xy)I$, where $\mu = \frac{1}{2}[(2\beta)^{-1} + q]$, so that $2\mu - q = (2\beta)^{-1} \in E$. Thus, as in part (3), A is isomorphic to an algebra with product $(2\beta)^{-1} x * y$ for $x,y \in A^-$. Also, if "$*$" is a product on A^- defined by (3.38), then

$$x * y - y * x = (2\mu - q)[x,y] \in A$$

for all $x,y \in A^-$, and so $2\mu - q \neq 0$ is an element of E .

Conversely, let A be an algebra over F with multiplication given by (3.37) or (3.38). It is easy to see that A^- is isomorphic to $[G,G]$ or to $[S,S]$ over F , and A is flexible since by (3.31) each ad_x is a derivation of A^+ . □

<u>Corollary 3.23</u>. Let A be a flexible Malcev-admissible algebra over a field F of characteristic 0 such that A^- is semisimple. Then, A is a direct sum of simple algebras described in Corollary 3.22.

<u>Proof</u>. By Theorem 1.7, A is a direct sum of ideals A_i of A

such that A_i^- is simple over F, and the result follows from Corollary 3.22. □

Corollary 3.24. Let A be the same as in Corollary 3.23. If A is power-associative, then A is a Malcev algebra.

Proof. Let K be the algebraic closure of F. Since the Killing form on A_K^- is nondegenerate, by (3.3) and (3.4) A_K^- is semisimple also. Thus, by Theorem 1.7 A_K is a direct sum of ideals A_i of A such that A_i^- is simple over K, and by Theorem 3.20 each A_i is either a Malcev algebra or an algebra with product defined by (3.36). In the latter case, A_i^+ is isomorphic to an algebra with product "#" on $\mathfrak{sl}(n+1,K)\,(n \geq 2)$ defined by (3.19), which must be power-associative, since A_i^+ is. But, it has been shown in Example 1.1 (Chapter 1) that the product "#" can not be power-associative for $n \geq 2$. □

Corollary 3.24 has been proved in Chapter 1 (Corollary 1.16) and Chapter 2 (Corollary 2.25) using different arguments. The following result is an extension of Theorem 2.33 to a non-split case.

Corollary 3.25. Let A be a Malcev-admissible algebra over a field F of characteristic 0 with product denoted by "$*$" such that A^- is central simple over F. Assume that "$*$" satisfies the third power identity $(x * x) * x = x * (x * x)$. Then

$$(3.40) \qquad x * y = \tfrac{1}{2}[x,y] + \tau(x)y + \tau(y)x + \beta x \# y$$

for some linear form $\tau : A \to F$ and some $\beta \in F$, where $\beta x \# y = 0$ for all $x,y \in A$ unless A^- is of type A_n for $n \geq 2$, and $x \# y$ is defined by (3.30) if A^- is of type A_n for $n \geq 2$. If A is power-associative, then $\beta x \# y = 0$ for all x, y.

Proof. If K denotes the algebraic closure of F, then $A_K = K \otimes_F A$ is third power-associative, since F has characteristic 0, and A_K^- is split simple over K. If A^- is a Lie algebra of type A_n for $n \geq 2$, then the product "#" is extended to a nonzero product on $s\ell(n+1,K)$ defined by (3.19). Thus, in view of Theorem 2.33, there exist a scalar $\beta \in K$ and a linear mapping τ of A into K such that (3.40) holds for all $x,y \in A$. In case A^- is a Lie algebra of type A_n for $n \geq 2$, there exist $x,y \in A$ such that x, y, $x \# y$ are linearly independent, because that is true in $A_K^- = s\ell(n+1,K)$ for the trace zero matrices e_{ij}, e_{jk} with $i < j < k$ and $e_{ij} \# e_{jk} = e_{ik}$. For that choice of x and y in A, equation (3.40) shows that $\beta \in F$, since $x * y + y * x$ belongs to A. But then τ must map A into F. It follows from Lemma 2.24 that if A is power-associative, then $\beta x \# y = 0$ for all $x,y \in A$. □

In view of Corollary 3.25, the following result for a non-split semi-simple case is immediate from Theorem 2.38 and Corollary 2.49.

Corollary 3.26. Let A be a Malcev-admissible algebra over a field F of characteristic 0 such that A^- is the direct sum $A^- = S_1^- \oplus \cdots \oplus S_m^-$ of simple ideals S_i^- of A^- which are central over F. Assume that multiplication "*" in A is third power-associative. Then, for each j there exists a scalar $\beta_j \in F$, and for each ordered pair (i,j) there exists a linear form $\tau_{ij} : S_j \to F$ such that when $x \in S_i$ and $y \in S_j$,

$$(3.41) \qquad x * y = \frac{1}{2}[x,y] + \tau_{ji}(x)y + \tau_{ij}(y)x + \beta_j\, x \# y ,$$

where $x \# y$ is defined by $x \# y = x \#_i y$ if $i = j$, and $x \# y = 0$ if $i \neq j$, and where $x \#_i y$ is defined on S_i by relation (3.30). Moreover, A is power-associative if and only if the product "$*$" is given by (3.41), where $\beta_j\, x \# y = 0$ for all $x,y \in A$ and for all j, and the linear forms τ_{ij} satisfy the conditions described in Theorem 2.48. □

We end this chapter with an application of Theorem 3.20 and Corollary 3.22 to real flexible Malcev-admissible algebras with A^- simple. Recall that any central simple, non-Lie, Malcev algebra over a field F of characteristic $\neq 2,3$ is one of the algebras $M(\alpha,\beta,\gamma)$ defined by Table 2.2, where α, β, γ are scalars in F with $\alpha\beta\gamma \neq 0$. There are exactly two nonisomorphic central simple, non-Lie, Malcev algebras over the field $\mathbb{R}$ of real numbers. The split one is $M(1,1,-1)$ while all non-split ones are isomorphic to $M(1,1,1)$ (Kuzmin [1]). Consider next central simple real Lie algebras L of type A_n. By the well known classification (Jacobson [2,p.316] and Sagle and Walde [1,p.323]), L is one of the following four types of algebras. For type A_I,

(3.42) $$L = \mathfrak{sl}(n + 1, \mathbb{R}),\ n \geq 1\ ,$$

(3.43) $$L = \mathfrak{sl}(\tfrac{n+1}{2}, \Delta),\ n \geq 1 \text{ odd}\ ,$$

where Δ denotes the division ring of real quaternions, $\mathfrak{sl}(\frac{n+1}{2}, \Delta)$ the Lie algebra of $\frac{n+1}{2} \times \frac{n+1}{2}$ trace zero matrices with entries in Δ, and for $x = (a_{ij}) \in M(m,\Delta)$, $\operatorname{tr} x = \sum_i \operatorname{tr} a_{ii}$ with $\operatorname{tr} a_{ii}$ being the trace of a_{ii} in Δ. For type A_{II},

(3.44) $$L = \mathfrak{su}(n + 1) = \{x \in \mathfrak{sl}(n + 1,\mathbb{C}) \mid \bar{x}^t = - x\}\ ,$$

$$L = S(n+1,r,\mathbb{C}) = \{x \in \mathfrak{sl}(n+1,\mathbb{C}) \mid T_r \bar{x}^t T_r = -x\}, \tag{3.45}$$

where $\mathbb{C}$ denotes the field of complex numbers, $\bar{x}$ the complex conjugate of x, "t" the transpose, and T_r is the diagonal matrix

$$T_r = \text{diag}\{\underbrace{1, \cdots, 1}_{n+1-r}, \underbrace{-1, \cdots, -1}_{r}\}$$

with $r = 1, 2, \cdots, [\frac{n+1}{2}]$.

Theorem 3.27. If A is a real flexible Malcev-admissible algebra with A^- simple over $\mathbb{R}$, then A is isomorphic to one of the following.

(1) The algebra $M(1,1,1)$.

(2) The algebra $M(1,1,-1)$.

(3) The 14-dimensional real Malcev algebra which is the realification of the complex algebra C_0^- defined by (3.24).

(4) A central simple real Lie algebra.

(5) The realification of a simple complex Lie algebra.

(6) An algebra with multiplication

$$x * y = \mu\, xy + (1-\mu)yx - \frac{1}{n+1}(\text{tr } xy)I \tag{3.46}$$

defined on the real vector space $\mathfrak{sl}(n+1,\mathbb{R})$ or $\mathfrak{sl}(\frac{n+1}{2},\Delta)$ for $n \geq 2$ given by (3.42) or (3.43), where $\mu \neq \frac{1}{2}$ is a fixed real number and xy denotes the matrix product.

(7) An algebra with multiplication

$$x * y = \mu\, xy + (i-\mu)yx - \frac{i}{n+1}(\text{tr } xy)I \tag{3.47}$$

defined on the real vector space $\mathfrak{su}(n+1)$ or $S(n+1,r,\mathbb{C})$ for $n \geq 2$ with $r = 1, \cdots, [\frac{n+1}{2}]$, given by (3.44) or (3.45), where μ is a fixed complex number such that $2\mu - i$ is a nonzero real number.

(8) The realification of a complex algebra with multiplication "$*$" given by (3.46) which is defined on $s\ell(n+1,\mathbb{C})$ for $n \geq 2$, where $\mu \neq \frac{1}{2}$ is a complex number.

Conversely, all algebras above are real flexible Malcev-admissible algebras with A^- simple, and no algebras from different lists are isomorphic.

Proof. We note that the centroid of A^- is either $\mathbb{R}$ or $\mathbb{C}$. Assume first that the centroid of A^- is $\mathbb{R}$, so that A^- is central simple over $\mathbb{R}$. If A^- is non-Lie, then by Corollary 3.22 A is non-Lie, Malcev, and so either $A \cong M(1,1,1)$ or $A \cong M(1,1,-1)$. Assume then that A^- is Lie. If A^- is not of type $A_n (n \geq 2)$, then A is Lie, to give case (4). Let A^- be of type A_n for $n \geq 2$. If A^- is of type A_I, then by Corollary 3.22 (3) A is isomorphic to an algebra described by (3.46) in case (6). If A^- is of type A_{II}, then using Corollary 3.22 (4) and relations (3.44) and (3.45) we have the algebra described by (3.47) in case (7). Consider the case that the centroid of A^- is $\mathbb{C}$, so that $\mathbb{C}$ is also the centroid of A by Corollary 3.21. If A^- is non-Lie, then by Corollary 3.22 (1) A is a non-Lie, Malcev algebra with centroid $\mathbb{C}$, and hence A is isomorphic to C_0^- as a $\mathbb{C}$-algebra. But then, A is isomorphic to the realification of the complex algebra C_0^- as a real algebra, to give case (3). If A^- is Lie but not of type A_n for $n \geq 2$ over the centroid $\mathbb{C}$, then by Corollary 3.22 (2) A is Lie, and simple over $\mathbb{C}$. Thus, A is an algebra described by case (5). Assume that A^- is of type A_n for $n \geq 2$ over $\mathbb{C}$. Since A^- is simple over $\mathbb{C}$ and $\mathbb{C}$ is algebraically closed, Corollary 3.22 (3) and (4) imply that A is isomorphic as a $\mathbb{C}$-algebra to an algebra defined by (3.46) with a fixed complex number $\mu \neq \frac{1}{2}$. Thus, A is isomorphic as a real algebra to the realification of that algebra, to give case (8).

Clearly, no algebras from different lists are isomorphic, and it follows from Corollary 3.18 that all of those algebras are flexible Malcev-admissible with A^- simple over $\mathbb{R}$. □

Corollary 3.28. Assume that A is a real flexible Malcev-admissible algebra with A^- simple. Then, the complexification $A_{\mathbb{C}} = \mathbb{C} \otimes_{\mathbb{R}} A$ of A is one of the following algebras.

(1) The algebra C_0^- over $\mathbb{C}$.

(2) The direct sum of two isomorphic copies of C_0^- over $\mathbb{C}$.

(3) A simple complex Lie algebra.

(4) A direct sum of two copies of a simple complex Lie algebra.

(5) An algebra with multiplication (3.46) defined on $\mathfrak{sl}(n+1,\mathbb{C})$ with a fixed $\mu \neq \frac{1}{2}$ in $\mathbb{C}$.

(6) The direct sum of an algebra given by part (5) and its conjugate under the standard complex conjugate.

Proof. Under the hypotheses, A is one of the algebras described in Theorem 3.27. Since $M(1,1,\pm 1)$ are central simple over $\mathbb{R}$, cases (1) and (2) of Theorem 3.27 give the algebra C_0^- over $\mathbb{C}$. If A is the realification of C_0^- over $\mathbb{C}$, then since C_0^- over $\mathbb{C}$ has a real form by (3.24), it follows from Lemma 3.15 that $A_{\mathbb{C}}$ is the direct sum of two copies of C_0^- over $\mathbb{C}$, to give case (2). For an algebra described by case (4), (6) or (7) of Theorem 3.27, A^- is central simple over $\mathbb{R}$, and hence $A_{\mathbb{C}}^-$ is a simple complex Lie algebra, which gives case (3) or (5). When A is the realification of a simple complex Lie algebra, by Lemma 3.15 $A_{\mathbb{C}}$ is the direct sum of two copies of that complex algebra, since every simple complex Lie algebra has a real form. Finally, if A is the algebra of case (8) of Theorem 3.27, then again by Lemma 3.15 $A_{\mathbb{C}}$ is an algebra described by case (6). □

Assume that A is a simple complex algebra with multiplication "$*$" on $\mathfrak{sl}(n+1,\mathbb{C})$ defined by (3.46) where $\mu \neq \frac{1}{2}$ is a complex number. If A has a real form, then by Lemma 3.15 the algebra of case (6) in Corollary 3.28 is the direct sum of two copies of A. It is shown in Okubo [2] that one can choose a basis $\{x_i \mid i = 1, \cdots, (n+1)^2 - 1\}$ of $A^- = \mathfrak{sl}(n+1,\mathbb{C})$ consisting of Hermitian matrices x_i such that

$$\text{tr}\, x_i x_j = 2\delta_{ij} ,$$

(3.48)

$$x_i x_j = \frac{2}{n+1}\delta_{ij} I + \sum_{k=1}^{n^2+2n} (\alpha_{ijk} + i\beta_{ijk}) x_k ,$$

where $\alpha_{ijk} = \frac{1}{4}\text{tr}[(x_i x_j + x_j x_i)x_k]$ and $\beta_{ijk} = \frac{-i}{4}\text{tr}[(x_i x_j - x_j x_i)x_k]$ are real numbers which are completely symmetric and skew-symmetric tensors, respectively. When $n = 2$, the eight trace zero Hermitian matrices of Gell-Mann [1] satisfy relations (3.48). It is easy to see from (3.48) that the product "$*$" is given by

$$x_i * x_j = \sum_k [\alpha_{ijk} + i(2\mu - 1)\beta_{ijk}] x_k . \tag{3.49}$$

Thus, if μ is a complex number such that $2\mu - 1$ is purely imaginary, then $\{x_i\}$ is a basis of a real form of A.

When $n = 2$ and μ satisfies the equation $3\mu(1-\mu) = 1$, so that $\mu = \frac{1}{2} \pm \frac{\sqrt{3}}{6} i$, the algebra defined by (3.46) is the complex pseudo-octonion algebra M_3 of Okubo (see Definition 2.3, Chapter 2). Since

$2\mu - 1 = \pm \frac{\sqrt{3}}{3} i$, $x_1, \cdots, x_8$ form a basis of a real form of M_3 which is shown to be a division algebra (see Section 2.2). We do not know whether A for $n \geq 2$ has a real form for any complex number $\mu \neq \frac{1}{2}$.

4

MALCEV-ADMISSIBLE ALGEBRAS WITH THE SOLVABLE RADICAL OF A^- NONZERO

4.1. Derivation decompositions

In this chapter we investigate the structure of those flexible Malcev-admissible algebras A for which A^- has a semisimple subalgebra S^- such that an S^--submodule R of A complementary to S^- is a subalgebra of A^-. In particular, we give a characterization of such algebras for which either R is a direct summand of A^- (i.e., [R,S] = 0), or A is simple and R is abelian (i.e., [R,R] = 0). Included in these classes of algebras are all simple flexible Malcev-admissible algebras such that the solvable radical of A^- is either a direct summand of A^- or abelian. The main technique is to regard A as a Lie module for a semisimple Lie algebra of derivations of A and to see what restrictions the module structure imposes on the multiplication of A. Our approach is based on the fact that if S^- is a semisimple subalgebra of A^-, then by (3.6) and (3.7) Der S^- is semisimple and inner, and hence by flexibility of A each element of Der S acts as a derivation on A.

Let A be an arbitrary algebra with multiplication denoted by xy over a field F, and let P be a subset of $\mathrm{Hom}_F\, A$ which acts as derivations on A, i.e., $\sigma(xy) = \sigma(x)y + x\sigma(y)$ for $\sigma \in P$ and $x,y \in A$. For a subset S of A which is stable under P, we put

$$\mathrm{Ker}_S\, P = \{x \in S \mid \sigma(x) = 0 \text{ for all } \sigma \in P\},$$

$$\mathrm{Im}_S\, P = \{\sigma(x) \mid \sigma \in P \text{ and } x \in S\}.$$

If S = A, then we write

$$\mathrm{Ker}\, P = \mathrm{Ker}_A\, P,\ \mathrm{Im}\, P = \mathrm{Im}_A\, P.$$

Lemma 4.1. Let P be a subset of $\mathrm{Hom}_F A$ which acts as derivations on A. If S is a subalgebra of A stable under P, then $\mathrm{Ker}_S P$ is a subalgebra of S, and $(\mathrm{Ker}_S P)(\mathrm{Im}_S P)$ and $(\mathrm{Im}_S P)(\mathrm{Ker}_S P)$ are contained in $\mathrm{Im}_S P$.

Proof. If $x, y \in \mathrm{Ker}_S P$, then $\sigma(xy) = \sigma(x)y + x\sigma(y) = 0$ for $\sigma \in P$, so that $\mathrm{Ker}_S P$ is a subalgebra of S. For $x \in \mathrm{Ker}_S P$ and $y \in \mathrm{Im}_S P$, let $y = \sigma(z)$ for some $z \in S$ and $\sigma \in P$. Then, $xy = x\sigma(z) = \sigma(xz) - \sigma(x)z = \sigma(xz) \in \mathrm{Im}_S P$. □

In the remainder of this chapter, unless otherwise stated, all algebras and modules are assumed to be finite-dimensional over a field of characteristic 0.

Suppose that A is an arbitrary algebra over a field F, and that D is a semisimple Lie algebra of derivations on A. Thus, D decomposes into the direct sum

$$D = D_1 \oplus \cdots \oplus D_n \tag{4.1}$$

of simple ideals D_i of D, and A is regarded as a D-module, so that each D-submodule S of A is a direct sum of irreducible D-submodules S_i of S. We focus on a special type of D-submodules of A. Let S be a D-submodule of A such that S decomposes into the direct sum

$$S = S_1 \oplus \cdots \oplus S_n \tag{4.2}$$

of irreducible D-submodules, which satisfy the properties

$$\mathrm{Ker}_S D_i = \sum_{j \neq i} S_j , \quad \mathrm{Im}_S D_i = S_i \tag{4.3}$$

for $i = 1, 2, \cdots, n$. Relation (4.3) clearly implies that

(4.4) $$\mathrm{Ker}_{S_i} D_i = 0,\ \mathrm{Im}_{S_i} D_i = \mathrm{Im}_{S_i} D = S_i\ ,$$

and in particular, each S_i is a non-trivial D-module.

Let A be a flexible Malcev-admissible algebra over F and S^- be a split semisimple Malcev subalgebra of A^-. Then, by (3.5) S^- is the direct sum of split simple ideals $S_1^-, \cdots, S_n^-$ of S^-, so that $\mathrm{Der}\, S^-$ is a direct sum of simple ideals $\mathrm{Der}\, S_i^-$. Since A is flexible, by (3.6) $\mathrm{Der}\, S^-$ acts as derivations on A and S is a $\mathrm{Der}\, S^-$-submodule of A which satisfies properties (4.3). If F is algebraically closed, then such a subalgebra S^- of A^- exists by Levi's theorem for Malcev algebras (Kuzmin [3] and Grishkov [1]), when A^- is a non-solvable Malcev algebra.

Assume that A is an arbitrary algebra over F such that A has a semisimple Lie algebra D of derivations and a D-submodule S, which satisfy properties (4.1) - (4.3). Then, by Weyl's theorem, there exists a D-submodule R of A complementary to S, so that

(4.5) $$A = S_1 \oplus S_2 \oplus \cdots \oplus S_n \oplus R\ ,$$

where the S_i are irreducible D-submodules of S. For an irreducible D-submodule W of R, recall from Definition 3.3 that the support Γ of W in the D-module R is the set

$$\Gamma = \{j \mid 1 \le j \le n \text{ and } D_j W \ne 0\}$$

and that R_Γ denotes the sum of all irreducible D-submodules of R with support Γ. If $\Gamma = \phi$, $\{i\}$, or $\{i,j\}$, then we write $R_\Gamma = R_0$, R_i, or R_{ij}, respectively, so that R_0 is the sum of one-dimensional trivial D-submodules. For simplicity, we denote

$$\operatorname{Ker} D_i = \operatorname{Ker}_i \ , \ \operatorname{Im} D_i = \operatorname{Im}_i \ .$$

It is convenient to express A as the module direct sum

$$A = \sum_{i=1}^{n} S_i + \sum_{\Gamma \neq \phi} R_\Gamma + R_0 \ , \tag{4.6}$$

where Γ ranges over all nonempty subsets of $\{1,2,\cdots,n\}$. From Lemma 4.1, (4.3) and (4.4) it is easy to see that

$$\operatorname{Ker}_i = \sum_{j \neq i} S_j + \sum_{i \notin \Gamma} R_\Gamma + R_0 \ , \tag{4.7}$$

$$\operatorname{Im}_i = S_i + \sum_{i \in \Gamma} R_\Gamma \ , \tag{4.8}$$

(4.9) Ker_i is a subalgebra of A ,

(4.10) $R_0 = \bigcap_{i=1}^{n} \operatorname{Ker}_i$ is a subalgebra of A .

<u>Lemma 4.2</u>. Let A, S, D, and R be the same as above, and assume that D and S have decompositions given by (4.1) and (4.2). For each $\Gamma \subseteq \{1,2,\cdots,n\}$, the D-submodule

$$B_\Gamma = \sum_{i \in \Gamma} S_i + \sum_{\Delta \subseteq \Gamma} R_\Delta + R_0 \tag{4.11}$$

is a subalgebra of A , and furthermore,

(1) $S_i^2 \subseteq S_i + R_i + R_0$,

(2) $S_i S_j \subseteq R_{ij}$ for $i \neq j$,

(3) $S_i R_0$ and $R_0 S_i$ are contained in $S_i + R_i$,

(4) $(S_i + R_i)(S_j + R_j) \subseteq R_{ij}$ for $i \neq j$.

<u>Proof</u>. If $\Delta(i) = \{1,2,\cdots,n\} - \{i\}$, then $B_{\Delta(i)} = \mathrm{Ker}_i$, which is a subalgebra of A by (4.9). It is easy to see that

$$B_\Gamma = \bigcap_{i \notin \Gamma} B_{\Delta(i)}$$

for each subset Γ of $\{1,\cdots,n\}$, so that B_Γ is a subalgebra of A . Since $S_i \subseteq B_i$, part (1) holds. For $i \neq j$, since by Lemma 4.1

$$S_iS_j \subseteq B_{ij} = S_i + S_j + R_{ij} + R_i + R_j + R_0 ,$$

$$S_iS_j \subseteq (\mathrm{Im}_i)(\mathrm{Ker}_i) \subseteq \mathrm{Im}_i ,$$

$$S_iS_j \subseteq (\mathrm{Ker}_j)(\mathrm{Im}_j) \subseteq \mathrm{Im}_j ,$$

from (4.8) we have $S_iS_j \subseteq (\mathrm{Im}_i) \cap (\mathrm{Im}_j) \cap B_{ij} = R_{ij}$, which is part (2). In light of Lemma 4.1 and (4.7), $S_iR_0 \subseteq (\mathrm{Im}_i)(\mathrm{Ker}_i) \subseteq \mathrm{Im}_i$ and $S_iR_0 \subseteq B_iB_i \subseteq B_i$, to show that $S_iR_0 \subseteq \mathrm{Im}_i \cap B_i = S_i + R_i$, and similarly $R_0S_i \subseteq S_i + R_i$. This proves part (3). For part (4), note that $(S_i + R_i)(S_j + R_j)$ is a trivial D_k-module for $k \neq i$ and $k \neq j$, since each element of D_k acts as a derivation on A . For $i \neq j$, we have

$$(S_i + R_i)(S_j + R_j) \subseteq (\mathrm{Ker}_j)(\mathrm{Im}_j) \cap (\mathrm{Im}_i)(\mathrm{Ker}_i)$$

$$\subseteq (\mathrm{Im}_j) \cap (\mathrm{Im}_i) \subseteq \sum_{i,j\in\Gamma} R_\Gamma$$

by (4.8), which gives part (4). □

4.2. THE CASE R IS A DIRECT SUMMAND OF A^-

Let M be a semisimple Malcev algebra over a field F and let $D = \text{Der } M$. Then, by (3.6) D is inner, and the invariance of the Killing form $K(\ ,\ )$ of M implies the relation

$$K(\sigma x,y) + K(x,\sigma y) = 0 \tag{4.12}$$

for all $x,y \in M$ and $\sigma \in D$. Thus, the mapping $x \otimes y \to K(x,y)$ is an element of $\text{Hom}_D(M \otimes M,F)$. In particular, if M is central simple over F, then from the decomposition of $V(\lambda_1) \otimes V(\lambda_1)$ in the proof of Lemma 3.13 and from Lemma 3.8 we have

$$\dim_F \text{Hom}_D(M \otimes M,F) = 1\ , \tag{4.13}$$

and hence the mapping $x \otimes y \to K(x,y)$ is a basis of $\text{Hom}_D(M \otimes M,F)$. It follows also from (3.11) that if Fa is a one-dimensional trivial D-module, then

$$\dim_F \text{Hom}_D(Fa \otimes M,M) = 1\ , \tag{4.14}$$

and the mapping $a \otimes x \to x$ is a basis of $\text{Hom}_D(Fa \otimes M,M)$.

In this section we determine the structure of those flexible algebras A for which A^- has a semisimple Malcev subalgebra S^- and there is a subalgebra W^- of A^- such that $A^- = S^- \oplus W^-$ and $[S,W] = 0$. In this part of investigation, A is not necessarily Malcev-admissible.

<u>Theorem 4.3.</u> Let $S_1, \cdots, S_n$ be flexible Malcev-admissible algebras over F where multiplication in S_i is denoted by "$\underset{i}{*}$" for each i, and let W be a flexible algebra over F. Assume that $a_1, \cdots, a_n$ are

elements of W satisfying $[a_i,W] = 0$, $1 \le i \le n$, and that $\tau_1, \cdots, \tau_n$ are linear forms on W such that $\tau_i([W,W]) = 0$, $1 \le i \le n$. Let $K_i(\ ,\ )$ be the Killing form of S_i^-. Then, the vector space direct sum $A = S_1 \oplus \cdots \oplus S_n \oplus W$ with multiplication xy defined by

$$xy = x \underset{i}{*} y + K_i(x,y)a_i \ , \ x,y \in S_i \ ,$$

$$(4.15) \qquad xy = 0 \ , \ x \in S_i \ , \ y \in S_j \ , \ i \neq j \ ,$$

$$xa = ax = \tau_i(a)x \ , \ x \in S_i \ , \ a \in W$$

and with product ab in A for $a,b \in W$ to be the same as in W, is a flexible algebra such that $S^- = S_1^- + \cdots + S_n^-$ is a Malcev subalgebra of A^- and such that $[S^-,W^-] = 0$.

Conversely, suppose that A is a flexible algebra over F and that S^- is a semisimple Malcev subalgebra of A^- such that $S^- = S_1^- \oplus \cdots \oplus S_n^-$ with S_i^- central simple over F. Let W^- be a subalgebra of A^- such that $A^- = S^- \oplus W^-$ and $[S,W] = 0$. Then, W is a subalgebra of A and multiplication in A is given by (4.15) for some flexible Malcev-admissible product "$\underset{i}{*}$" on S_i for each i.

Proof. We show first that the algebra A constructed by (4.15) is flexible and that $S^- = S_1^- + \cdots + S_n^-$ is a Malcev subalgebra of A^- with $[S,W] = 0$. It follows from the first relation of (4.15) that $[x,y] = xy - yx = x \underset{i}{*} y - y \underset{i}{*} x$ for $x,y \in S_i$, so that the commutator in S_i under the product in A is the same as that under the original product in S_i. Hence, by the second relation of (4.15) $S^- = S_1^- + \cdots + S_n^-$ is a Malcev algebra under the product in A^-. To show flexibility of A, we verify that the linearized form of the flexible identity,

$$(x,y,z) + (z,y,x) = 0 \tag{4.16}$$

holds for all possible choices of x, y, z from $S_1, \cdots, S_n$ and W. If x and y lie in the same subspace, then it suffices to show the special case $z = x$, since the more general case will follow from this by linearization.

If $x,y,z \in W$, then (4.16) is valid since W is a flexible subalgebra of A. For $a,b \in W$ and $x,y \in S_i$, using (4.15) we have

$$(a,x,a) = (ax)a - a(xa) = \tau_i(a)xa - a(\tau_i(a)x) = 0 ,$$

$$\begin{aligned}(a,b,x) + (x,b,a) &= (ab)x - a(bx) + (xb)a - x(ba)\\ &= \tau_i(ab)x - \tau_i(b)ax + \tau_i(b)xa - \tau_i(ba)x\\ &= \tau_i(ab - ba)x = 0 ,\end{aligned}$$

$$\begin{aligned}(a,x,y) + (y,x,a) &= (\tau_i(a)x)y - a(xy) + (yx)a - y(\tau_i(a)x)\\ &= \tau_i(a)(xy - yx) - a(x \underset{i}{*} y + K_i(x,y)a_i)\\ &\quad + (y \underset{i}{*} x + K_i(y,x)a_i)a\\ &= \tau_i(a)(xy - yx) - \tau_i(a)(x \underset{i}{*} y - y \underset{i}{*} x)\\ &\quad - K_i(x,y)(aa_i - a_ia) = 0 .\end{aligned}$$

If $a \in W$, $x \in S_i$ and $y \in S_j$ for $i \neq j$, then we have

$$(x,a,y) + (y,a,x) = (\tau_i(a)x)y - x(\tau_j(a)y) + (\tau_j(a)y)x - y(\tau_i(a)x) = 0 ,$$

$$(a,x,y) + (y,x,a) = (\tau_i(a)x)y - a0 + 0a - y\tau_i(a)x = 0 .$$

Thus, (4.16) holds if at least one of x, y, z is in W .

If $x,y \in S_i$, then using the invariance of the Killing form $K_i(\, , \,)$ and flexibility of "$\underset{i}{*}$" in S_i , we compute

$$\begin{aligned}(x,y,x) &= (x \underset{i}{*} y + K_i(x,y)a_i)x - x(y \underset{i}{*} x + K_i(y,x)a_i) \\ &= (x \underset{i}{*} y) \underset{i}{*} x + K_i(x \underset{i}{*} y,x)a_i + K_i(x,y)a_i x \\ &\quad - x \underset{i}{*} (y \underset{i}{*} x) - K_i(x,y \underset{i}{*} x)a_i - K_i(x,y)xa_i \\ &= K_i(x,x \underset{i}{*} y)a_i - K_i(y \underset{i}{*} x,x)a_i \\ &= K_i(x,[x,y])a_i = K_i([x,x],y)a_i = 0 .\end{aligned}$$

If $x,y \in S_i$ and $z \in S_j$ for $i \neq j$, then we have

$$(x,z,x) = (xz)x - x(zx) = 0 ,$$

$$\begin{aligned}(x,y,z) + (z,y,x) &= (x \underset{i}{*} y + K_i(x,y)a_i)z - z(y \underset{i}{*} x + K_i(y,x)a_i) \\ &= K_i(x,y)(a_i z - za_i) = 0 .\end{aligned}$$

Identity (4.16) is clearly satisfied when $x \in S_i$, $y \in S_j$ and $z \in S_k$ for i, j, k distinct. We have shown that the algebra A is a flexible algebra over F for which A^- has a semisimple Malcev subalgebra $S^- = S_1^- \oplus \cdots \oplus S_n^-$ with S_i^- central simple over F such that $A^- = S^- \oplus W^-$ for a subalgebra W^- of A^- with $[W,S] = 0$. If we let $D = \text{Der } S^-$ and $D_i = \text{Der } S_i^-$, then $D = \sum_{i=1}^{n} \oplus D_i$ where each D_i is a central simple Lie algebra over F, and every element of D acts as a derivation on A^+, since D is inner and all ad_x are derivations of A^+ by flexibility of A. We also note that S, D and A^+ satisfy the properties in (4.3), since each S_i is an irreducible D-submodule. By the assumption, W is a trivial D-submodule of A complementary to S^-, and hence $W_\Gamma = 0$ for any nonempty subset Γ of $\{1,2,\cdots,n\}$. Let $x \circ y = \frac{1}{2}(xy + yx)$ denote the product in A^+. It follows from (4.10) and Lemma 4.2 (1) - (3) that $S_i \circ S_i \subseteq S_i + W$, $S_i \circ S_j = 0$ for $i \neq j$, $S_i \circ W \subseteq S_i$, and W is a subalgebra of A^+. Since $[S_i,S_i] \subseteq S_i$, $[S_i,S_j] = 0$ for $i \neq j$, and $[S_i,W] = 0$, we have that

$$S_iS_i \subseteq S_i + W, \quad i = 1, 2, \cdots, n,$$

(4.17) $$S_iS_j = 0 \quad \text{for } i \neq j,$$

$$S_iW = WS_i \subseteq S_i,$$

and that W is a subalgebra of A.

For $0 \neq a \in W$, the mapping $a \otimes x \to a \circ x = ax$ is clearly an element of $\mathrm{Hom}_{D_i}(Fa \otimes S_i^-, S_i^-)$, and since S_i^- is central simple over F, by (4.14) there exists a linear form τ_i on W such that $ax = xa = \tau_i(a)x$ for all $x \in S_i$ and $a \in W$. If $a,b \in W$ and $x \in S_i$, then since

$$\begin{aligned} 0 &= (a,b,x) + (x,b,a) \\ &= (ab)x - a(\tau_i(b)x) + (\tau_i(b)x)a - x(ba) \\ &= \tau_i(ab)x - \tau_i(ba)x = \tau_i([a,b])x, \end{aligned}$$

we have $\tau_i([W,W]) = 0$.

For $x,y \in S_i$, by (4.17) we can let $x \underset{i}{*} y$ and $Q(x,y)$ be the projections of xy onto S_i and W, respectively, so that

$$xy = x \underset{i}{*} y + Q(x,y) .$$

Letting $\{u_1,\cdots,u_r\}$ be a basis of W, we have

$$Q(x,y) = \sum_{k=1}^{r} Q_k(x,y)u_k$$

for $Q_k(x,y) \in F$. Since $[S_i,S_i] \subseteq S_i$, $Q_k(,)$ is a symmetric bilinear form on S_i. Noting that each element of D acts as a derivation on S_i^- and on $S_i \circ S_i$, and since $D_iW = 0$, we have for $\sigma \in D_i$

$$\sigma(x \underset{i}{*} y) = \sigma(xy) = \sigma(x)y + x\sigma(y)$$

$$= \sigma(x) \underset{i}{*} y + x \underset{i}{*} \sigma(y) + \sum_{k=1}^{r} [Q_k(\sigma x,y) + Q_k(x,\sigma y)]u_k ,$$

which implies $Q_k(\sigma x,y) + Q_k(x,\sigma y) = 0$ for all $x,y \in S_i$, $k = 1, \cdots, r$. Hence, $x \otimes y \to Q_k(x,y)$ is an element of $\mathrm{Hom}_{D_i}(S_i^- \otimes S_i^-,F)$, and by (4.13) $Q_k(,)$ is a multiple of the Killing form $K_i(,)$. Thus, we can write

$$xy = x \underset{i}{*} y + \sum_{k=1}^{r} \alpha_k K_i(x,y)u_k = x \underset{i}{*} y + K_i(x,y)a_i ,$$

where $a_i = \sum_k \alpha_k u_k \in W$. To see that $[a_i,b] = 0$ for all $b \in W$, we compute

$$0 = (b,x,y) + (y,x,b) = (\tau_i(b)x)y - b(xy) + (yx)b - y(\tau_i(b)x)$$

$$= \tau_i(b)(x \underset{i}{*} y + K_i(x,y)a_i) - b(x \underset{i}{*} y + K_i(x,y)a_i)$$

$$+ (y \underset{i}{*} x + K_i(y,x)a_i)b - \tau_i(b)(y \underset{i}{*} x + K_i(y,x)a_i)$$

$$= - K_i(x,y)ba_i + K_i(y,x)a_ib = K_i(x,y)[a_i,b] .$$

Finally, we show that S_i is Malcev-admissible and flexible under the product "$\underset{i}{*}$" . For the former, we note that $x \underset{i}{*} y - y \underset{i}{*} x = xy - yx$, and since S_i^- is a Malcev subalgebra of A^- , so is the product "$\underset{i}{*}$" in S_i . For flexibility, we have that

$$(x \underset{i}{*} y) \underset{i}{*} x - x \underset{i}{*} (y \underset{i}{*} x) = (x \underset{i}{*} y)x - K_i(x \underset{i}{*} y,x)a_i - x(y \underset{i}{*} x)$$

$$+ K_i(x,y \underset{i}{*} x)a_i$$

$$= (xy - K_i(x,y)a_i)x - x(yx - K_i(y,x)a_i) - K_i(x \underset{i}{*} y - y \underset{i}{*} x,x)a_i$$

$$= - K_i(x,y)(a_ix - xa_i) - K_i(xy - yx,x)a_i$$

$$= K_i([y,x],x)a_i = K_i(y,[x,x])a_i = 0 . \qquad \Box$$

In the second part of Theorem 4.3, if, in addition, F is algebraically closed and W^- is also a Malcev algebra, then the multiplication between S and W can be further refined and $a_1, \cdots, a_n$ can be chosen from the solvable radical of A^- . In this case, since A^- is a Malcev algebra and the linear mapping of the form (3.1) is a derivation of A^- , every element of D acts as a derivation on A . Let

$$W^- = \sum_{k=n+1}^{m} \tilde{S}_k^- + R^-$$

be a Levi decomposition of W^- , where each $\tilde{S}_k^-$ is a simple Malcev algebra over F and R^- is the solvable radical of W^- . Since $A^- = \Sigma S_i^- + \Sigma \tilde{S}_k^- + R^-$ and $[S,W] = 0$, R^- is the radical of A^- . We note that $R_\Gamma = 0$ for any subset Γ of $\{1,\cdots,n,\cdots,m\}$ such that

$\Gamma \cap \{1,\cdots,n\} \neq \phi$. Thus, by Lemma 4.2 we have

$$S_i S_i \subseteq S_i + R_0 \subseteq S_i + R \ , \ 1 \le i \le n \ ,$$

$$S_i \tilde{S}_k \subseteq R_{ik} = 0 \ , \ 1 \le i \le n < k \le m \ .$$

Hence, the elements $a_1, \cdots, a_n$ can be chosen from R_0 . Also, if $1 \le i \le n$ and $k \in \Gamma \subseteq \{n+1,\cdots,m\}$, then by Lemma 4.1, (4.7) and (4.8), we have

$$S_i R_\Gamma \subseteq (\text{Im } D_i)(\text{Ker } D_i) \subseteq \text{Im } D_i \ ,$$

$$S_i R_\Gamma \subseteq (\text{Ker } \tilde{D}_k)(\text{Im } \tilde{D}_k) \subseteq \text{Im } \tilde{D}_k \ ,$$

showing that

$$S_i R_\Gamma = (\text{Im } D_i) \cap (\text{Im } \tilde{D}_k) = \sum_{\{i,k\} \subseteq \Gamma} R_\Gamma = 0 \ ,$$

where $\tilde{D}_k = \text{Der } \tilde{S}_k^-$.

Since Theorem 4.3 essentially determines the multiplication between the radical R^- of A^- and a simple subalgebra S of A^- which centralizes R^- , Theorem 4.3 may be viewed as a reduction theorem in the sense that it reduces the study of the structure of flexible Malcev-admissible algebras of characteristic 0 to the subclass of algebras A such that no simple subalgebra of A^- centralizes the radical of A^- . However, a general structure theory for that subclass of algebras is not known.

<u>Corollary 4.4</u>. Let A be a flexible algebra over F with multiplication denoted by xy , and let $S^- = S_1^- \oplus \cdots \oplus S_n^-$ and W be the same as in the second part of Theorem 4.3. Then, W is a subalgebra of A , and there is a flexible Malcev-admissible product "$\underset{i}{*}$" on S_i such that either

(1) $x \underset{i}{*} y = \frac{1}{2}[x,y]$, or

(2) S_i^- is a Lie algebra of type $A_n (n \geq 2)$ over F and for some fixed $\beta \in F$, $x \underset{i}{*} y = \frac{1}{2}[x,y] + \beta x \# y$ where $x \# y$ is defined by (3.30).

Furthermore, multiplication xy in A is given by relation (4.15), where $x \# y$ is as in (1) or (2), and $a_1, \cdots, a_n$ and $\tau_1, \cdots, \tau_n$ are as in Theorem 4.3.

Proof. By the assumption, $[S,W] = 0$ and by Theorem 4.3 W is a subalgebra of A and the product "$\underset{i}{*}$" on S_i described there is flexible Malcev-admissible such that $(S_i, \underset{i}{*})^- = S_i^-$, being central simple over F . Thus, the remainder of the results follows from Theorems 3.20 and 4.3. □

Recall that a non-solvable Malcev algebra is called *reductive* if the solvable radical equals its center. The following characterizes flexible Malcev-admissible algebras A with A^- reductive.

Corollary 4.5. Let R be any commutative algebra over F , and let $S_1, \cdots, S_n$ be any flexible Malcev-admissible algebras such that S_i^- is semisimple for each i . Then, for any choice of a_i and τ_i , the vector space direct sum $A = S_1 + \cdots + S_n + R$ with multiplication xy defined by (4.15) is a flexible Malcev-admissible algebra with A^- reductive. Conversely, if A is a flexible Malcev-admissible algebra over F with subspaces $S_1, \cdots, S_n, R$ such that $A^- = S_1^- \oplus \cdots \oplus S_n^- \oplus R^-$, where each S_i^- is a central simple Malcev algebra over F and where R^- is the center of A^- , then A has the multiplication described by (4.15).

Proof. Since R is commutative, the elements a_i of R and the linear forms τ_i on R in Theorem 4.3 can be arbitrarily chosen. Also, since each S_i^- has center zero, R^- is the center of A^-. The remainder of the corollary is immediate from Theorem 4.3. □

<u>Theorem 4.6</u>. Let A, $S^- = S_1^- \oplus \cdots \oplus S_n^-$, and W be the same as in Corollary 4.4, and let B be an ideal of A. Then, there is an ideal I of W and a subset $\Gamma \subseteq \{1,2,\cdots,n\}$ such that $B = \Sigma_{j\in\Gamma} S_j + I$. Moreover, if $j \in \Gamma$, then $a_j \in I$ and $\tau_k(I) = 0$ for all $k \notin \Gamma$. Conversely, any subspace B of A constructed in this manner is an ideal of A.

<u>Proof</u>. Assume that B is a nonzero ideal of A, and let $0 \neq b = \Sigma x_i + a$ be an element of B, where $x_i \in S_i$ and $a \in W$. If $x_j \neq 0$, then there is some $y \in S_j$ with $[x_j,y] \neq 0$. But then $0 \neq [x_j,y] = [b,y] \in B \cap S_j$ since $[S,W] = 0$, and hence $B \cap S_j = S_j$ since $B \cap S_j$ is a nonzero ideal of the simple algebra S_j under the product $[\ ,\ ]$. Thus, there is a subset $\Gamma \subseteq \{1,\cdots,n\}$ such that $B \cap S_j = S_j$ for all $j \in \Gamma$ and $B \cap S_k = 0$ for $k \notin \Gamma$. This implies that $B = \Sigma_{j\in\Gamma} S_j \oplus (W \cap B)$, and the subspace $I = W \cap B$ is an ideal of W. For $j \in \Gamma$ and $x,y \in S_j$, xy and $x \underset{j}{*} y$ lie in B, so that $K_j(x,y)a_j = xy - x \underset{j}{*} y \in B$, to give $a_j \in I$. To show $\tau_k(I) = 0$ for all $k \notin \Gamma$, suppose that $\tau_k(a) \neq 0$ for some $a \in I$. Then, by (4.15) $xa = \tau_k(a)x \in B$ for all $x \in S_k$, to show $S_k \subseteq B$. But this contradicts the choice of k, and so $\tau_k(I) = 0$ for all $k \notin \Gamma$. The converse to show that such a subspace B is an ideal of A is straightforward from relation (4.15). □

<u>Corollary 4.7</u>. Let A, S and W be the same as in Theorem 4.6.

(1) A is simple if and only if for each proper subset $\Gamma \subseteq \{1,\cdots,n\}$ there is no ideal I of W with $a_j \in I$ for all $j \in \Gamma$ and with $\tau_k(I) = 0$ for all $k \notin \Gamma$.

(2) If A is simple and $W \neq 0$, then $a_i \neq 0$ and $\tau_i \neq 0$ for all i.

(3) If W is simple and $a_i \neq 0$ and $\tau_i \neq 0$ for all i, then A is simple.

Proof. (1) If for some proper subset Γ of $\{1,\cdots,n\}$, there is an ideal I of W with $a_j \in I$ for all $j \in \Gamma$ and with $\tau_k(I) = 0$ for all $k \notin \Gamma$, then $B = \Sigma_{j\in\Gamma} S_j + I$ is a proper ideal of A. Conversely, if B is a proper ideal of A, then it follows from Theorem 4.6 that B has the form $B = \Sigma_{j\in\Gamma} S_j + I$ for some proper subset $\Gamma \subseteq \{1,\cdots,n\}$ and an ideal I of W such that $a_j \in I$ for all $j \in \Gamma$ and $\tau_k(I) = 0$ for all $k \notin \Gamma$.

(2) If $a_i = 0$, then S_i is an ideal of A, while if $\tau_i = 0$, then $\Sigma_{j\neq i} S_j + W$ is an ideal of A.

(3) If B is an ideal of A, then B has the structure described in Theorem 4.6 with $I = 0$ or $I = W$, since W is simple. If $I = 0$, then since $a_j \in I$ for all $j \in \Gamma$, it must be that $\Gamma = \phi$ and $B = 0$. On the other hand, if $I = W$, then $\tau_k(I) = 0$ for all $k \notin \Gamma$ implies $\Gamma = \{1,\cdots,n\}$. Hence $B = A$ in this case, and the only ideals in A are 0 and A. □

Corollary 4.8. Let $M = M_1 \oplus \cdots \oplus M_n \oplus Z$ be any reductive Malcev algebra with center $Z \neq 0$, where each M_i is a simple Malcev algebra over F. Then there is a simple flexible Malcev-admissible algebra A with A^- isomorphic to M.

Proof. Suppose that $S_i = M_i$ for $i = 1, 2, \cdots, n$ and $Z = W$ in Theorem 4.6, and let $u_1, \cdots, u_m$ be a basis of W. Since a_i and τ_i for $i = 1, \cdots, n$ can be arbitrarily chosen in this case, we can let

$a_i = u_i$ for each i and define the linear form τ_i on W by $\tau_i(u_i) = 1$ and $\tau_i(u_j) = 0$ for $j \neq i$, $i = 1, \cdots, n$, $j = 1, \cdots, m$. Let ab be any commutative product defined on W for which W is a simple algebra. Then, since $a_i \neq 0$ and $\tau_i \neq 0$ for each i, by Corollary 4.7 (3) and Theorem 4.3 the vector space direct sum $A = S_1 \oplus \cdots \oplus S_n \oplus W$ with multiplication xy defined by the following relations is a simple flexible Malcev-admissible algebra with A^- isomorphic to M:

$$xy = \tfrac{1}{2}[x,y] + K_i(x,y)a_i \ , \quad x, y, \in S_i \ ,$$

$$xy = 0 \ , \ x \in S_i \ , \ y \in S_j \ , \ i \neq j \ ,$$

$$ax = xa = \tau_i(a)x \ , \ x \in S_i \ , \ a \in W \ ,$$

and ab for $a,b \in W$ is the same as the commutative product in W. Such a commutative product always exists. In fact, let W_0 be the linear span of $u_2, \cdots, u_m$, so that $W = Fu_1 \oplus W_0$. Let $f(\, , \,)$ be a bilinear form on W_0 defined by $f(u_i,u_j) = \delta_{ij}$, $i, j = 2, \cdots, m$. Then, $f(\, , \,)$ is symmetric and nondegenerate. Associated with $f(\, , \,)$ is the commutative product on W defined by

$$(\alpha u_1 + a)(\beta u_1 + b) = (\alpha\beta + f(a,b))u_1 + (\beta a + \alpha b)$$

for $\alpha,\beta \in F$ and $a,b \in W_0$. It is well known that W with this product becomes a Jordan algebra (Jacobson [3,p.13]), and is simple, since $f(\, , \,)$ is nondegenerate. □

Corollary 4.8 indicates that the number of simple flexible Malcev-admissible algebras is large. The following example shows also that the simplicity of A need not impose any restriction on W.

Example 4.1. Let W be any commutative algebra with basis u_1, $\cdots$, u_n , and assume $\tau_1, \cdots, \tau_n$ are linear forms on W defined by $\tau_i(u_j) = \delta_{ij}$. Let $a_i = u_{i+1}$ for $1 \le i \le n-1$ and $a_n = u_1$. Assume also that $S_1, \cdots, S_n$ are any simple Malcev algebras and $A = S_1 + \cdots + S_n + W$ is a vector space direct sum. We show that A with multiplication given by (4.15) with $x \underset{i}{*} y = \frac{1}{2}[x,y]$ is a simple flexible Malcev-admissible algebra. To verify that A is simple, let B be an ideal of A . By Theorem 4.6 there is a subset Γ of $\{1,\cdots,n\}$ and an ideal I of W with $B = \Sigma_{j\varepsilon\Gamma} S_j + I$. If $\Gamma = \phi$ or $\{1,\cdots,n\}$, then $B = 0$ or A , respectively. Thus, it suffices to assume that there is some $j \in \Gamma$ with $j + 1 \notin \Gamma$ where we interpret the subscripts mod n . In this case, $\tau_{j+1}(I) = 0$ by Theorem 4.6, but then, $0 = \tau_{j+1}(a_j) = \tau_{j+1}(u_{j+1}) = 1$, and this contradiction shows that A is simple. Therefore, the simplicity of W is not restricted by that of A . □

We will provide in Section 4.5 more examples of simple flexible Malcev-admissible algebras to show that widely diversified classes of Malcev algebras can occur as the attached algebra A^- of a simple flexible Malcev-admissible algebra A . We end this section with an application of these results to a flexible Malcev-admissible algebra A of dimension 4 or 8 for which A^- has a simple Lie subalgebra or a simple, non-Lie, Malcev subalgebra.

Theorem 4.9. Let A be a flexible Malcev-admissible algebra over an algebraically closed field F of dimension ≤ 4 such that A^- has a simple subalgebra. Then,

(1) A is a Lie algebra isomorphic to $\mathfrak{sl}(2)$, or

(2) A has a basis e, f, h, a , and there are scalars $\alpha,\beta,\gamma \in F$

such that multiplication in A is given by

$$he = -eh = e,\ hf = -fh = -f,\ ef = \frac{1}{2}h + \alpha a ,$$

$$\text{(4.18)} \qquad fe = -\frac{1}{2}h + \alpha a,\ h^2 = 2\alpha a,\ a^2 = \beta a ,$$

$$ax = xa = \gamma x \quad \text{for all } x \text{ in the span of } e, h, f ,$$

and all other products are 0. Moreover, A is Lie-admissible such that $A^- = s\ell(2) \oplus Fa$ is reductive with center Fa, where $\{e,h,f\}$ is a standard basis of $s\ell(2)$, and A is simple if and only if $\gamma \neq 0$.

Proof. Let S denote a simple Malcev subalgebra of A^-. Since $\dim S \leq 4$, by a dimension argument we find that $S = s\ell(2)$. When A is regarded as an $s\ell(2)$-module under the adjoint action, $s\ell(2)$ is an irreducible $s\ell(2)$-submodule of A. Thus, if $\dim A = 3$, then by Theorem 3.20 A is a Lie algebra isomorphic to $s\ell(2)$. Assume that $\dim A = 4$. Then, a submodule complementary to $s\ell(2)$ is a one-dimensional trivial $s\ell(2)$-module Fa, and hence we are in the situation of Theorem 4.3 with $x * y = \frac{1}{2}[x,y]$ for $x,y \in s\ell(2)$. Let e, h, f be a standard basis of $s\ell(2)$ such that $[e,f] = h,\ [h,e] = 2e,\ [h,f] = -2f$. Then, the Killing form $K(\ ,\ )$ of $s\ell(2)$ relative to this has matrix

$$\text{(4.19)} \qquad \begin{bmatrix} 0 & 0 & 4 \\ 0 & 8 & 0 \\ 4 & 0 & 0 \end{bmatrix}.$$

Now by the results of Corollary 4.4, $xy = \frac{1}{2}[x,y] + K(x,y)a_1$ for some $a_1 \in Fa$. By letting $4a_1 = \alpha a$ for $\alpha \in F$, we have multiplication (4.18), since Fa is a subalgebra of A. The remainder of the proof is immediate from Corollary 4.7. □

Corollary 4.10. Let A be the same as in Theorem 4.9. If A^- is not solvable, then A is a Lie algebra isomorphic to $s\ell(2)$ or an algebra with multiplication given by (4.18).

Proof. Since A^- is not solvable, by Levi's theorem A^- has a simple subalgebra. Thus, the result follows from Theorem 4.9. □

Theorem 4.11. Let A be a flexible Malcev-admissible algebra over an algebraically closed field F of dimension ≤ 8 such that A^- has a simple, non-Lie, Malcev subalgebra. Then,

(1) A is a Malcev algebra isomorphic to C_0^-, or

(2) A has a basis $e_0, e_i, e_{-i} (i = 1,2,3)$, a, and there are scalars α, β, γ in F such that multiplication in A is given by

$$e_0 e_{\pm i} = - e_{\pm i} e_0 = \pm e_{\pm i}, \quad i = 1, 2, 3 ,$$

$$e_{\pm i} e_{\pm j} = - e_{\pm j} e_{\pm i} = \pm e_{\mp k}, \quad (ijk) = (123), (231), (312) ,$$

$$e_i e_{-i} = \frac{1}{2} e_0 + \alpha a, \quad e_{-i} e_i = - \frac{1}{2} e_0 + \alpha a, \quad i = 1, 2, 3 , \tag{4.20}$$

$$e_0^2 = 2\alpha a, \quad a^2 = \gamma a ,$$

$$ax = xa = \beta x \quad \text{for all } x \text{ in the span of } e_0, e_i, e_{-i}, i = 1,2,3 ,$$

and all other products are 0. Moreover, A is non-Lie, Malcev-admissible such that $A^- = C_0^- \oplus Fa$ is reductive with center Fa, and A is simple if and only if $\gamma \neq 0$.

Proof. Since C_0^- is the unique, non-Lie, simple Malcev algebra over F, it must be that $\dim A = 7$ or 8. In the former case, $A^- \cong C_0^-$ and by Theorem 3.20 A is a Malcev algebra isomorphic to C_0^-. Suppose now that $\dim A = 8$. Under the assumption, C_0^- is a subalgebra of A^-. When A is regarded as a Malcev module for C_0^- under the adjoint action or as a Lie module for $\mathrm{Der}\, C_0^- = G_2$, a submodule of A complementary to C_0^- is a one-dimensional trivial module Fa for C_0^-, and hence we are in the situation of Theorem 4.3 with $x \underset{i}{*} y = \frac{1}{2}[x,y]$ for $x,y \in C_0^-$. Let e_0, e_i, e_{-i}, $i = 1, 2, 3$ be a standard basis of C_0^- for which C_0^- has multiplication given by (3.24)(Chapter 3). If $x = \alpha_0 e_0 + \Sigma_{i=1}^3 (\alpha_i e_i + \alpha_{-i} e_{-i})$ and $y = \beta_0 e_0 + \Sigma_{i=1}^3 (\beta_i e_i + \beta_{-i} e_{-i})$ are elements of C_0^-, then from matrix (3.26) one can easily compute the Killing form $K(x,y)$ as

$$K(x,y) = 12[2\alpha_0\beta_0 + \sum_{i=1}^{3} (\alpha_i\beta_{-i} + \alpha_{-i}\beta_i)] ,$$

which gives the matrix of $K(e_i,e_j)$:

$$(4.21) \qquad K(e_i,e_j) = \begin{cases} 24 , & \text{if } i = j = 0 , \\ 12 , & \text{if } j = -i,\ i = 1, 2, 3 , \\ 0 , & \text{otherwise} . \end{cases}$$

It follows from Corollary 4.4 that if $x,y \in C_0^-$, then $xy = \frac{1}{2}[x,y] + K(x,y)a_1$ for some $a_1 \in Fa$. Using (4.21) and letting $12a_1 = \alpha a$ for $\alpha \in F$, we have multiplication (4.20) for A. The remainder of the proof is immediate. □

4.3. MULTIPLICATION RELATIONS BETWEEN IRREDUCIBLE SUMMANDS

In this section we assume that A is a flexible Malcev-admissible algebra over a field F for which A^- has a split semisimple subalgebra S^-. Then, A can be regarded as a Malcev module for S^- under the adjoint action, and by Weyl's theorem for Malcev modules, there is an S^--submodule R of A complementary to S. We further assume that R is a subalgebra of A^-, so that R is an ideal of A^- also. When A^- is not solvable, by Levi's theorem, a Levi factor S^- of A^- and the solvable radical R of A^- satisfy these properties. Let $S^- = S_1^- \oplus \cdots \oplus S_n^-$ be the decomposition of S^- into simple ideals, and let $D = \text{Der}\, S^-$ and $D_i = \text{Der}\, S_i^-$. Then, $D = D_1 \oplus \cdots \oplus D_n$, where each D_i is a split simple Lie algebra over F. Since D is inner by (3.6) (Chapter 3) and all ad_x are derivations on A^+ by flexibility, D acts as derivations on A, and furthermore satisfies properties (4.1) - (4.3). Recall from Lemma 3.12 that the decomposition of R into the irreducible S^--submodules coincides with that into the irreducible D-submodules and that the support of an irreducible S^--submodule W of R is the same as the support of W as a D-module. Thus, relations (4.6) - (4.10) and Lemma 4.2 hold for the present case. We use the decomposition

$$A = \sum_{i=1}^{n} S_i + \sum_{\Gamma\neq\phi} R_\Gamma + R_0 \tag{4.22}$$

as in (4.6) and denote

$$R_\star = \sum_{\Gamma\neq\phi} R_\Gamma . \tag{4.23}$$

We retain the same notations as in Section 4.1. Throughout, S^- denotes a fixed split semisimple subalgebra of A^- for which A is decomposed as in (4.22).

Lemma 4.12. Let A, S^-, and R_* be the same as in (4.22) and (4.23). Then $SR_* \subseteq R$ and $R_*S \subseteq R$.

Proof. Since $[S,R_*] \subseteq R_* \subseteq R$, it suffices to show that $S \circ R_* \subseteq R$, where "$\circ$" denotes the product in A^+. Let Γ be any non-empty subset of $\{1,2,\cdots,n\}$. Assume first that i is an index with $i \notin \Gamma$. Then, by (4.7) and (4.8), we have $S_i \subseteq Im_i$, $R_\Gamma \subseteq Ker_i$, $S_i \subseteq Ker_j$, $R_\Gamma \subseteq Im_j$ for $j \in \Gamma$, and $S_i, R_\Gamma \subseteq \bigcap_{k \notin \Gamma \cup \{i\}} Ker_k$. Thus, it follows from Lemma 4.1 and (4.9) that

(4.24) $$S_i \circ R_\Gamma \subseteq \left(\bigcap_{k \notin \Gamma \cup \{i\}} Ker_k\right) \cap \left(\bigcap_{j \in \Gamma} Im_j\right) \cap Im_i .$$

Since $\left(\bigcap_{j \in \Gamma} Im_j\right) \cap Im_i \subseteq \Sigma_{\Gamma \cup \{i\}} \subseteq {}_\Lambda R_\Lambda$ by (4.8), (4.24) and (4.7) imply

(4.25) $$S_i \circ R_\Gamma \subseteq R_{\Gamma \cup \{i\}} \subseteq R ,$$

and we are done in this case. Assume then that $i \in \Gamma$. Let $s,t \in S_i$, $v \in R_\Gamma$, and for an element w of A, let w_i denote the image of w under the projection of A onto S_i. Using the linearized form $[x,y \circ z] = [x \circ y,z] + [x \circ z,y]$ of the third power identity $x^2x = xx^2$, we have

$$[s,v \circ t]_i = [s \circ v,t]_i + [s \circ t,v]_i = [s \circ v,t]_i ,$$

since $[s \circ t,v]_i = 0$ by Lemma 4.2(1). This identity is equivalent to the relation

$$[s,(v \circ t)_i] = [(s \circ v)_i,t] .$$

Hence, for a fixed $v \in R_\Gamma$, if we define a linear mapping $\theta : S_i \to S_i$ by

$\theta(s) = (s \circ v)_i$ for $s \in S_i$, then since S_i^- is a split simple Malcev algebra over F , by Lemma 2.37 θ is a scalar mapping on S_i , i.e., $\theta(s) = \alpha s$ for some $\alpha \in F$. Since ad_s is a derivation of A^+ ,

$$(t \circ [s,v])_i = [s,t \circ v]_i - ([s,t] \circ v)_i$$

$$= \alpha[s,t] - \alpha[s,t] = 0 ,$$

which shows that $(S_i \circ R_\Gamma)_i = 0$, since by Lemma 3.12 the elements of the form $[s,v]$ span R_Γ .

We finish our proof by showing that $(S_i \circ R_\Gamma)_j = 0$ also. Suppose that $(S_i \circ R_\Gamma)_j \neq 0$ and hence $(s \circ v)_j \neq 0$ for some $s \in S_i$ and $v \in R_\Gamma$. Then, there is an element $x \in S_j$ such that

$$0 \neq [x,(s \circ v)_j] = [x,s \circ v]_j = [x \circ s,v]_j + [x \circ v,s]_j = 0 .$$

This contradiction shows that $(S_i \circ R_\Gamma)_j = 0$. Hence, $S_i \circ R_\Gamma \subseteq R$ for all $i = 1, 2, \cdots, n$, to give $S \circ R_* \subseteq R$. □

Lemma 4.13. Let Γ and Δ be nonempty subsets of $\{1,\cdots,n\}$. Then

(1) if $i \notin \Gamma$, then $S_iR_\Gamma \subseteq R_{\Gamma\cup\{i\}}$,

(2) if $i \in \Gamma$, then $S_iR_\Gamma \subseteq R_\Gamma + R_{\Gamma-\{i\}}$,

(3) $R_0R_\Gamma \subseteq R_\Gamma$,

(4) $R_\Gamma R_\Delta \subseteq [R,R] + R_0$.

Proof. (1) If $i \notin \Gamma$, then $[S_i,R_\Gamma] = 0$, and since $S_i \circ R_\Gamma \subseteq R_{\Gamma\cup\{i\}}$ by (4.25), we have $S_iR_\Gamma \subseteq R_{\Gamma\cup\{i\}}$.

(2) Assume first $\Gamma = \{i\}$. From (4.11) it follows that $B_i = S_i + R_i + R_0 = \bigcap_{j\neq i} Ker_j$ is a subalgebra. Hence, in view of Lemma

4.12, we have

$$S_iR_i \subseteq R \cap B_i = R_i + R_0 = R_\Gamma + R_{\Gamma-\{i\}} ,$$

as desired. To complete the proof, it suffices to treat the case of $i,j \in \Gamma$ with $i \neq j$. Since $S_i \subseteq \mathrm{Ker}_j$ and $R_\Gamma \subseteq \mathrm{Im}_j$ by (4.7) and (4.8), it follows from Lemma 4.1 that

$$S_iR_\Gamma \subseteq S_j + \sum_{j\varepsilon\Delta} R_\Delta ,$$

which combines with Lemma 4.12 to show that $S_iR_\Gamma \subseteq \sum_{j\varepsilon\Delta} R_\Delta$ for every $j \neq i$ in Γ . But, since B_Γ of (4.11) is a subalgebra of A , we have

$$S_iR_\Gamma \subseteq B_\Gamma = \sum_{k\varepsilon\Gamma} S_k + \sum_{\Delta\subseteq\Gamma} R_\Delta + R_0 .$$

This implies that $S_iR_\Gamma \subseteq R_\Gamma + R_{\Gamma-\{i\}}$.

(3) For $i \in \Gamma$, since $R_0 \subseteq \mathrm{Ker}_i$ and $R_\Gamma \subseteq \mathrm{Im}_i$, by Lemma 4.1 and (4.8) we have $R_0R_\Gamma \subseteq S_i + \Sigma_{i\varepsilon\Delta} R_\Delta$. Since this is true for all $i \in \Gamma$, if Γ has at least two elements, then

$$R_0R_\Gamma \subseteq \bigcap_{i\varepsilon\Gamma} (S_i + \sum_{i\varepsilon\Delta} R_\Delta) \subseteq \bigcap_{i\varepsilon\Gamma} (\sum_{i\varepsilon\Delta} R_\Delta) = R_\Gamma .$$

Thus, it suffices to assume that $\Gamma = \{i\}$. From (4.7) and (4.8), we see that

$$[R_0,R_i] \subseteq \mathrm{Im}_i \cap (\bigcap_{k\neq i} \mathrm{Ker}_k) \cap R = R_i . \tag{4.26}$$

The proof is finished by showing $R_0 \circ R_i \subseteq R_i$. From (4.26) and the fact that ad_x is a derivation of A^+ for each $x \in A$, it follows that

$$R_0 \circ R_i = R_0 \circ [S_i,R_i] \subseteq [R_0 \circ S_i,R_i] + [R_0,R_i] \circ S_i . \tag{4.27}$$

Since $R_0 \circ S_i \subseteq S_i + R_i$ by Lemma 4.2(3) and R_i, S_i are contained in the subalgebra B_i , the right side of (4.27) is contained in $R \cap B_i = R_i + R_0$ by (4.11), so that $R_0 \circ R_i \subseteq R_i + R_0$. But, $R_0 \circ R_i \subseteq Im_i$ by Lemma 4.1, and hence by (4.8) we conclude that $R_0 \circ R_i \subseteq R_i$.

(4) Since $[S,R_\Gamma \circ R_\Delta] \subseteq [S,R_\Gamma] \circ R_\Delta + R_\Gamma \circ [S,R_\Delta] \subseteq R_\Gamma \circ R_\Delta$, $R_\Gamma \circ R_\Delta$ is a submodule. From $[S,R_\Gamma \circ R_\Delta] \subseteq [S \circ R_\Gamma,R_\Delta] + [S \circ R_\Delta,R_\Gamma] \subseteq [R,R]$ by parts (1) and (2), we have $R_\Gamma \circ R_\Delta \subseteq [R,R] + R_0$ and this proves part (4). □

Corollary 4.14. (1) R is a subalgebra of A .

(2) $S[R,R]$ and $[R,R]S$ are contained in R for $S = S_1 + \cdots + S_n$.

(3) If A is simple, then $R_0 \nsubseteq [R,R]$.

Proof. (1) Since R_0 is a subalgebra of A by (4.10), it is immediate from Lemma 4.13 (3) and (4) that R is a subalgebra of A .

(2) Since $[S,[R,R]] \subseteq [S,R] \subseteq R$, $[R,R]$ is an ideal of A^- . Thus, it suffices to show $S \circ [R,R] \subseteq R$. Let Γ, Δ be nonempty subsets of $\{1,\cdots,n\}$. We complete the proof by showing that $S \circ [R_\Gamma,R_\Delta]$ and $S \circ [R_0,R]$ are contained in R . Since $R_\Delta \circ R_\Gamma \subseteq R$ by Lemma 4.13 (4), $S \circ R_\Gamma \subseteq R$ by Lemma 4.13 (1),(2), and $S_i \circ R_0 \subseteq S_i + R_i$ by Lemma 4.2 (3), the desired containments follow from the calculation

$$S \circ [R_\Gamma,R_\Delta] \subseteq [S \circ R_\Gamma,R_\Delta] + [S,R_\Delta] \circ R_\Gamma \subseteq R + R_\Delta \circ R_\Gamma \subseteq R \ ,$$

$$S \circ [R_0,R] \subseteq [R_0,S \circ R] + [R_0,S] \circ R = [R_0,S \circ R] \subseteq R \ .$$

(3) We prove this by arguing that $R_0 \subseteq [R,R]$ implies that R is an ideal of A . By Lemma 4.13 (1) and (2) we have $S \circ R_\Gamma \subseteq R$, and by part (2) of this corollary $S \circ [R,R] \subseteq R$. Hence, if $R_0 \subseteq [R,R]$, then $S \circ R_0 \subseteq R$, which implies that R is an ideal of A , since R is a subalgebra of A and an ideal of A^- . □

Lemmas 4.2 and 4.13 establish our fundamental containment relations between submodules for the product in A. Our principal result in this section is to establish a more stringent containment relation:

$$S' \circ W \subseteq [S' \circ S', W] + R_0 , \tag{4.28}$$

where $S' = S_{i_1} + \cdots + S_{i_q}$ is a split semisimple subalgebra of A^- and W is a nontrivial irreducible S'-submodule of A^- such that $[S_j, W] = 0$ for $j \notin \{i_1, \cdots, i_q\}$. Relation (4.28) plays a main role for the characterization of simple flexible Malcev-admissible algebras with the solvable radical of A^- abelian. The proof of (4.28) involves the consideration of the weight decomposition.

Definition 4.1. Let M be a nilpotent Malcev algebra over a field F and let V be an M-module over F. For a mapping $\mu : M \to F$, define

$$V_\mu = \{w \in V \mid (\rho(x) - \mu(x))^{n(x)} w = 0 \quad \text{for} \quad x \in M\} ,$$

where ρ is the representation afforded by the M-module V and $n(x) > 0$ is an integer depending on x. If $V_\mu \neq 0$, then μ is called a *weight* of M and V_μ the *weight space* of weight μ. □

Lemma 4.15. Let M be a split semisimple Malcev algebra over F with a split Cartan subalgebra H, and let V be an M-module over F. Then, V is the direct sum of weight spaces of H in V, and H acts diagonally on each weight space.

Proof. In view of Weyl's theorem, it suffices to assume that V is a nontrivial irreducible M-module. Let $M = S_1 + \cdots + S_m$ be the direct sum of simple ideals. If V is a Lie module for M, then by Corollary 3.11 V is annihilated by all non-Lie summands of M, and hence V is regarded as a module for a split semisimple Lie algebra. The result in this case is well known. Assume that V is a non-Lie module for M. It follows from Corollary 3.11 that V is either isomorphic to the two-dimensional module $\tilde{M}_2$ for $s\ell(2)$ defined by (3.25), or to the adjoint module C_0^- given by (3.24). In either case, $s\ell(2)$ or C_0^- is a direct summand of M, and V is annihilated by all other summands of M. If $V \cong \tilde{M}_2$, then by relation (3.25) V is the direct sum of two one-dimensional weight spaces. Suppose that V is isomorphic to C_0^-, and e_0, e_i, e_{-i} $(i=1,2,3)$ is a standard basis of C_0^- with multiplication given by (3.24). Then, we can choose a basis u_0, u_i, u_{-i} $(i=1,2,3)$ of V such that

$$(4.29)\qquad \begin{aligned} &[e_0,u_{\pm i}] = \pm 2u_{\pm i},\ [e_{\pm i},u_0] = \mp u_{\pm i},\ [e_{\mp i},u_{\pm i}] = \mp u_0 , \\ &[e_i,u_j] = 2u_{-k},\ [e_{-i},u_{-j}] = -2u_k,\ (ijk) = (123),\ (231),\ (312) \end{aligned}$$

and all other module products are 0. Since $H = Fe_0$ is a Cartan subalgebra, (4.29) in particular shows that V is a direct sum of weight spaces V_0, V_2, and V_{-2} of H where $V_0 = Fu_0$, and V_2 and V_{-2} are respectively the linear spans of weight vectors u_i and u_{-i} $(i=1,2,3)$. □

As a consequence of Lemma 4.15, we note that a nontrivial, non-Lie, irreducible module for a split semisimple, non-Lie, Malcev algebra is isomorphic to the adjoint module C_0^- and is determined by (4.29), where ± 2 are the nonzero weights and their corresponding weight spaces are the linear spans of u_1, u_2, u_3 and u_{-1}, u_{-2}, u_{-3}. Consider next the Lie case. Let L be a split semisimple Lie algebra over F and let V be a nontrivial irreducible L-module with highest weight λ relative to the set Φ^+ of positive roots. Thus, V is the direct sum $V = \sum_\nu V_\nu$ of weight spaces. If v_λ denotes a maximal vector of V and $\Phi^+ = \{\beta_1,\cdots,\beta_m\}$, then, as well known (Humphreys [1,p.108]), V is the linear span of weight vectors $x_{-\beta_1}^{i_1} x_{-\beta_2}^{i_2} \cdots x_{-\beta_m}^{i_m}(v_\lambda)$, where $x_{-\beta_k}$ is a basis element of $\mathfrak{sl}(2)$ corresponding to β_k and $x_{-\beta_1}^{i_1} x_{-\beta_2}^{i_2} \cdots x_{-\beta_m}^{i_m}$ is an element of the universal enveloping algebra of L . It follows from this that for weight $\nu \neq \lambda$, each weight vector $v_\nu \in V_\nu$ is a linear combination of elements of the form $x_{-\alpha} v_{\nu+\alpha}$, where $\alpha \in \Phi^+$, $x_{-\alpha}$ is in the root space $L_{-\alpha}$, and $v_{\nu+\alpha} \in V_{\nu+\alpha}$.

Theorem 4.16. Assume that A is a flexible Malcev-admissible algebra over F for which A^- has a split semisimple subalgebra $S = S_1 \oplus \cdots \oplus S_n$, where the S_i are simple ideals of S . Suppose that A is decomposed as in (4.22). If M is the semisimple Malcev subalgebra $S_{i_1} + \cdots + S_{i_q}$ of A^- and W is a nontrivial irreducible M-submodule of A^- such that $[S_j,W] = 0$ for $j \notin \{i_1,\cdots,i_q\}$, then

$$(4.30) \qquad M \circ W \subseteq [M \circ M,W] + R_0 .$$

In particular, (4.30) holds for $M = S$.

Proof. Under the assumption, we note that W is an irreducible S-submodule of A^- . Assume first that $[S_i,W] = 0$ for some i . Since W is nontrivial, there is an index $k \in \{i_1,\cdots,i_q\}$ such that $[S_k,W] \neq 0$, hence $[S_k,W] = W$ by Lemma 3.12. Since ad_A acts as derivations on A^+ ,

$$S_i \circ W = S_i \circ [S_k,W] \subseteq [S_i \circ S_k,W] + [S_i,W] \circ S_k \subseteq [M \circ M,W] ,$$

and hence

$$(4.31) \qquad S_i \circ W \subseteq [M \circ M,W] .$$

Consider first the case that W is a non-Lie M-module. Then, by

Lemma 3.10 and Corollary 3.11, we may assume that either $W \cong \tilde{M}_2$ as an $s\ell(2)$-module with $\tilde{M}_2$ defined by (3.25), or $W \cong \bar{C}_0$ for the adjoint module $\bar{C}_0$. In this case, there is an index $i \in \{i_1,\cdots,i_q\}$ such that $S_i = s\ell(2)$ or $S_i = \bar{C}_0$ and $[S_j,W] = 0$ for all $j \neq i$. Suppose that $W \cong \tilde{M}_2$ and $S_i = s\ell(2)$. Let $\{u,v\}$ and $\{e,h,f\}$ be the standard bases of $\tilde{M}_2$ and $s\ell(2)$ given by (3.25). Denoting the module action in $\tilde{M}_2$ by $[\ ,\ ]$, it follows from (3.25) that

$$u \circ e = \frac{1}{2}[f \circ e,v],\ u \circ h = \frac{1}{4}[h \circ h,u],\ u \circ f = \frac{1}{2}[h \circ f,u]\ ,$$

$$v \circ e = -\frac{1}{2}[h \circ e,v],\ v \circ h = -\frac{1}{4}[h \circ h,v],\ v \circ f = \frac{1}{2}[e \circ f,u]\ .$$

Hence, $S_i \circ W \subseteq [S_i \circ S_i,W]$, and by (4.31) we have $M \circ W \subseteq [M \circ M,W]$ in this case.

Assume that $W \cong \bar{C}_0$ and $S_i = \bar{C}_0$. Using the basis u_0, u_i, u_{-i} $(i=1,2,3)$ of W and relation (4.29), it is easy to compute

$$
\begin{aligned}
&[e_0 \circ e_0,u_{\pm i}] = \pm\ 4(u_{\pm i} \circ e_0)\ ,\\
(4.32)\qquad &[e_0 \circ e_{\pm i},u_{\pm i}] = \pm\ 2(u_{\pm i} \circ e_{\pm i})\ ,\\
&[e_{\pm i} \circ e_{\pm i},u_{\mp i}] = \pm\ 2(u_0 \circ e_{\pm i})
\end{aligned}
$$

for $i = 1, 2, 3$. Let $j,k = \pm 1, \pm 2, \pm 3$ with $j + k \neq 0$. Then,

$$
\begin{aligned}
[e_j \circ e_0,u_k] &= [e_j,u_k] \circ e_0 + e_j \circ [e_0,u_k]\\
&= [e_j,u_k] \circ e_0 \pm 2e_j \circ u_k\ .
\end{aligned}
$$

Since by (4.32) $[e_j,u_k] \circ e_0 \in [S_i \circ S_i,W]$, $e_j \circ u_k \in [S_i \circ S_i,W]$. Therefore, we have $S_i \circ W \subseteq [S_i \circ S_i,W]$ by (4.32), except possibly for elements $e_0 \circ u_0$, $e_i \circ u_{-i}$ and $e_{-i} \circ u_i (i=1,2,3)$ which are weight vectors of $H = Fe_0$ of weight 0. Since $S_i \circ W$ is an S_i-submodule, it is a direct sum of irreducible S_i-submodules, and by Lemma 3.10 any non-trivial irreducible summand is an isomorphic copy of C_0^- which must be contained in the submodule $[S_i \circ S_i, W]$, since any weight vector in $S_i \circ W$ of nonzero weight is a linear combination of weight vectors of the form $e_i \circ u_j$ with $i + j \neq 0$. Thus, letting N_0 denote the sum of trivial S_i-submodules of $S_i \circ W$, we have $S_i \circ W \subseteq [S_i \circ S_i, W] + N_0$. But then, since $[S_j, S_i \circ W] = 0$ for all $j \neq i$, it must be that $N_0 \subseteq R_0$, which combines with (4.31) to show that (4.30) holds in this case also.

Assume that W is a Lie module for M. Since all non-Lie simple summands of M annihilate W, in view of (4.31) we may further assume that M is a Lie algebra L. Let λ be the highest weight of W relative to the set Φ^+ of positive roots of L. Let $W = \Sigma W_\mu$ be the weight space decomposition and $L = H + \Sigma L_\alpha$ be the Cartan decomposition of L relative to a Cartan subalgebra H. Letting $Q = [L \circ L, W]$, we first verify the relations

(4.33) $$H \circ W_\mu \subseteq Q \quad \text{for} \quad \mu \neq 0 ,$$

(4.34) $$L_\alpha \circ W_\mu \subseteq Q \quad \text{for} \quad \alpha + \mu \neq 0 .$$

If $h_1, h_2 \in H$ and $v_\mu \in W_\mu$, then

(4.35) $$[h_1 \circ h_2, v_\mu] = \mu(h_1)h_2 \circ v_\mu + \mu(h_2)h_1 \circ v_\mu .$$

The left side of (4.35) is an element of Q for all $h_1, h_2 \in H$. Suppose $\mu \neq 0$ and h_1 is any element of H such that $\mu(h_1) \neq 0$. Let h_2 be arbitrary. If $\mu(h_2) = 0$, then (4.35) gives $h_2 \circ v_\mu \in Q$. If $\mu(h_2) \neq 0$, then we can choose $h_1 = h_2$ to give $h_2 \circ v_\mu \in Q$ again by (4.35). This proves relation (4.33).

Assume now that $\alpha + \mu \neq 0$ and let $x_\alpha \in L_\alpha$, $v_\mu \in W_\mu$. Then, if $\mu \neq 0$, we have for $h \in H$

$$[x_\alpha \circ h, v_\mu] = [x_\alpha, v_\mu] \circ h + x_\alpha \circ [h, v_\mu]$$

$$= v_{\alpha+\mu} \circ h + \mu(h) x_\alpha \circ v_\mu$$

for some $v_{\alpha+\mu} \in W_{\alpha+\mu}$. The left side of this is in Q as is $v_{\alpha+\mu} \circ h$, and hence $x_\alpha \circ v_\mu \in Q$ since $\mu \neq 0$. Suppose then that $\mu = 0$ and $\alpha \in \Phi^+$. Since $\alpha \neq \mu$ and the module action of L on W is $[\ ,\ ]$ in this case, each weight vector v_μ in W_μ is a linear combination of elements of the form $[x_{-\beta}, v_\beta]$ for $\beta \in \Phi^+$, $x_{-\beta} \in L_{-\beta}$, and $v_\beta \in W_\beta$. For each $v_\beta \in W_\beta$, there exists a $v_{\alpha+\beta} \in W_{\alpha+\beta}$ such that

$$[x_{-\beta} \circ x_\alpha, v_\beta] = x_{-\beta} \circ v_{\alpha+\beta} + [x_{-\beta}, v_\beta] \circ x_\alpha .$$

Since we have already shown that (4.34) holds for $\mu \neq 0$, we deduce from this that $[x_{-\beta}, v_\beta] \circ x_\alpha \in Q$. Therefore, $x_\alpha \circ v_\mu \in Q$ in this case also. The case of $\mu = 0$ and $-\alpha \in \Phi^+$ can be shown from this by taking $-\lambda$ to be the highest weight and $-\Phi^+$ to be the set of positive roots.

Given (4.33) and (4.34), we see that $L \circ W \subseteq Q$, except possible for some vectors in $L \circ W$ of weight zero. Since $L \circ W$ and Q are L-submodules, this implies that all nontrivial irreducible L-submodules of $L \circ W$ are contained in Q, and hence $L \circ W \subseteq Q + N_0$ for some trivial L-submodule N_0 of $L \circ W$. Since L and W are contained in B_Γ given by (4.11) for $\Gamma = \{i_1, \cdots, i_q\}$ and B_Γ is a subalgebra of A, $L \circ W \subseteq B_\Gamma$, and hence $N_0 \subseteq B_\Gamma$. Since B_Γ is a trivial S_k-module for $k \notin \Gamma$, we find that $N_0 \subseteq R_0$ and thus $L \circ W \subseteq Q + R_0$. □

If the M-submodule W of Theorem 4.16 is not isomorphic to any summand S_{i_j} of M as an M-module, then we can strengthen relation (4.30). For this, we need

Lemma 4.17. Assume that $S_1, \cdots, S_q$ are split simple Malcev algebras over F, and let $M = S_1 \oplus \cdots \oplus S_q$ and $D = \mathrm{Der}\, M$. Suppose W is an irreducible M-module. If the D-module $M \otimes W$ has a nonzero trivial D-submodule, then $S_i \cong W$ as a D-module for some $i \in \{1,\cdots,q\}$.

Proof. Since $M \otimes W$ is isomorphic to the direct sum $S_1 \otimes W + \cdots + S_q \otimes W$ as a D-module, one of the $S_i \otimes W$ must contain a nonzero trivial D-submodule. Let $S_1 \otimes W$ be this module. When $S_1 \otimes W$ decomposes into the direct sum of irreducible D-submodules, it contains a trivial D-module which we identify with F. If $x \in S_1$ and $v \in W$, then let $(x \otimes v)'$ denote the projection of $x \otimes v$ onto F. Define a mapping $\rho : W \to S_1^*$, the dual module for D, by $\rho(v) = v^* : x \to (x \otimes v)'$ for $x \in S_1$. Clearly, ρ is linear and nonzero. For $x \in S_1$, $v \in W$ and $\sigma \in D$, we have $(\sigma v^*)(x) = -v^*(\sigma x) = -(\sigma x \otimes v)'$. But then $(\sigma v)^*(x) = (x \otimes \sigma v)'$ $= -(\sigma x \otimes v)'$ since F is a trivial D-module. This shows that ρ is a D-module homomorphism of W into S_1^*. Since W and S_1^* are irreducible D-modules (Lemma 3.12(1)), ρ is an isomorphism and since S_1 is self-dual, the proof is complete. □

Corollary 4.18. Let A, M and W be as in Theorem 4.16. If W is not isomorphic to S_{i_j} as an M-module for any $1 \le j \le q$, then

(4.36) $$M \circ W \subseteq [M \circ M, W] .$$

Proof. Note by Lemma 3.12(3) that $W \not\cong S_{i_j}$ as D-modules for any $1 \le j \le q$, where $D = \mathrm{Der}\, M$. By (4.30) we have either $M \circ W \subseteq [M \circ M, W]$, or $M \circ W$ contains a nonzero trivial D-submodule (also see the proof of Theorem 4.16). In the latter case, since $x \otimes v \to x \circ v$ induces a D-module homomorphism of $M \otimes W$ into $M \circ W$, it must be that $M \otimes W$ contains a nonzero trivial D-submodule also. But this is impossible by Lemma 4.17, since $W \not\cong S_{i_j}$ as D-modules for any $1 \leq j \leq q$. □

4.4. Flexible Malcev-admissible algebras with abelian radical

We assume that A is a flexible Malcev-admissible algebra over an algebraically closed field F for which the solvable radical R of A^- is abelian. In this section, we classify all such simple algebras. Let $S = S_1 \oplus \cdots \oplus S_n$ denote a Levi factor of A^-, where each S_i is a simple ideal of S. Thus, A decomposes as in (4.22) with R_* given by (4.23). Throughout this section, we let R denote the abelian radical of A^-. We begin with

<u>Lemma 4.19</u>. $R_*R_* = 0$.

<u>Proof</u>. If $x \in S$ and $u,v \in R_*$, then

$$[x,u] \circ v = [x \circ v,u] - x \circ [v,u] .$$

Both terms on the right side vanish since $S \circ R_* \subseteq R$ by Lemma 4.12 and $[R,R] = 0$. Since the elements $[x,u]$ span R_* by (4.23), $R_* \circ R_* = 0$, and hence $R_*R_* = 0$. □

<u>Lemma 4.20</u>. Let U be the sum of all irreducible S-submodules in R_* which are not isomorphic to S_i as an S-module for any $i = 1, 2, \cdots, n$.

Then, U is an ideal of A and contains R_Γ for any subset Γ of $\{1,\cdots,n\}$ with $|\Gamma| \geq 2$.

Proof. Since the module action is the adjoint action, $[S,U] \subseteq U$ and $[R,U] = 0$. Thus, U is an ideal of A^- . In view of Lemmas 4.13(3) and 4.19, we have $R_0 \circ U \subseteq U$ and $R_* \circ U = 0$. Hence, it suffices to to show that $S \circ U \subseteq U$. But, by Corollary 4.18, $S \circ U \subseteq [S \circ S, U] \subseteq [S,U] + [R,U] \subseteq U$, which shows that $A \circ U \subseteq U$ and U is an ideal of A . If W denotes an irreducible summand in R_Γ such that $W \cong S_i$ for some i as an S-module, then $[S_k, W] = [S_k, S_i] = 0$ for all $k \neq i$, to show that the support Γ of W is $\{i\}$. □

In view of Lemma 4.20, if A is simple, then the ideal U is not present, and hence any irreducible summand in R_* is isomorphic to some S_i as an S-module and so as a D-module. In the remainder of the discussion, we assume that A is simple, so that A decomposes as

$$(4.37) \qquad A = \sum_{j=1}^{m_1} S_1^j + \sum_{j=1}^{m_2} S_2^j + \cdots + \sum_{j=1}^{m_n} S_n^j + R_0$$

where $S_i^j \cong S_i$ as S-modules for $j = 1, 2, \cdots, m_i$, and we have set $S_i^1 = S_i$, $i=1, \cdots, n$. Hence

$$(4.38) \qquad R_* = \sum_{i=1}^{n} \sum_{j \geq 2} S_i^j \ , \ S = \sum_{i=1}^{n} S_i^1 \ .$$

We note that 2-dimensional irreducible submodules are not present and the only non-Lie summands are isomorphic copies of the adjoint module C_0^- .

Let $\tilde{S}_i = \Sigma_{j=1}^{m_i} S_i^j$ and denote by ω_j the module isomorphism of S_i to the j th copy S_i^j . If $s \in S_i$, then it is convenient to write

$$(4.39) \qquad \omega_j(s) = s \otimes q_{ij} \ .$$

Let P_i be the vector space over F with basis $q_{i1}, \cdots, q_{im_i}$, and we make P_i into a trivial S-module, and so a trivial D-module, where $D = \mathrm{Der}\, S$. If $s = \Sigma_{j=1}^{m_i} \omega_j(s_j)$ for $s_j \in S_i$, $j = 1, \cdots, m_i$, then it is easy to see that the mapping $s \to \Sigma_j s_j \otimes q_{ij}$ induces an S-module isomorphism of $\tilde{S}_i$ to the tensor product $S_i \otimes P_i$, which can be extended to an S-module isomorphism of A to the direct sum $S_1 \otimes P_1 + \cdots + S_n \otimes P_n + R_0$. The use of tensor products as indexing sets has been a useful tool in the study of Lie algebras, especially in the Tits' construction of the exceptional Lie algebras (cf. Jacobson [5]). It is convenient to identify A with $S_1 \otimes P_1 + \cdots + S_n \otimes P_n + R_0$ under relation (4.39), and this enables us to transfer products in A to this sum, where the results in Chapter 3 can be readily applied. Thus, we write

(4.40) $$A = S_1 \otimes P_1 + \cdots + S_n \otimes P_n + R_0$$

for decomposition (4.37) of A. Then, the actions of the S-module and D-module on each irreducible summand $S_i \otimes q_{ij}$ in $S_i \otimes P_i$ are given by

(4.41) $$[x, s \otimes q_{ij}] = [x,s] \otimes q_{ij}, \quad \sigma(s \otimes q_{ij}) = \sigma(s) \otimes q_{ij}$$

for $x \in S$, $s \in S_i$, and $\sigma \in D$.

The projection of the product $(S_i \otimes q_{ij})(S_i \otimes q_{ik})$ onto $S_i \otimes q_{i\ell}$ induces a D_i-module homomorphism of $(S_i \otimes q_{ij}) \otimes (S_i \otimes q_{ik})$ into $S_i \otimes q_{i\ell}$ which, by (4.41), acts on the first components, where $D_i = \mathrm{Der}\, S_i$. This is given by an element of $\mathrm{Hom}_{D_i}(S_i \otimes S_i, S_i)$, and by Theorem 3.19, is a linear combination of the products $[\ ,\ ]$ and "#", where $x \# y$ is defined by (3.19), and $x \# y = 0$ for $x,y \in S_i$ if $S_i \not\cong s\ell(m+1)$ for $m \geq 2$. Therefore,

$$\mathrm{Proj}[(s \otimes q_{ij})(t \otimes q_{ik})] = (\alpha_{jk\ell}[s,t] + \beta_{jk\ell}\, s \# t) \otimes q_{i\ell}$$

for $s,t \in S_i$ and for some scalars $\alpha_{jk\ell}, \beta_{jk\ell} \in F$. We define two products "*" and "•" in P_i by extending the relations

$$(4.42) \qquad q_{ij} * q_{ik} = \Sigma_{\ell=1}^{m_i} \alpha_{jk\ell} q_{i\ell} ,$$

$$(4.43) \qquad q_{ij} \cdot q_{ik} = \Sigma_{\ell=1}^{m_i} \beta_{jk\ell} q_{i\ell}$$

bilinearly to P_i. Then, for all $s \otimes a, t \otimes b$ in $S_i \otimes P_i$, the projection of the product $(s \otimes a)(t \otimes b)$ onto $S_i \otimes P_i$ can be written as

$$(4.44) \qquad \text{Proj}[(s \otimes a)(t \otimes b)] = [s,t] \otimes (a * b) + (s \# t) \otimes (a \cdot b) .$$

On the other hand, this projection onto $S_j \otimes P_j$ is zero if $i \neq j$, since $B_i = S_i + R_i + R_0$ is a subalgebra of A by Lemma 4.2, while it can have a nonzero projection onto R_0. Thus, let $z_1, \cdots, z_r$ be a basis of R_0. Then, the projection of $(S_i \otimes q_{ij})(S_i \otimes q_{ik})$ onto Fz_ℓ induces a D_i-module homomorphism of $(S_i \otimes q_{ij}) \otimes (S_i \otimes q_{ik})$ into the trivial D_i-module Fz_ℓ, which is given by an element of $\text{Hom}_{D_i}(S_i \otimes S_i, Fz_\ell)$ and hence is, by (4.13), a multiple of the Killing form $K_i(\ ,\ )$ of S_i. Thus,

$$(4.45) \qquad \text{Proj}[(s \otimes q_{ij})(t \otimes q_{ik})] = \gamma_{jk\ell} K_i(s,t) z_\ell .$$

We define a bilinear mapping $\phi : P_i \times P_i \to R_0$ by

$$(4.46) \qquad \phi(q_{ij}, q_{ik}) = \sum_\ell \gamma_{jk\ell} z_\ell .$$

Since $(S_i + R_i)(S_j + R_j) \subseteq R_{ij} \subseteq U = 0$ for $i \neq j$ by Lemmas 4.2(4) and 4.20, $S_i^k S_j^\ell \subseteq R_{ij} = 0$ and hence $(S_i \otimes P_i)(S_j \otimes P_j) = 0$ for $i \neq j$. To simplify the multiplication in A, we extend the "*" and "•" products and the bilinear mapping ϕ to the vector space direct sum $P = P_1 + \cdots + P_n$ by letting

$$(4.47) \qquad a * b = a \cdot b = 0, \quad \phi(a,b) = 0$$

for all $a \in P_i$, $b \in P_j$ and $i \neq j$. If $K(\; , \;)$ denotes the Killing form on S , then relations (4.44)-(4.47) can combine to give

$$(4.48) \qquad (s \otimes a)(t \otimes b) = [s,t] \otimes (a * b) + (s \# t) \otimes (a \cdot b) + K(s,t)\phi(a,b)$$

for all $s \otimes a \in S_i \otimes P_i$ and $t \otimes b \in S_j \otimes P_j$.

To determine the general form of products in A , it remains to consider products of the type $z(S_i \otimes q_{ij})$ where $z \in R_0$. By (3.11) and (4.14) $z(S_i \otimes P_i) \subseteq S_i \otimes P_i$, and, as above, the projection of these products onto $S_i \otimes q_{i\ell}$ arises from an element in $\mathrm{Hom}_{D_i}(Fz \otimes S_i, S_i)$. Hence, by (4.14) $\mathrm{Proj}\; z(s \otimes q_{ij}) = \zeta_{\ell j} s \otimes q_{i\ell}$ for some $\zeta_{\ell j} \in F$ and for all $s \in S_i$. If we let

$$T_i(z, q_{ij}) = \sum_{\ell} \zeta_{\ell j} q_{i\ell} ,$$

then $T_i(z, \;)$ can be extended to P_i by linearity. Thus, we can write

$$(4.49) \qquad (s \otimes a)z = z(s \otimes a) = s \otimes T_i(z,a)$$

for all $s \otimes a \in S_i \otimes P_i$, since $[R,R_0] = [S,R_0] = 0$.

We have described the general form of products in A in relations (4.48) and (4.49). The Malcev and flexible identities together with the fact that R is abelian impose further restrictions on the products "$*$" and "$\cdot$" in P , and on the mappings ϕ and T_i . These conditions on "$*$" , "$\cdot$" , and ϕ are stated in the following lemma.

<u>Lemma 4.21</u>. Let A be a simple flexible Malcev-admissible algebra over an algebraically closed field F such that the solvable radical R of A^- is abelian, and assume that A is decomposed as in (4.40), where

each P_i is a trivial S-module with basis $\{q_{i1},\cdots,q_{im_i}\}$. Let $e_i = q_{i1}$ for $i = 1, \cdots, n$, and let $P_i = Fe_i + Q_i$, where Q_i is the subspace of P_i spanned by $q_{i2}, \cdots, q_{im_i}$. Define two products "$*$" and "$\bullet$" on the vector space direct sum $P = P_1 + \cdots + P_n$ by (4.42), (4.43), and (4.47). Then, "$*$" and "$\bullet$" are commutative products and for each i, there exist a scalar $\lambda_i \in F$ and an element $q_i \in Q_i$ such that

(4.50) $$a * b = a \bullet b = 0 \quad \text{for all} \quad a \in P_i,\ b \in P_j, \ \text{and} \ i \neq j,$$

(4.51) $$e_i * a = \frac{1}{2}a \qquad \text{for all} \quad a \in P_i ,$$

(4.52) $$a * b = a \bullet b = 0 \quad \text{for all} \quad a,b \in Q_i ,$$

(4.53) $$e_i \bullet e_i = \frac{1}{2}\lambda_i e_i + q_i ,$$

(4.54) $$e_i \bullet a = \frac{1}{2}\lambda_i a \qquad \text{for all} \quad a \in Q_i .$$

Furthermore, if we define a bilinear mapping $\phi : P \times P \to R_0$ by (4.46) and (4.47), then ϕ is symmetric and satisfies the relation

(4.55) $$\phi(a * c,b) = \phi(a,c * b)$$

for all $a,b,c \in P$.

Proof. Relation (4.50) holds by definition in (4.47). Under the assumption, we have relation (4.48) holding for products of the form $(s \otimes a)(t \otimes b)$, and by Lemmas 4.19 and 4.20, and by (4.38) $(s \otimes a)(t \otimes b) = 0$ whenever $a,b \in Q_i$. This implies

(4.56) $$\phi(Q_i,Q_i) = 0, \ i = 1, \cdots, n .$$

If $S_i \not\cong \mathfrak{sl}(m + 1)$ for $m \geq 2$, then since $s \# t = 0$, we obtain $Q_i * Q_i = 0$ from (4.48). If $S_i \cong \mathfrak{sl}(m + 1)$ for $m \geq 2$, then by choosing

$s = t = e_{11} - e_{22}$ in $(s \otimes a)(t \otimes b) = 0$ we have $Q_i \cdot Q_i = 0$, and hence $Q_i * Q_i = 0$ as well. This proves (4.52).

For $s \otimes e_i$, $t \otimes a \in S_i \otimes P_i$, the product

$$[s \otimes e_i, t \otimes a] = [s,t] \otimes (e_i * a + a * e_i) + (s \# t) \otimes (e_i \cdot a - a \cdot e_i)$$
$$+ K(s,t)[\phi(e_i,a) - \phi(a,e_i)]$$

should be the same as the module action of $S_i = S_i^1$ on $S_i^1 + \cdots + S_i^{m_i}$, and hence by (4.41), should equal $[s,t] \otimes a$. From this, it follows that

(4.57) $$\phi(e_i,a) = \phi(a,e_i)$$

for all $a \in P_i$, and that if $S_i \not\cong \mathfrak{sl}(m+1)$ for $m \geq 2$, then

(4.58) $$e_i * a + a * e_i = a .$$

If $S_i \cong \mathfrak{sl}(m+1)$ for $m \geq 2$, then by choosing $s = t = e_{11} - e_{22}$ and equating the two expressions for $[s \otimes e_i, t \otimes a]$, we obtain

(4.59) $$e_i \cdot a = a \cdot e_i$$

in addition to (4.58). Symmetry of ϕ now follows from (4.47) and (4.56), and since $Q_i \cdot Q_i = 0$, (4.59) shows that "$\cdot$" is a commutative product. These combine with (4.48) to give

(4.60) $$[s \otimes a, t \otimes b] = [s,t] \otimes (a \Delta b)$$

for all $s \otimes a \in S_i \otimes P_i$, $t \otimes b \in S_j \otimes P_j$, where we have set

$$a \Delta b = a * b + b * a .$$

Let $\{a,b\} = \frac{1}{2}(a * b - b * a)$. If $r \otimes c \in S_k \otimes P_k$, then we use (4.48) and (4.60) to compute

$$(4.61)\quad (s \otimes a) \circ (t \otimes b) = [s,t] \otimes \{a,b\} + (s \# t) \otimes (a \cdot b) + K(s,t)\phi(a,b) ,$$

$$(4.62)\quad \begin{aligned} ad_{r\otimes c}[(s \otimes a) \circ (t \otimes b)] &= [r,[s,t]] \otimes (c \Delta \{a,b\}) \\ &+ [r,s \# t] \otimes (c \Delta (a \cdot b)) , \end{aligned}$$

$$(4.63)\quad \begin{aligned} &[ad_{r\otimes c}(s \otimes a)] \circ (t \otimes b) + (s \otimes a) \circ ad_{r\otimes c}(t \otimes b) \\ &= [[r,s],t] \otimes \{c \Delta a,b\} + ([r,s] \# t) \otimes ((c \Delta a) \cdot b) \\ &+ K([r,s],t)\phi(c \Delta a,b) + [s,[r,t]] \otimes \{a,c \Delta b\} \\ &+ (s \# [r,t]) \otimes (a \cdot (c \Delta b)) + K(s,[r,t])\phi(a,c \Delta b) . \end{aligned}$$

Since $ad_{r\otimes c}$ is a derivation on A^+, the two expressions in (4.62) and (4.63) are equal. If $r,s,t \in S_i$ and $S_i \not\cong s\ell(m+1)$ for $m \geq 2$, then by choosing $r = s$ with $[r,[r,t]] \neq 0$, and equating (4.62) and (4.63), we have

$$(4.64)\quad c \Delta \{a,b\} = \{a,c \Delta b\}$$

for all $a,b,c \in P_i$. If we let $a = b = e_i$ in (4.64), then $0 = \{a,c \Delta b\} = \{e_i,c \Delta e_i\} = \{e_i,c\}$ for all $c \in P_i$ by (4.58), and $e_i * c = c * e_i$. Since we have already proven that $Q_i * Q_i = 0$, it follows that

$$(4.65)\quad a * b = b * a$$

for all $a,b \in P$. In the case that $S_i \cong s\ell(m+1)$ for $m \geq 2$, setting $r = e_{11} - e_{22} = s$ and $t = e_{22}$ in (4.62) and (4.63) gives

$$4e_{21} \otimes (c \Delta \{a,b\}) = 4e_{21} \otimes \{a,c \Delta b\} ,$$

since $s \# t = s \# [r,t] = [r,s] \# t = 0$. Thus, (4.64) holds in this case as well, and hence by the same argument as above we find that the product

"$*$" is commutative. Noting that $\{a,b\} = 0$, so $a \Delta b = 2a * b$, it follows from (4.62), (4.63), and the invariance of the Killing form $K(\, , \,)$ that $\phi(a * c,b) = \phi(a,c * b)$ for all $a,b,c \in P_i$. Since $P_i * P_j = \phi(P_i,P_j) = 0$ for $i \neq j$, equation (4.55) holds for all $a,b,c \in P$. Commutativity of "$*$" combines with (4.58) to give (4.51).

To derive the remaining relations (4.53) and (4.54), we again equate (4.62) and (4.63), and use (4.55) and the fact that the product $\{ \, , \, \}$ is identically zero. Thus, (4.62) and (4.63) reduce to

$$(4.66) \qquad [r,s \# t] \otimes (c \Delta (a \cdot b)) = ([r,s] \# t) \otimes ((c \Delta a) \cdot b) + (s \# [r,t]) \otimes (a \cdot (c \Delta b)) .$$

The "#" product is nonzero only when $S_i \cong \mathfrak{sl}(m + 1)$ for $m \geq 2$, so we let $r = e_{12}$, $s = e_{23}$, $t = e_{33} - e_{11}$. Then, since $[r,s] = e_{13} = [r,s \# t] = s \# [r,t]$, (4.66) becomes

$$e_{13} \otimes (c \Delta (a \cdot b)) = e_{13} \otimes (a \cdot (c \Delta b)) .$$

Therefore,

$$(4.67) \qquad c * (a \cdot b) = \frac{1}{2}c \Delta (a \cdot b) = \frac{1}{2}a \cdot (c \Delta b) = a \cdot (c * b) ,$$

since $a \Delta b = 2a * b$. Let $e_i \cdot e_i = \frac{1}{2}\lambda_i e_i + q_i$ for some $q_i \in Q_i$ and $\lambda_i \in F$. Then, letting $c \in Q_i$ and $a = b = e_i$ in (4.67), we have by (4.51) and (4.52)

$$\frac{1}{4}\lambda_i c = c * (\frac{1}{2}\lambda_i e_i + q_i) = c * (e_i \cdot e_i) = e_i \cdot (c * e_i) = \frac{1}{2}e_i \cdot c .$$

Hence, (4.53) and (4.54) hold, and this completes the proof. □

The following theorem characterizes simple flexible Malcev-admissible algebras over an algebraically closed field for which the solvable radical R of A^- is abelian.

Theorem 4.22. Let $S_1, \cdots, S_n$ be simple Malcev algebras over an algebraically closed field F and let R_0 be a commutative algebra over F. Assume that $K(\, , \,)$ is the Killing form of $S = S_1 \oplus \cdots \oplus S_n$ and $s \# t$ is defined by (3.19) for $s,t \in S_i$ if $S_i \cong \mathfrak{sl}(m+1)$ for $m \geq 2$, and $s \# t = 0$, otherwise. For each $i = 1, 2, \cdots, n$, let $P_i = Fe_i + Q_i$ be a vector space over F, and for fixed $\lambda_i \in F$, $q_i \in Q_i$, define two commutative products "$*$" and "$\cdot$" on the vector space direct sum $P = P_1 + \cdots + P_n$ by relations (4.50)-(4.54). Suppose that T is a linear mapping from R_0 to P and ϕ is a symmetric bilinear mapping from $P \times P$ to R_0 satisfying (4.55). Define a multiplication on the vector space direct sum $A = S_1 \otimes P_1 + \cdots + S_n \otimes P_n + R_0$ as follows: The product $(s \otimes a)(t \otimes b)$ is defined by (4.48) for $s \otimes a \in S_i \otimes P_i$, $t \otimes b \in S_j \otimes P_j$, the product $z(s \otimes a) = (s \otimes a)z$ by the relation

$$z(s \otimes a) = (s \otimes a)z = s \otimes (a * (Tz)) \tag{4.68}$$

for $z \in R_0$, and yz is the same as in R_0 for $y,z \in R_0$. Then, A is a flexible Malcev-admissible algebra such that $\Sigma S_i \otimes e_i$ is a Levi factor of A^- and the radical of A^- is $\Sigma S_i \otimes Q_i + R_0$, which is abelian. Conversely, if A is a simple flexible Malcev-admissible algebra over an algebraically closed field F such that the radical of A^- is abelian, then A arises from this construction.

Proof. We first prove that the algebra constructed above is Malcev-admissible and flexible by showing that A^- is a Malcev algebra and ad_x is a derivation of A^+ for any $x \in A$. If $u = \Sigma s_i \otimes a_i + x$ and

$v = \Sigma\, t_i \otimes b_i + y$ for $x,y \in R_0$ are any elements of A, then by (4.48) and (4.50)-(4.54) we have

$$[u,v] = 2\,\Sigma\,[s_i,t_i] \otimes (a_i * b_i)\ ,$$

since $K(\ ,\ )$, ϕ are symmetric and R_0 lies in the center of A^-. Thus, A^- satisfies the same identities as the S_i^-, and so A is Malcev-admissible.

For the flexibility of A, it suffices to show that ad_x for $x \in A$ acts as a derivation on the products $(S_i \otimes P_i) \circ (S_j \otimes P_j)$, $R_0 \circ (S_i \otimes P_i)$ and $R_0 \circ R_0$. If $s \otimes a \in S_i \otimes P_i$, $t \otimes b \in S_j \otimes P_j$, $r \otimes c \in S_k \otimes P_k$, and $x,y \in R_0$, then it follows from (4.48) and (4.68) that

$$(s \otimes a) \circ (t \otimes b) = (s \# t) \otimes (a \cdot b) + K(s,t)\phi(a,b)\ , \tag{4.69}$$

$$x \circ (s \otimes a) = (s \otimes a) \circ x = s \otimes (a * (Tx))\ , \tag{4.70}$$

$$[r \otimes c,\ (s \otimes a) \circ (t \otimes b)] = 2[r,s \# t] \otimes (c * (a \cdot b))\ . \tag{4.71}$$

Since $s \otimes t \to s \# t$ is an S-module homomorphism of $S \otimes S$ into S, the right side of (4.71) equals

$$2([r,s] \# t) \otimes (a \cdot b) * c + 2(s \# [r,t]) \otimes (a \cdot b) * c\ . \tag{4.72}$$

On the other hand, by (4.69)

$$\begin{aligned}(4.73)\qquad & [r \otimes c,\ s \otimes a] \circ (t \otimes b) + (s \otimes a) \circ [r \otimes c, t \otimes b] \\ &= 2([r,s] \# t) \otimes ((a * c) \cdot b) + 2\,K([r,s],t)\phi(a * b,c) \\ &+ 2(s \# [r,t]) \otimes (a \cdot (b * c)) + 2\,K(s,[r,t])\phi(a,b * c) \\ &= 2([r,s] \# t) \otimes ((a * c) \cdot b) + 2(s \# [r,t]) \otimes (a \cdot (b * c))\ ,\end{aligned}$$

since $K(\ ,\ )$ is invariant and ϕ satisfies (4.55). To show $\mathrm{ad}_{r\otimes c}$ acts as a derivation on the product $(s \otimes a)\circ(t \otimes b)$, by comparing the right side of (4.71) with (4.72) and (4.73), it suffices to verify the identity

$$(4.74) \qquad c * (a \cdot b) = (c * a) \cdot b = a \cdot (c * b)$$

for all $a,b,c \in P$. By (4.50) we may assume that a, b, c lie in one particular P_i. If $c = e_i$, then (4.74) clearly holds by (4.51)-(4.54). Thus, we may suppose that $c \in Q_i$. If both a and b lie in Q_i, then all terms of (4.74) vanish, so we may assume that $a = e_i$ or $b = e_i$. By symmetry of (4.74) in a and b, we can let $a = e_i$. Then, for $b = e_i$, by (4.51)-(4.54)

$$c * (e_i \cdot e_i) = c * (\tfrac{1}{2}\lambda_i e_i + q_i) = \tfrac{1}{4}\lambda_i c = \tfrac{1}{2}c \cdot e_i$$

$$= (c * e_i) \cdot e_i = c \cdot (e_i * e_i) ,$$

and for $b \in Q_i$,

$$c * (e_i \cdot b) = 0 = (c * e_i) \cdot b = c \cdot (e_i * b) .$$

Hence, (4.74) is valid, and so $\mathrm{ad}_{r\otimes c}$ acts as a derivation on $(s \otimes a) \circ (t \otimes b)$.

Consider now the action of $\mathrm{ad}_{r\otimes c}$ on $x \circ (s \otimes a)$. By (4.70) we have

$$[r \otimes c, x \circ (s \otimes a)] = 2[r,s] \otimes (a * (Tx) * c)$$

$$= 2[r,s] \otimes ((a * c) * (Tx))$$

$$= 2x([r,s] \otimes (a * c)) , \text{ by } (4.68)$$

$$= x \circ [r \otimes c, s \otimes a]$$

since "$\star$" is a commutative associative product. Thus, $ad_{r\otimes c}$ acts as a derivation on the product $x \circ (s \otimes a)$. If $z \in R_0$, then ad_z is clearly a derivation of A^+ . Finally, note that $ad_{r\otimes c}$ acts trivially on R_0 and $R_0 \circ R_0$. Therefore, A is flexible. It follows from (4.48) and (4.50)-(4.54) that $\Sigma\, S_i \otimes e_i$ is a Levi factor of A^- and the radical of A^- is $\Sigma\, S_i \otimes Q_i + R_0$, which is abelian.

For the converse, all but relation (4.68) have already been proved in (4.48) and Lemma 4.21. To verify (4.68), recall that for each $z \in R_0$ there is a linear transformation $T_i(z,\ )$ on P_i satisfying (4.49). We apply $ad_{t\otimes b}$ to (4.49) to obtain

$$[t,s] \otimes (b \star T_i(z,a)) = [t,s] \otimes T_i(z,b \star a) .$$

Thus, if $a = 2e_i$ and $b \in P_i$, then it follows from this and (4.51) that

$$b \star T_i(z,2e_i) = T_i(z,b) . \tag{4.75}$$

Define a linear mapping $T : R_0 \to P = P_1 + \cdots + P_n$ by

$$T(z) = T_1(z,2e_1) + \cdots + T_n(z,2e_n) . \tag{4.76}$$

Then, (4.49) and (4.75) imply that

$$z(s \otimes a) = (s \otimes a)z = s \otimes T_i(z,a) = s \otimes (a \star Tz) ,$$

to give (4.68), and the proof of Theorem 4.22 is complete. □

Ideals of the algebra constructed in Theorem 4.22 are determined by

<u>Corollary 4.23</u>. Let A be a flexible Malcev-admissible algebra constructed by the process described in Theorem 4.22. Then, J is an ideal of A if and only if J is given by

(4.77) $$J = S_1 \otimes I_1 + \cdots + S_n \otimes I_n + I_0 ,$$

where

(1) I_i is an ideal of P_i under "*" and "•" ,

(2) I_0 is an ideal of R_0 ,

(3) $\phi(P_i, I_i) \subseteq I_0$ for all $i = 1, 2, \cdots, n$,

(4) $T(I_0) \subseteq I_1 + \cdots + I_n$.

Proof. Let J be an ideal of A constructed in Theorem 4.22. Then, since J is an S-submodule for $S = S_1 \otimes e_1 + \cdots + S_n \otimes e_n$ and the $S_i \otimes q_{ij}$ are irreducible S-submodules, J is described by (4.77) for some subspace I_i of P_i . Since R_0 is a subalgebra of A , I_0 is an ideal of R_0 . It follows from (4.48) that I_i is an ideal of P_i under "*" and $\phi(P_i, I_i) \subseteq I_0$ for all $i = 1, \cdots, n$. If $z \in I_0$, then by (4.49) $z(S_i \otimes e_i) \subseteq S_i \otimes T_i(z, 2e_i) \subseteq S_i \otimes I_i$ for all $i = 1, 2, \cdots, n$, and hence $T_i(z, 2e_i) \in I_i$ for all $z \in I_0$, which by (4.76) proves part (4). If I_i contains e_i , then since I_i is an ideal of P_i under "*" , by (4.51) $I_i = P_i$. Thus, we may assume that $I_i \subseteq Q_i$. But then, by (4.52) and (4.54) I_i is an ideal of P_i under "•" also. The converse is straightforward. □

Corollary 4.24. Let A be the same as in Corollary 4.23. If A is simple, then we have

(1) $\dim R_0 \geq \dim Q_i = \dim P_i - 1$ for all $i = 1, \cdots, n$,

(2) the ideal of R_0 generated by $\phi(P, P)$ is R_0 .

Proof. Let $I_i = \{a \in Q_i \mid \phi(a, e_i) = 0\}$. Then, I_i is clearly an ideal of P_i under "*" and "•" , and by Corollary 4.23 $J = \Sigma\, S_i \otimes I_i$

is an ideal of A . Thus, $I_i = 0$ for all i and $\phi(\, ,e_i)$ is an injective linear mapping from Q_i to R_0 , which implies that $\dim R_0 \geq \dim Q_i = \dim P_i - 1$, as desired in (1). For part (2), let Z denote the ideal of R_0 generated by $\phi(P,P)$. Then, by Corollary 4.23 $\Sigma S_i \otimes P_i + Z$ is an ideal of A , and hence $Z = R_0$. □

We end this section with an example which shows that if A is a simple flexible Malcev-admissible algebra such that the radical of A^- is abelian, then the assumptions on A impose no constraints on multiplication in R_0 except for commutativity.

Example 4.2. Let S be a simple Malcev algebra over a field F and let P be a vector space over F with basis $e, q_1, \cdots, q_k$. Let Q be the subspace spanned by $q_1, \cdots, q_k$ and let q_0 be an arbitrary element of Q . For a fixed $\lambda \in F$, define two commutative products "$*$" and "$\cdot$" on P satisfying $e \cdot e = \frac{1}{2}\lambda e + q_0$ and (4.50)-(4.52), (4.54). Choose R_0 to be any commutative algebra with basis $\{z_0, z_1, \cdots, z_k\}$, and define a symmetric bilinear mapping $\phi : P \times P \to R_0$ by

$$\phi(e,e) = z_0, \ \phi(e,q_i) = z_i = \phi(q_i,e), \ \phi(q_i,q_j) = 0$$

for all $i, j = 1, \cdots, k$. Let $T : R_0 \to P$ be a linear mapping defined by

$$T(z_0) = e, \ T(z_i) = i(e + q_i)$$

for $i = 1, \cdots, n$. It is easy to verify that ϕ satisfies condition (4.55). We claim that $A = S \otimes P + R_0$ with multiplication given in Theorem 4.22 is a simple flexible Malcev-admissible algebra.

By the construction, it follows from Theorem 4.22 that A is flexible and Malcev-admissible. To show that A is simple, let J be an ideal of A. By Corollary 4.23 J has the form $S \otimes I + I_0$ where conditions (1)-(4) of Corollary 4.23 hold. If $I = P$, then $\phi(P,P) = R_0 \subseteq I_0$, so $J = A$ in this case. Assume that $I = 0$, and $z = \Sigma_{j=0}^{k} \beta_j z_j$ is an element of I_0. Then, $T(z) = \beta_0 e + (\Sigma_{i=1}^{k} i\beta_i)e + \Sigma_{i=1}^{k} \beta_i q_i \in I = 0$ by Corollary 4.23(4), and hence $\beta_i = 0$ for all $i = 0, 1, \cdots, k$. This shows that $z = 0$ and $J = 0$. Suppose now that $I \neq 0, P$. If $\gamma_0 e + \Sigma \gamma_i q_i \in I$ and $\gamma_0 \neq 0$, then by multiplying by q_j under "*" we find that $q_j \in I$ for all $j = 1, \cdots, k$, and hence $\gamma_0 e \in I$. Thus, $I = P$, whenever any element of I involves a multiple of e, and so we may assume that $0 \neq I \subseteq Q$. If $0 \neq q = \Sigma_{j=1}^{k} \alpha_j q_j$ is an element of I, then $z = \Sigma_{j=1}^{k} \alpha_j z_j = \phi(e,q) \in I_0$ and

$$T(z) = (\Sigma_{j=1}^{k} j\alpha_j)e + \Sigma_{j=1}^{k} j\alpha_j q_j \in I .$$

Then, we have $\Sigma_{j=1}^{k} j\alpha_j = 0$ for any element $q = \Sigma \alpha_j q_j \in I$. Let $q \in I$ be one with a minimal number of nonzero coefficients α_j. Then, for some $\ell \in \{1,\cdots,k\}$, the element $\ell q - T(z)$ will have fewer nonzero coefficients, unless $q = \alpha_j q_j \neq 0$. But then $j\alpha_j = 0$. This contradiction shows that A is simple. □

4.5. Quasi-classical Malcev algebras

In this section we construct examples of simple flexible Malcev-admissible algebras over fields of characteristic $\neq 2$. These examples can be used to illustrate the diversity of Malcev algebras which can occur as the algebra A^- of a simple flexible Malcev-admissible algebra A. Our construction is based on the construction of a generalized form of a quadratic algebras, introduced by Okubo [8,14], and on the so-called quasi-classical Malcev algebras. For the latter algebras, we give the following general definition.

Definition 4.2. An algebra A over a field of arbitrary characteristic is termed a *quasi-classical algebra* if there is a symmetric non-degenerate bilinear form (,) on A satisfying the invariant condition

$$(xy,z) = (x,yz) \tag{4.78}$$

for all $x,y,z \in A$, where xy denotes the multiplication in A . □

A quasi-classical algebra has been called a symmetric algebra by Winter [1,p.31]. For Lie algebras, Okubo [1] first used the term "classical", which was later modified to the present term by Okubo and Myung [1] to distinguish it from "classical" Lie algebras of Seligman [1]. It is well known that the semisimple alternative and Jordan algebras of characteristic ≠ 2 as well as reductive Lie algebras of characteristic 0 are quasi-classical. Included in the class of quasi-classical algebras are the para-Hurwitz and the pseudo-octonion algebras of characteristic ≠ 2 (Theorem 2.2, Lemma 2.3 and Definition 2.4). Quasi-classical algebras also play important roles in the work of Sagle [4,6]. In the remainder of this section, we retain the assumption that all algebras and modules are finite-dimensional. It is clear that if A is quasi-classical, then so are A^- and A^+. It is a theorem of Dieudonné that a semisimple quasi-classical algebra of arbitrary characteristic is a direct sum of simple quasi-classical algebras. Conversely, if $A_i (i=1,\cdots,n)$ are quasi-classical algebras with associated bilinear forms $(,)_i$, then the direct sum $A = A_1 \oplus \cdots \oplus A_n$ is quasi-classical with bilinear form (,) on A defined by

$$(\Sigma\, u_i, \Sigma\, v_i) = \Sigma\, (u_i, v_i)_i \,, \tag{4.79}$$

where $u_i, v_i \in A_i$. Any trivial algebra, i.e., $A^2 = 0$, is automatically quasi-classical, since $(\ ,\ )$ can be chosen to be any nondegenerate symmetric bilinear form. Thus, by (3.5) every reductive Malcev algebra of characteristic 0 is quasi-classical.

Let $x_1, \cdots, x_n$ be a basis of an algebra A over a field F and let A have the multiplication table

$$x_i x_j = \sum_{k=1}^{n} \gamma_{ij}^k x_k \tag{4.80}$$

where $\gamma_{ij}^k \in F$ are the structure constants of A . If (β_{ij}) denotes the matrix of $(\ ,\ )$ relative to this basis, then it is readily seen that condition (4.78) is equivalent to

$$\sum_{p=1}^{n} \gamma_{ij}^p \beta_{pk} = \sum_{p=1}^{n} \gamma_{jk}^p \beta_{ip} \tag{4.81}$$

for all $i, j, k = 1, \cdots, n$. Hence, A is quasi-classical if and only if there is a symmetric nonsingular matrix (β_{ij}) satisfying (4.81) .

Consider next the tensor product $A = A_1 \otimes \cdots \otimes A_n$ of quasi-classical algebras A_i over F with associated bilinear forms $(\ ,\ )_i$. Define a multiplication on A and a bilinear form $(\ ,\ )$ on A by extending

$$(u_1 \otimes \cdots \otimes u_n)(v_1 \otimes \cdots \otimes v_n) = u_1 v_1 \otimes \cdots \otimes u_n v_n \,, \tag{4.82}$$

$$(u_1 \otimes \cdots \otimes u_n, v_1 \otimes \cdots \otimes v_n) = (u_1, v_1)_1 \cdots (u_n, v_n)_n \tag{4.83}$$

bilinearly to A , where the u_i and v_i are elements of A_i .

<u>Lemma 4.25</u>. The tensor product A of quasi-classical algebras $A_1, \cdots, A_n$ is quasi-classical under definitions (4.82) and (4.83).

Proof. The invariance and symmetry of $(\ ,\ )$ are immediate from (4.83) and those of $(\ ,\ )_i$. Let $x_1, \cdots, x_m$ be a basis of A_1. Since every element of A is uniquely expressed as $\Sigma_{i=1}^m x_i \otimes w_i$ for $w_i \in A_2 \otimes \cdots \otimes A_n$, for the nondegeneracy of $(\ ,\ )$, it suffices to assume $n = 2$. Let $y_1, \cdots, y_p$ denote a basis of A_2. For each $1 \leq k \leq p$, consider the equations

$$0 = (\sum_i x_i \otimes w_i, x_j \otimes y_k) = \sum_i (x_i, x_j)_1 (w_i, y_k)_2$$

for $j = 1, \cdots, m$. Since the matrix of $(\ ,\ )_1$ relative to the basis x_i is nonsingular, $(w_1, y_k)_2 = \cdots = (w_m, y_k)_2 = 0$ for all $k = 1, \cdots, p$, and hence $\Sigma x_i \otimes w_i = 0$, which shows that $(\ ,\ )$ is nondegenerate. □

Using a Casimir invariant argument, Okubo [8] has constructed an infinite class of indecomposable quasi-classical solvable Lie algebras L, i.e., L cannot be a direct sum of nontrivial ideals. Okubo's construction starts from a 4-dimensional indecomposable, non-nilpotent, solvable Lie algebra, denoted by $A_{4,8}$, whose multiplication table is given by

$$[x_2, x_3] = x_1, \ [x_2, x_4] = x_2, \ [x_3, x_4] = -x_3,$$

and all other products are zero. It can be routinely checked that the matrix (β_{ij}) with $\beta_{14} = \beta_{41} = -1 = -\beta_{23} = -\beta_{32}$ and all other $\beta_{ij} = 0$ satisfies relations (4.81) for $A_{4,8}$. The algebra $A_{4,8}$ originally arose from the classification of algebraically independent Casimir invariants of all indecomposable Lie algebras over the real field of dimensions 3, 4 and 5 as well as of all nilpotent Lie algebras of dimension 6 (Patera et al [1]). It is shown that the only non-Lie, Malcev algebra of dimension ≤ 4 over F of characteristic 0 is the 4-dimensional non-nilpotent solvable algebra given by

$$[e_1,e_2] = e_3, [e_1,e_4] = e_1, [e_2,e_4] = e_2, [e_3,e_4] = - e_3 ,$$

and all other products are zero (Kuzmin [2]). It can be directly checked that for this algebra, there is no a nonsingular symmetric matrix (β_{ij}) satisfying (4.81).

For Okubo's construction of a generalized form of quadratic algebras, let M denote any algebra over an arbitrary field F of characteristic $\neq 2$ with a nondegenerate symmetric bilinear form $(,)$ on M, and let A be an algebra over F with a linear form $t \neq 0$. Let $e \neq 0$ be a fixed element of A, and define a multiplication on the vector space direct sum $M \oplus A$ by

$$(4.84) \qquad (x + a) * (y + b) = xy + t(a)y + t(b)x + ab + (x,y)e$$

for $x,y \in M$ and $a,b \in A$, where xy and ab denote the products in M and A. We denote the resulting algebra by $M(A,e)$. When M is an anticommutative algebra, $A = F$, $e = 1$ and t is the identity mapping on F, (4.84) reduces to

$$(4.85) \qquad (x + \alpha 1) * (y + \beta 1) = xy + \alpha y + \beta x + [\alpha\beta + (x,y)]1$$

for $\alpha\ \beta \in F$. It is well known that any quadratic algebra B with 1 over F satisfying

$$(4.86) \qquad x^2 - 2t(x)x + n(x)1 = 0$$

for all $x \in B$ has a multiplication of the form (4.85), where $M = \{x \in B \mid t(x) = 0\}$, so $B = M \oplus F1$, xy is an anticommutative product on M, and xy and $(x,y)1$ for $x,y \in M$ denote the projections of the product of x and y in B onto M and $F1$, respectively (Osborn [1]). Furthermore, B is flexible if and only if $(,)$ is symmetric and satisfies

$$(x,xy) = 0 \tag{4.87}$$

for all $x,y \in M$. Note that relation (4.87) for a symmetric bilinear form $(\, , \,)$ on an anticommutative algebra is equivalent to the invariant condition (4.78).

Lemma 4.26. Let $M(A,e)$ be the algebra defined by (4.84), where A, M, $(\, , \,)$ and t satisfy the conditions preceding (4.84).

(1) $M(A,e)^- = M^- \oplus A^-$ and $[x + a, y + b] = [x,y] + [a,b]$ for all $x,y \in M$ and $a,b \in A$.

(2) If A has an identity element 1 and $t(1) = 1$, then 1 is the identity element of $M(A,1)$.

(3) If A is a quadratic algebra satisfying equation (4.86) for all $a \in A$ and M is anticommutative, then $M(A,1)$ is a quadratic algebra satisfying (4.86) under the product "$*$" , where $t(x + a) = t(a)$ and $n(x + a) = n(a) - (x,x)$.

(4) $M(A,e)$ is flexible if and only if $(\, , \,)$ is an invariant bilinear form on M^- , both M and A are flexible, t is symmetric, i.e., $t(ab) = t(ba)$ for all $a,b \in A$, and $[A,e] = 0$.

Proof. Parts (1)-(3) are immediate from (4.84) and (4.86). For part (4), we compute the associator $(\, , \, , \,)^*$ in $M(A,e)$ as

$$\begin{aligned}(x + a, y + b, z + c)^* &= (x,y,z) + [t(ab) - t(a)t(b) + (x,y)t(e)]z \\ &\quad - [t(bc) - t(b)t(c) + (y,z)t(e)]x + (a,b,c) \\ &\quad + (x,y)ec - (y,z)ae + [(xy,z) - (x,yz) + t(a)(y,z) - t(c)(x,y)]e \, .\end{aligned}$$

for $x,y,z \in M$ and $a,b,c \in A$. This implies

$$
\begin{aligned}
&(x + a,y + b,z + c)^* + (z + c,y + b,x + a)^* \\
&= (x,y,z) + (z,y,x) + [t(ab) - t(ba)]z + [t(bc) - t(cb)]x \\
&+ (a,b,c) + (c,b,a) + (y,z)(ea - ae) + (x,y)(ec - ce) \\
&+ [([z,y],x) - (z,[y,x])]e .
\end{aligned} \tag{4.88}
$$

Assume that M and A are flexible, $(\, , \,)$ is invariant on M^- , t is symmetric, and $[A,e] = 0$. Then, flexibility of $M(A,e)$ immediately follows from (4.88). Conversely, suppose that $M(A,e)$ is flexible, so that the left side of (4.88) is identically zero. When $a = b = c = 0$, (4.88) gives flexibility of M and the invariance of $(\, , \,)$ on M^- , which in turn implies flexibility of A . Setting $x = 0$ and choosing $y,z \in M$ with $(y,z) \neq 0$ in (4.88), we have $t(ab) = t(ba)$ and $ae = ea$ for all $a,b \in A$. □

<u>Theorem 4.27</u>. Let $M(A,e)$ be the same as in Lemma 4.26. If A is simple and if either M is anticommutative or M^- is simple, then $M(A,e)$ is simple.

<u>Proof</u>. Let J denote an ideal of $M(A,e)$ and let

$$M_0 = \{x \in M \mid x + a \in J \text{ for some } a \in A\} ,$$

$$A_0 = \{a \in A \mid x + a \in J \text{ for some } x \in M\} .$$

Then, M_0 and A_0 are clearly subspaces of M and A . We first show that A_0 is an ideal of A . If $b \in A_0$, then $y + b \in J$ for some $y \in M$, and hence by (4.84) $a * (y + b) = t(a)y + ab \in J$ for all $a \in A$, to give $ab \in A_0$. Similarly, $ba \in A_0$ and A_0 is an ideal of A . Since A is simple, $A_0 = 0$, or $A_0 = A$. If $A_0 = 0$, then for each $y \in M_0$, $x * y = (x,y)e \in J$ for all $x \in M$ and so $(x,y)e = 0$, which implies $y = 0$ by the nondegeneracy of $(\, , \,)$. Hence, $M_0 = 0$ and $J = 0$ in this case. Assume then that $A_0 = A$. Suppose M^- is simple. It follows from (4.84) that M_0 is an ideal of M^- and hence $M_0 = 0$ or $M_0 = M$. If $M_0 = 0$, then $J = A$, and consequently $x * b = t(b)x = 0$ for all $x \in M$ and all $b \in A$. Since $t \neq 0$, this case cannot occur. But then, the case $M_0 = M$ implies $J = M(A,e)$. Thus, it remains to treat the case that M is anticommutative, when $A_0 = A$. Then, the Jordan product "$\circ$" in $M(A,e)^+$ is given by

$$(4.89) \qquad (x + a) \circ (y + b) = t(a)y + t(b)x + \frac{1}{2}(ab + ba) + (x,y)e \ .$$

Since $A_0 = A$, for each $b \in A$ there is a $y \in M_0$ such that $y + b \in J$. Since J is an ideal of $M(A,e)$, it follows from (4.89) that $t(a)y + t(b)x$ lies in M_0 for all $x \in M$ and $a \in A$. Thus, $t(b)x \in M_0$ for each $b \in A$, since $y \in M_0$, and so $x \in M_0$, showing $M = M_0$. Therefore, $M(A,e)$ is simple. □

Corollary 4.28. Assume that M is a quasi-classical Malcev algebra over F under $(\, , \,)$ and A is a simple flexible Malcev-admissible algebra with nonzero symmetric linear form t . For any nonzero element $e \in A$ with $[A,e] = 0$, the algebra $M(A,e)$ constructed by (4.84) is a simple flexible Malcev-admissible algebra over F such that $M(A,e)^- \cong M \oplus A^-$ as an algebra. Suppose, in addition, that A is a quadratic algebra with identity element 1 satisfying equation (4.86), and $t(1) = 1$. Then, $M(A,1)$ is a quadratic algebra with identity element 1 , where $t(x + a) = t(a)$ and $n(x + a) = n(a) - (x,x)$. Furthermore, $M(A,1)$ is a noncommutative Jordan algebra with a symmetric linear form t , and the bilinear form $(,)$ associated with $M(A,1)$ is given by

$$(x + a, y + b) = (x,y) + (a,b)_0 \tag{4.90}$$

for $x,y \in M$ and $a,b \in A$ with $t(a) = t(b) = 0$, where $(,)_0$ is the bilinear form associated with the quadratic algebra A , so that $(,)$ is symmetric and nondegenerate on $M(A,1)_0 = \{x + a \mid t(x + a) = t(a) = 0\}$.

<u>Proof</u>. The first part immediately follows from Lemma 4.26 and Theorem 4.27. Assume A is a quadratic algebra with identity element 1 and let $X = x + a \in M(A,1)$. Then, by Lemma 4.26(3)

$$X * X - 2t(X)X + n(X)1 = 0 \tag{4.91}$$

and hence the Jordan identity $(X * X) * (Y * X) = [(X * X) * Y] * X$ follows from (4.91) and flexibility of "$*$" . Since $t(X * Y) = t(ab) + (x,y)t(1) = t(ba) + (y,x) = t(Y * X)$, t is symmetric on $M(A,1)$. Since A and $M(A,1)$ are quadratic over F , by (4.84) the projection of $(x + a) * (y + b)$ onto $F1$ is given by $[(x,y) + (a,b)_0]1$, and (4.90) holds. We note that the nondegeneracy of $(,)$ on $M(A,1)_0$ is equivalent to the simplicity of $M(A,1)$. □

Theorem 4.27 and Corollary 4.28 provide a general method of constructing simple flexible Malcev-admissible algebras from a simple flexible Malcev-admissible algebra A with a symmetric linear form and a quasi-classical Malcev algebra M . In particular, A can be chosen to be a simple commutative algebra, e.g., a simple Jordan algebra. We also note that any simple flexible quadratic Malcev-admissible algebra B over F can be constructed by (4.85) for some quasi-classical Malcev algebra M

over F. In fact, letting $M = \{x \in B \mid t(x) = 0\}$, this immediately follows from (4.86) and (4.87).

In the remainder of this section, we focus on the construction of examples of quasi-classical Malcev algebras. Consequently, each of these Malcev algebras may be used in the construction given in Corollary 4.28 to yield an example of a simple flexible Malcev-admissible algebra. We begin with finite-dimensional commutative associative algebras over F of arbitrary characteristic. For a positive integer m, let $N(m)$ denote the commutative associative nilalgebra generated by a nilpotent element a of nil-index $m + 1$, i.e., $a^m \neq 0$ and $a^{m+1} = 0$, so that $a, a^2, \cdots, a^m$ form a basis of $N(m)$. Let $C(m)$ be the algebra over F obtained by adjoining the identity element 1 to $N(m)$, so that $C(m) = F1 + N(m)$. Then, $N(m)$ and $C(m)$ are indecomposable, since if $N(m)$ or $C(m)$ is a direct sum of nontrivial ideals, then a would be of nil-index $< m + 1$ in $N(m)$, or there would be an idempotent $\neq 1$, 0 in $C(m)$. Define a bilinear form $(\, , \,)$ on $C(m)$ and $N(m)$ by

$$(4.92) \qquad (a^i, a^j) = \begin{cases} 1, & \text{if } i,j \geq 0 \text{ and } i + j = m \text{ for } C(m), \\ 1, & \text{if } i,j > 0 \text{ and } i + j = m + 1 \text{ for } N(m), \\ 0, & \text{otherwise, for both } C(m) \text{ and } N(m). \end{cases}$$

It can be routinely checked that $(\, , \,)$ is a symmetric nondegenerate invariant bilinear form for both cases.

We call a Malcev algebra M *nilpotent of class* m if $M^{m+1} = 0$ and $M^m \neq 0$, where $M^1 = M$ and M^m is defined inductively by $M^m = [M^{m-1}, M]$.

<u>Proposition 4.29</u>. Let S be a semisimple quasi-classical Malcev algebra over a field F of arbitrary characteristic. For each positive

integer m ,

(1) there exists a quasi-classical Malcev algebra M over F with radical which is nilpotent of class m ,

(2) there exists a quasi-classical nilpotent Malcev algebra over F of class m .

Proof. Let $C(m)$ and $N(m)$ be the commutative associative quasi-classical algebras over F defined by (4.92). Then, $N(m)$ is the nilpotent radical of $C(m)$ of codimension 1 which is nilpotent of class exactly m . By Lemma 4.25, the tensor products $M = S \otimes C(m)$ and $S \otimes N(m)$ are quasi-classical Malcev algebras over F . It is immediate that the radical of M is $S \otimes N(m)$, which is nilpotent of class exactly m , and that $M / S \otimes N(m) \cong S \otimes 1 \cong S$. □

If A is a simple commutative algebra over F and M is an algebra constructed by Proposition 4.29(1), then the algebra $M(A,e)$ defined by (4.84) is a simple flexible Malcev-admissible algebra such that $M(A,e)^-$ has a radical which is nilpotent of class m . When $m = 1$, $M(A,e)$ is one of the algebras described in Theorem 4.22. For case (2) of Proposition 4.29, $M(A,e)$ is a simple flexible Malcev-admissible algebra with $M(A,e)^-$ nilpotent of class m .

Examples constructed in Proposition 4.29 and each algebra described in Theorem 4.22 are modules for a semisimple Malcev algebra $S(= S \otimes 1)$, and as an S-module, are direct sums of trivial S-modules and S-modules isomorphic to the simple components of S . We next construct examples of quasi-classical Lie algebras to illustrate the great diversity of irreducible S-modules that can occur in the radical of A^- for a simple flexible Lie-admissible algebra A .

Proposition 4.30. Let $S = [S,S]$ be a semisimple Lie algebra over F, and let $\{U_1, \cdots, U_n, V_1, \cdots, V_n\}$ be any set of irreducible S-modules such that no nonzero element of S annihilates all the U_i's and V_i's. Assume that for each i the S-module $U_i \otimes V_i$ has an S-submodule of codimension one. Then, there exists a quasi-classical Lie algebra L such that the radical R of L is nilpotent of class 2, the semisimple part of L is isomorphic to S, and each of the modules U_i and V_i $(i=1,\cdots,n)$ is isomorphic to an S-submodule of R.

Proof. We denote the S-module action by $[s,w]$ for $s \in S$ and $w \in U_i$ or $w \in V_i$. Fix i between 1 and n, and let Z_i denote an S-submodule of $U_i \otimes V_i$ of codimension one. Let π be the natural module homomorphism of $U_i \otimes V_i$ onto the quotient module $U_i \otimes V_i/Z_i$ which we identify with F. Then, F is a trivial S-module since $S = [S,S]$. Consider the S-module direct sum $U_i \otimes V_i + V_i \otimes U_i$ and define a bilinear mapping ϕ from $S \times (U_i \otimes V_i + V_i \otimes U_i)$ into F by

$$\phi(s, u \otimes v + v' \otimes u') = \pi([s,u] \otimes v + u' \otimes [s,v']) \tag{4.93}$$

for $s \in S$, $u, u' \in U_i$ and $v, v' \in V_i$. Since π is an S-module homomorphism of $U_i \otimes V_i$ onto the trivial module F,

$$\pi([s,u] \otimes v) + \pi(u \otimes [s,v]) = 0 . \tag{4.94}$$

Using this, we show that

$$\phi([s,t],w) = \phi(s,[t,w]) \tag{4.95}$$

for all $s,t \in S$ and $w \in U_i \otimes V_i + V_i \otimes U_i$. It suffices to verify (4.95) when $w = u \otimes v$ or $v \otimes u$. For $w = u \otimes v$, we have

$$\phi(s,[t,u \otimes v]) = \pi([s,[t,u]] \otimes v + [s,u] \otimes [t,v])$$

$$= \pi([s,[t,u]] \otimes v - [t,[s,u]] \otimes v) \text{ , by (4.94)}$$

$$= \pi([[s,t],u] \otimes v) = \phi([s,t],u \otimes v) \text{ .}$$

Similarly, $\phi(s,[t,v \otimes u]) = \phi([s,t],v \otimes u)$, and hence (4.95) holds.

Let W_i denote the set of elements w in $U_i \otimes V_i + V_i \otimes U_i$ such that $\phi(s,w) = 0$ for all $s \in S$. It follows from (4.93)-(4.95) that W_i is an S-submodule containing all the symmetric elements $\Sigma_i(u_i \otimes v_i + v_i \otimes u_i)$. Denoting the quotient module $(U_i \otimes V_i + V_i \otimes U_i)/W_i$ by Q_i, we see that ϕ induces a nondegenerate bilinear form ϕ^* on $S \times Q_i$ into F. Let $M_i = U_i \oplus V_i \oplus Q_i$ be the S-module direct sum. We want to make $S + M_i$ into a Lie algebra. Let multiplication of S on M_i be the module action of S on M_i, and let multiplication on S be the same as in the Lie algebra S. Define the product $[\ ,\ ]$ between U_i and V_i to be an element of Q_i given by

$$\text{(4.96)} \qquad \begin{aligned} [u,v] &= u \otimes v + W_i \text{ ,} \\ [v',u'] &= v' \otimes u' + W_i \end{aligned}$$

for $u,u' \in U_i$, and $v,v' \in V_i$. Then, $[u,v] = -[v,u]$ for all $u \in U_i$, $v \in V_i$, since $u \otimes v + v \otimes u \in W_i$. We define all other products in $S + M_i$ to be zero. It is immediate that $S + M_i$ is a Lie algebra under these products, and that the radical M_i of $S + M_i$ is nilpotent of class 2. In fact, since M_i is an S-module, the radical M_i acts as derivations on S, and hence the product on $S + M_i$ is the same as $S + M_i$, viewed as the holomorph of S by M_i.

We define a symmetric bilinear form on $S + M_i$ by letting

(4.97)
$$(u,v) = \pi(u \otimes v),\ (s,q) = \phi^*(s,q) ,$$
$$(U_i,U_i) = (V_i,V_i) = (Q_i,Q_i) = (S,S) = (S,U_i) = (S,V_i) = 0 ,$$

for $u \in U_i$, $v \in V_i$, $q \in Q_i$, and $s \in S$. To show that $(,)$ is invariant, let $q = [u,v]$ for $u \in U_i$ and $v \in V_i$. From (4.93), (4.95) and (4.96), we compute

$$(s,[u,v]) = \phi^*(s,[u,v]) = \phi(s,u \otimes v) = \pi([s,u] \otimes v) = ([s,u],v) ,$$

$$(s,[v,u]) = \phi^*(s,[v,u]) = \phi(s,v \otimes u)$$
$$= \pi(u \otimes [s,v]) = (u,[s,v]) = ([s,v],u) ,$$

$$(u,[s,v]) = ([s,v],u) = (s,[v,u]) = -(s,[u,v])$$
$$= -([s,u],v) = ([u,s],v) ,$$

$$(s,[t,q]) = \phi^*(s,[t,q]) = \phi(s,[t,u \otimes v])$$
$$= \phi([s,t],u \otimes v) = \phi^*([s,t],q) = ([t,s],q) ,$$

$$(s,[q,t]) = -(s,[t,q]) = -([s,t],q) = ([t,s],q)$$
$$= (t,[s,q]) = ([s,q],t) .$$

The remaining possibilities of associating 3 elements from S, V_i, U_i, Q_i will be valid because both sides of the desired equation will vanish.

Consider now the algebra $L' = S \oplus \Sigma_{i=1}^{n} M_i$, where products in each subalgebra $S + M_i$ have been defined, and where products between different M_i's are zero. The symmetric bilinear form on each $S + M_i$ can be extended to all of L' by letting $(M_i,M_j) = 0$ for $i \neq j$. This form is also invariant, but may be degenerate. If L_0 denotes the radical of the form, then L_0 is an ideal of L', and $L = L'/L_0$ is a quasi-classical Lie algebra over F. Since U_i, V_i, Q_i are not contained in L_0 and U_i, V_i are irreducible, $L_0 \cap U_i = L_0 \cap V_i = 0$. Hence, L has a radical of nilpotent class 2, and a submodule of its radical isomorphic to U_i and one isomorphic to V_i. Suppose that $S \cap L_0 \neq 0$, and let $S_0 = S \cap L_0$. Since $[S_0,U_i]$ and $[S_0,V_i]$ are submodules and S_0 cannot annihilate all the U_i's and V_i's, $[S_0,U_i] = U_i$ or $[S_0,V_i] = V_i$ for some i. If $[S_0,U_i] = U_i$, then $0 = (S_0,[U_i,V_i]) = ([S_0,U_i],V_i) = (U_i,V_i)$. This contradiction shows $S_0 = 0$. Thus, L has a semisimple part isomorphic to S. □

When A is a simple commutative algebra and L is the same as in Proposition 4.30, the algebra $L(A,e)$ constructed by (4.84) is a simple flexible Lie-admissible algebra and $L(A,e)^-$ has the same property as L. Using the above construction, one can obtain for each subset $\Gamma \subseteq \{1,\cdots,n\}$ a simple flexible Lie-admissible algebra in which $R_\Gamma \neq 0$. The special case of $\Gamma = \{1,\cdots,n\}$ is given by

Corollary 4.31. Let $S_1, \cdots, S_n$ be simple quasi-classical Lie algebras over a field F. Then, there exists a quasi-classical Lie algebra L whose semisimple part is isomorphic to $S = S_1 \oplus \cdots \oplus S_n$, and whose radical contains an irreducible S-module isomorphic to $S_1 \otimes \cdots \otimes S_n$.

Proof. In view of Proposition 4.30, it suffices to show that the irreducible S-module $U = V = S_1 \otimes \cdots \otimes S_n$ has the property that $U \otimes V$ has an S-module of codimension one. Denoting the bilinear form on each S_i by $(\ ,\ )$, we define a linear mapping $\phi : U \otimes V \to F$ by

$$\phi((s_1 \otimes \cdots \otimes s_n) \otimes (t_1 \otimes \cdots \otimes t_n)) = (s_1,t_1) \cdots (s_n,t_n)$$

for $s_i, t_i \in S_i$. Then, the invariance of each $(\, , \,)$ implies that ϕ is an S-module homomorphism. Since ϕ is not identically zero (cf. Lemma 4.25), the kernel of ϕ is a submodule of codimension one. □

Remark 4.1. Let U and V be irreducible modules over F for a semisimple Lie algebra $S = [S,S]$. Then, the following are equivalent.

(1) $U \otimes V$ has an S-submodule of codimension one.

(2) There is an S-module homomorphism of $U \otimes V$ onto the trivial module F.

(3) $V \cong U^*$ and $U \cong V^*$ as an S-module.

It is immediate that (1) and (2) are equivalent. Assume that π is an S-module homomorphism of $U \otimes V$ onto F. Define a mapping $\rho : V \to U^*$ by $\rho(v)(u) = \pi(u \otimes v)$ for $u \in U, v \in V$. It is easily seen that ρ is a nonzero S-module homomorphism, and must be injective, since V is irreducible (see the proof of Lemma 4.17). Similarly, the mapping $\rho' : U \to V^*$ defined by $\rho'(u)(v) = \pi(u \otimes v)$ is an injective S-module homomorphism. Thus, $V \cong U^*$ and $U \cong V^*$ as an S-module. If $V \cong U^*$ as an S-module, then the mapping : $u \otimes v^* \to v^*(u)$ for $u,v \in U$ is an S-module homomorphism of $U \otimes U^*$ onto F. □

Remark 4.2. Let U and V be irreducible modules over F for a semisimple Lie algebra $S = [S,S]$ such that no nonzero element of S annihilates both U and V, and such that $U \otimes V$ has an S-module homomorphism onto the trivial S-module F. Let $Q = (U \otimes V + V \otimes U)/W$ be the quotient S-module constructed in Proposition 4.30. Then, $Q = U \otimes V/W = V \otimes U/W$, since $u \otimes v + W = - v \otimes u + W$ for $u \in U, v \in V$. Let

$(\, , \,)$ denote the symmetric invariant bilinear form defined by (4.97). By the construction of ϕ^* in (4.97) and the invariance of ϕ^*, we see that the mapping $q \to \phi^*(\, ,q)$ is an injective S-module homomorphism of Q into the dual module S^* of S. Similarly, from the assumption that no nonzero element of S annihilates both U and V, it follows that the mapping $s \to \phi^*(s, \,)$ is also an injective S-module homomorphism of S into Q^*. Thus, Q is contragredient to S, i.e., $S \cong Q^*$ and $S^* \cong Q$ as an S-module. For characteristic zero, this says that if no nonzero element of S annihilates U, then by Weyl's theorem $U \otimes U^*$ has a summand isomorphic to S, since $U \otimes U^*$ has an S-module homomorphism onto Q and S is self-dual. □

The following result illustrates that the radical can also be solvable but not nilpotent when the semisimple part is nonzero.

Proposition 4.32. For any quasi-classical semisimple Lie algebra S over F, there exists a quasi-classical Lie algebra L such that the semisimple part of L is isomorphic to S, and the radical of L is not nilpotent.

Proof. Let U, V, Q be S-modules isomorphic to S under isomorphisms ρ, σ, τ, respectively, and let a, d span trivial S-modules. Consider the vector space $L = S \oplus Fa \oplus U \oplus V \oplus Q \oplus Fd$, and let $R = Fa + U + V + Q + Fd$. Let $(\, , \,)$ denote a nondegenerate symmetric invariant form on S which exists by hypothesis. For $s, t \in S$, and $r \in R$, and $u \in U, v \in V$, define a multiplication in L by:

$[s,t]$ is the same as in S,

$[s,r] = - [r,s]$ is the module action of S on R,

$$[a,u] = u = -[u,a] , [v,a] = v = -[a,v] ,$$

$$[\rho(s),\sigma(t)] = \tau([s,t]) + (s,t)d ,$$

$$[a,Q] = [L,d] = [U,U] = [V,V] = [Q,Q] = [U,Q] = [V,Q] = 0 .$$

The subalgebra $U + V + Q + Fd$ is clearly a Lie algebra since all double commutators are zero. Since L is an S-module and all products are S-bilinear mappings, any triple with an element of S satisfies the Jacobi identity. For example, if $s, t, x \in S$, then

$$[[x,\rho(s)],\sigma(t)] + [[\sigma(t),x],\rho(s)] + [[\rho(s),\sigma(t)],x] = [\rho([x,s]),\sigma(t)]$$

$$- [\sigma([x,t]),\rho(s)] + [\tau([s,t]),x]$$

$$= \tau([[x,s],t]) + ([x,s],t)d + \tau([s,[x,t]]) + (s,[x,t])d + \tau([[s,t],x])$$

$$= 0 .$$

To show that L is a Lie algebra, it suffices to verify the Jacobi identity for those triples of basis elements which contain a, and which contain no elements of S. The only triples of this type for which the Jacobi identity is not trivial consist of a, u, v for $u \in U$ and $v \in V$. But then, we have

$$[[a,u],v] + [[u,v],a] + [[v,a],u] = [u,v] + 0 + [v,u] = 0 ,$$

showing that L is a Lie algebra.

It follows that R is an ideal of L, and that $[R,R] = U + V + Q + Fd$ and $[[R,R],[R,R]] \subseteq Q + Fd$. Since $[Q + Fd, Q + Fd] = 0$, R is solvable, but the relation $[u,a] = u$ for $u \in U$ shows that R is not nilpotent. Since $L/R \cong S$, R is the radical of L. Define a symmetric bilinear form $(\ ,\ )'$ on L by

$$(\rho(s),\sigma(t))' = (s,t), \quad (s,\tau(t))' = (s,t), \quad (a,d)' = 1$$

for $s,t \in S$, and by letting the form vanish on any other pairs of elements from the subspaces. The form is clearly nondegenerate.

To verify the invariance of $(\ ,\ )'$, for $r,s,t \in S$, we have

$$(r,[\rho(s),\sigma(t)])' = (r,\tau([s,t]))' = (r,[s,t]) = ([r,s],t)$$

$$= (\rho([r,s]),\sigma(t))' = ([r,\rho(s)],\sigma(t))' ,$$

$$(r,[\sigma(t),\rho(s)])' = -\ (r,[\rho(s),\sigma(t)])' = -\ (r,[s,t]) = ([r,t],s)$$

$$= (\sigma([r,t]),\rho(s))' = ([r,\sigma(t)],\rho(s))' ,$$

$$(a,[\rho(s),\sigma(t)])' = (a,\tau([s,t]) + (s,t)d)' = (s,t)(a,d)' = (s,t)$$

$$= (\rho(s),\sigma(t))' = ([a,\rho(s)],\sigma(t))' ,$$

$$(r,[s,\tau(t)])' = (r,\tau([s,t]))' = (r,[s,t]) = ([r,s],t) = ([r,s],\tau(t))' .$$

The remaining cases follow either from these or from the fact that both sides of the desired equation are zero. □

4.6. Wedderburn - type theory

We retain in this section the assumption that all algebras are finite-dimensional over an arbitrary field F. We discuss for flexible Malcev-admissible algebras a possibility of developing any structure theory of the type that works so well for such classes of algebras as associative, more generally alternative, Jordan, Lie, or Malcev algebras. In order to develop a structure theory, one first needs to introduce an appropriate concept of the radical of an algebra A . Generally, a radical of an algebra A is defined by certain property (P), called a *radical property*, which A satisfies. An algebra A is called a (P)-*algebra* if it possesses the property (p), and an ideal B of A is called a (P)-*ideal* if B is a (P)-algebra. An algebra without nonzero (P)-ideals is said to be (P)-*semisimple*. The property (P) is called a radical property if the following conditions hold (Divinsky [1]):

(a) Any homomorphic image of a (P)-algebra is a (P)-algebra.

(b) Every algebra A contains a (P)-ideal R which contains all (P)-ideals of A .

(c) The quotient algebra A/R is (P)-semisimple.

It is well known that the solvable property is a radical property for any algebra and that the nilpotent property is not in general a radical property, since condition (b) is not always satisfied. The maximal (P)-ideal R of A guaranteed by condition (b) is called the (P)-*radical* of A . The concept of radical originated from the hope that a suitable radical property (P) can be defined in such a way that the quotient algebra A/R by the (P)-radical R gives rise to a useful structure theory. Thus, it is expected that a radical property (P) should not allow either too many (P)-semisimple algebras, or too many radical algebras to be very useful. For example, all ideals that square to zero ought to be contained in the (P)-radical, or equivalently, a (P)-semisimple algebra should not contain such ideals.

Since Jordan and alternative algebras are power-associative, the nil radical is well defined, and in fact coincides with the solvable radical of those algebras. The nil radical is also known to work so well for many well known algebras, such as flexible strictly power-associative algebras including noncommutative Jordan algebras (see Oehmke [1] and Schafer [3]). For these algebras, the nil property is a radical property (P) satisfying the additional condition:

(d) Any homomorphic image of every (P)-semisimple algebra is also (P)-semisimple.

When (P) is a radical property such that any algebra A is (P)-semisimple if and only if A is a direct sum of simple ideals of A , then condition (d) automatically holds. This is the case for the nil radical of any Jordan or alternative algebra over F , and of any flexible strictly power-associative algebra over F of characteristic ≠ 2, 3 (Oehmke [1]). On the other hand, the nil radical does not seem useful for flexible Malcev-admissible algebras, since too many useful algebras are radical algebras, and power-associativity discards too many useful algebras (see Chapter 2). A class of algebras closed under any homomorphism (and each algebra in that class) will be called of *Wedderburn-type* if it possesses a radical property (P) satisfying condition (d). In this case, the radical property (P) may also be called Wedderburn-type for that class of algebras. Thus, any Lie algebra or Malcev algebra of characteristic 0 is of Wedderburn-type with respect to the solvable radical property, since semisimple algebras are direct sums of simple ideals. However, as we shall see below, the class of flexible Malcev-admissible algebras is not of Wedderburn-type with respect to the solvable radical property, even for characteristic 0 . In fact, we give an example which seems to indicate that there is no radical property which leads to a well-behaved Wedderburn-type theory for flexible Malcev-admissible algebras. Our example is based on Theorems 4.3 and 4.6.

Example 4.3. Let W be any flexible Malcev-admissible algebra over F such that $[W,W]$ contains no nonzero ideals of W (for example, W could be any commutative algebra). Let $b_1, \cdots, b_n$ be elements of W such that the cosets $b_i + [W,W]$ for $i = 1, \cdots, n$ form a basis of the quotient space $W/[W,W]$, and let $\tau_i (i=1,\cdots,n)$ be the linear form on W defined by

$$\tau_i([W,W]) = 0, \ \tau_i(b_j) = \delta_{ij}$$

for $i, j = 1, \cdots, n$. Suppose that $S_1, \cdots, S_n$ are flexible Malcev-admissible algebras with $S_1^-, \cdots, S_n^-$ simple. We let $A = \Sigma S_i + W$ be the algebra defined by

$$xy = 0, \ x \in S_i, \ y \in S_j \quad \text{for} \quad i \neq j ,$$

$$ax = \tau_i(a)x = xa, \ a \in W, \ x \in S_i ,$$

by letting the product of any two elements in the same S_i be the same in A as in S_i, and by letting any two elements of W have the same product in A as in W. It follows from Theorem 4.3 that A is a flexible Malcev-admissible algebra.

If B is a nonzero ideal of A, then by Theorem 4.6 either $S_j \subseteq B$ for some j or B is an ideal of W. In the latter case, $B \nsubseteq [W,W]$ by the choice of W, and hence there is an element $b \in B$ but $b \notin [W,W]$. Then, there exists $i \in \{1,\cdots,n\}$ such that $\tau_i(b) \neq 0$, so that $S_i = \tau_i(b)S_i = bS_i \subseteq B$. This shows that every nonzero ideal of A

contains one of the S_i's . Conversely, each S_i is clearly an ideal of A which is simple as an algebra. Thus, the minimal ideals of A are exactly the S_i's and they are all simple. Also, we have $A/\Sigma S_i \cong W$.

Suppose that there exists a radical property (P) of Wedderburn-type for A . Assume first that A is (P)-semisimple. Then, the quotient algebra $A/\Sigma S_i \cong W$ must be a (P)-semisimple flexible Malcev-admissible algebra, which can contain all commutative algebras including all nilpotent commutative algebras as well as algebras squaring to zero. Assume then that A is not (P)-semisimple, and let R be the (P)-radical of A , which exists by condition (b). Since $R \neq 0$, it must contain at least one of the S_i's . Since the S_i's could have been made isomorphic to each other and since any simple algebra S with S^- simple could have been used in the construction of A , we see that all simple algebras S with S^- simple must be in the class of (P)-radical algebras, including the algebras described in part (2) of Corollary 4.4. Therefore, we can conclude that any Wedderburn-type theory for flexible Malcev-admissible algebras would have either too many semisimple algebras or too many radical algebras to be very useful. □

Example 4.3 shows also that the solvable radical cannot be of Wedderburn-type for the class of flexible Malcev-admissible algebras. For characteristic 0 , many well known algebras of Wedderburn-type possess a symmetric nondegenerate invariant bilinear form which essentially characterizes the (nil or solvable) semisimplicity of the algebra. In the terminology of Definition 4.2, we have called such algebras quasi-classical. If A is a noncommutative Jordan algebra of characteristic 0 , then Schafer [1] has shown that the bilinear form (,) on A defined by $(x,y) = \frac{1}{2}\mathrm{tr}(R_{xy} + L_{xy})$ is a symmetric invariant form on A , and the nil radical of A coincides with the radical of (,) and hence (,) is nondegenerate if and only if A is semisimple. Thus, A is a direct sum of simple ideals if and only if (,) is nondegenerate. The semisimplicity of Lie and Malcev algebras is similarly characterized by the nondegeneracy of the Killing form. In contrast, we give an example of a simple flexible Malcev-admissible algebra which does not possess such a bilinear form, even for characteristic 0 .

Example 4.4. Let $A = S \otimes P + R_0$ be the simple flexible Malcev-admissible algebra over F constructed in Example 4.2. Assume further that $k = 2$ and S is a simple Malcev algebra of dimension $n > 3$. Thus, $\dim R_* = 2n$, $\dim R_0 = 3$, and $\dim A = 3n + 3$. If (,) denotes an invariant bilinear form on A and if $s \in S$ and $u,v \in R_*$, then $([s,u],v) = (s,[u,v]) = 0$ since $[R_*,R_*] = 0$. Thus, $(R_*,R_*) = 0$, since R_* is spanned by elements of the form $[s,u]$. But then R_* is totally isotropic, and since $\dim R_* > \frac{1}{2} \dim A$, the form must be degenerate. □

In Section 4.5, we focused on the construction of simple flexible Malcev-admissible algebras of the form $M(A,e)$, using a simple commutative algebra and a quasi-classical Malcev algebra M , so that $M(A,e)^-$ is quasi-classical also. Example 4.4 illustrates that not every simple flexible Malcev-admissible algebra arises from this construction.

5

MALCEV-ADMISSIBLE ALGEBRAS OF LOW DIMENSION

5.1. Basic results

In this chapter, we construct some flexible Malcev-admissible algebras of low dimension over an algebraically closed field F of characteristic 0. Under some restrictions, we classify all flexible Malcev-admissible algebras A of dimension ≤ 8 over F, when A^- is not solvable. Since the case of dimension ≤ 4 has been determined in Theorem 4.9, we begin with dimension 5. Specifically, we determine completely the case of dimension 5. For dimension 6, we put the restriction that the radical R of A^- is nilpotent, when R is not a direct summand of A^-. For dimensions 7 and 8, we assume that either the radical R is abelian, or A^- has a Levi factor $\mathfrak{sl}(2) \oplus \mathfrak{sl}(2)$ or C_0^-.

When dim $A = 7$, we also investigate the structure of algebras A which are acted on by the Lie algebra $\mathfrak{sl}(3)$ as derivations and which have an irreducible $\mathfrak{sl}(3)$-submodule direct sum $A = U + V + Fz$, where U and V are the two nonisomorphic irreducible $\mathfrak{sl}(3)$-submodules of dimension 3, and Fz is a trivial module. This problem is motivated by the 7-dimensional color algebra constructed by Domokos and Kövesi-Domokos [1], where triplets in U represent quarks, antitriplets in the dual space V represent antiquarks, and leptons correspond to singlets Fz under the $\mathfrak{sl}(3)$-action. As we shall see below, the $\mathfrak{sl}(3)$-module decomposition $A = U + V + Fz$ naturally arises from a split octonion algebra C over F, since the derivations of C vanishing on an idempotent $e \neq 1$ induce two irreducible $\mathfrak{sl}(3)$-submodules of C of dimension 3. The color algebra of Domokos and Kövesi-Domokos is obtained as a special case of our construction.

We construct two classes of algebras of dimension 15. The first class consists of algebras A with $s\ell(3)$ as an algebra of derivations, which have an irreducible $s\ell(3)$-submodule decomposition $A = S + U + V + Fz$ where S is isomorphic to $s\ell(3)$ as a module, U, V are the same as above, and Fz is a trivial module. This construction is also motivated by physics and includes the 7-dimensional algebra mentioned above as a submodule and the simple Lie algebra G_2 used by Günaydin and Gürsey [1]. In the latter case, the standard basis of S represents gluons. The decomposition of A also arises from the fact that the derivation algebra G_2 of a split octonion algebra C has the decomposition $G_2 = S + U + V$ as an S-module under the adjoint action, where S is the subalgebra of derivations vanishing on an idempotent $e \neq 1$ of C, and U, V are two irreducible S-submodules of G_2. We give conditions for such algebras to be Lie-admissible as well as flexible. The second class is composed of algebras with G_2 as an algebra of derivations. Since the only nontrivial irreducible G_2-modules of dimension ≤ 15 are the 7-dimensional module $V(\lambda_1)$ and the 14-dimensional module $V(\lambda_2)$ where λ_1 and λ_2 are the fundamental weights of G_2, we consider in this case the decompositions $A = V(\lambda_1) + V(\lambda_1) + Fz$ and $A = V(\lambda_2) + Fz$.

As in Chapters 3 and 4, our fundamental techniques are based on Lemma 3.1, Theorem 3.5, and the fact that the products between irreducible summands of A are determined by module homomorphisms of $V_i \otimes V_j$ to V_k. Modules involved in our discussion are those irreducible Malcev or Lie modules of dimension ≤ 8 for $s\ell(2)$, $s\ell(3)$, G_2, or C_0^-. Thus, our main effort is focused on the determination of the tensor product $V_i \otimes V_j$ for those modules V_i, V_j of dimension ≤ 8 and of module homomorphisms of $V_i \otimes V_j$ to V_k.

Consider first the irreducible $s\ell(2)$-module $V(m)$ of highest weight

$m \geq 0$, and let $\{e,h,f\}$ be the standard basis of $\mathfrak{sl}(2)$ such that

(5.1) $$[e,f] = h, \ [h,e] = 2e, \ [h,f] = -2f .$$

Then, there exists a basis $\{v_0, v_1, \cdots, v_m\}$ of $V(m)$ such that

(5.2) $$\begin{aligned} hv_i &= (m - 2i)v_i , \\ fv_i &= (i + 1)v_{i+1}, \\ ev_i &= (m - i + 1)v_{i-1}, \ i = 0, 1, \cdots, m , \end{aligned}$$

where $v_{-1} = v_{m+1} = 0$ (Humphreys [1]). The Clebsch-Gordan formula provides the decomposition of the $\mathfrak{sl}(2)$-module $V(m) \otimes V(n)$:

(5.3) $$V(m) \otimes V(n) = V(m + n) \oplus V(m + n - 2) \oplus \cdots \oplus V(|m - n|) ,$$

and hence by Lemma 3.1

(5.4) $$\dim_F \mathrm{Hom}_{\mathfrak{sl}(2)} (V(m) \otimes V(n), V(s)) = \begin{cases} 1, & \text{if } s = m + n, m + n - 2, \cdots, |m - n| , \\ 0, & \text{otherwise} . \end{cases}$$

For the Lie algebra $\mathfrak{sl}(3)$, by Weyl's formula (Humphreys [1]), the dimension of the irreducible module $V(m_1\lambda_1 + m_2\lambda_2)$ $(m_i \geq 0)$ is given by

$$\dim V(m_1\lambda_1 + m_2\lambda_2) = \frac{1}{2}(m_1 + 1)(m_2 + 1)(m_1 + m_2 + 2) .$$

Using this we find that the only modules of dimension ≤ 8 are

	$V(0)$	$V(\lambda_1)$	$V(\lambda_2)$	$V(2\lambda_1)$	$V(2\lambda_2)$	$V(\lambda_1 + \lambda_2)$
dimension	1	3	3	6	6	8

where $V(\lambda_2) = V(\lambda_1)^*$, $V(2\lambda_2) = V(2\lambda_1)^*$ and $\mathfrak{sl}(3) \cong V(\lambda_1 + \lambda_2)$.

Consider the tensor products of these modules. For any irreducible module $V(\lambda)$, we have $V(0) \otimes V(\lambda) \cong V(\lambda)$ by (3.11). Of the remaining products we list only those needed for our investigation.

$$
\begin{aligned}
V(\lambda_1) \otimes V(\lambda_1) &= V(2\lambda_1) + V(\lambda_2) , \\
V(\lambda_1) \otimes V(\lambda_2) &= V(\lambda_1 + \lambda_2) + V(0) , \\
V(\lambda_2) \otimes V(\lambda_2) &= V(2\lambda_2) + V(\lambda_1) , \\
V(\lambda_1) \otimes V(\lambda_1 + \lambda_2) &= V(2\lambda_1 + \lambda_2) + V(2\lambda_2) + V(\lambda_1) , \\
V(\lambda_2) \otimes V(\lambda_1 + \lambda_2) &= V(\lambda_1 + 2\lambda_2) + V(2\lambda_1) + V(\lambda_2) , \\
V(\lambda_1+\lambda_2) \otimes V(\lambda_1+\lambda_2) &= V(2\lambda_1+2\lambda_2)+V(3\lambda_1)+V(3\lambda_2)+V(\lambda_1+\lambda_2)+V(\lambda_1+\lambda_2)+V(0) ,
\end{aligned} \tag{5.5}
$$

where $\dim V(2\lambda_1 + \lambda_2) = \dim V(\lambda_1 + 2\lambda_2) = 15$, $\dim V(3\lambda_1) = \dim V(3\lambda_2) = 10$, and $\dim V(2\lambda_1 + 2\lambda_2) = 27$. The last decomposition of (5.5) is given by Table 3.3, and the presence of $V(\lambda_1)$ only once in the decomposition of $V(\lambda_1) \otimes V(\lambda_1 + \lambda_2)$ also follows from Corollary 3.3 and Theorem 3.5.

For the Lie algebra G_2, Weyl's dimension formula reads:

$$\dim V(m_1\lambda_1+m_2\lambda_2) = \frac{1}{5!}(m_1+1)(m_2+1)(m_1+m_2+2)(m_1+2m_2+3)(m_1+3m_2+4)(2m_1+3m_2+5) .$$

Thus, the only modules of dimension ≤ 8 are the 1-dimensional module $V(0)$ and the 7-dimensional module $V(\lambda_1)$. The decomposition of $V(\lambda_1) \otimes V(\lambda_1)$ is given by

$$V(\lambda_1) \otimes V(\lambda_1) = V(2\lambda_1) + V(\lambda_2) + V(\lambda_1) + V(0) , \tag{5.6}$$

where $\dim V(2\lambda_1) = 27$ and $\dim V(\lambda_2) = 14$.

The relations between $s\ell(3)$, $s\ell(3)$-modules $V(\lambda_1)$, $V(\lambda_2)$, and G_2 can be best described in terms of a split octonion algebra. Let C denote a split octonion algebra over a field F of characteristic $\neq 2,3$. Thus, C may be given by a Zorn vector-matrix algebra (Sagle [4]). Let $F^3 = \{(\alpha_1,\alpha_2,\alpha_3) \mid \alpha_i \in F\}$ be the 3-dimensional space over F, and let $(,)$ and "$\times$" denote the inner and cross products on F^3. Then, C consists of the elements

$$x = \begin{bmatrix} \alpha & a \\ b & \beta \end{bmatrix}, \quad y = \begin{bmatrix} \gamma & c \\ d & \delta \end{bmatrix},$$

where $\alpha, \beta, \gamma, \delta \in F$ and $a, b, c, d \in F^3$. The addition $x + y$ and multiplication xy in C are defined by

$$x + y = \begin{bmatrix} \alpha + \gamma & a + c \\ b + d & \beta + \delta \end{bmatrix},$$

$$(5.7) \qquad xy = \begin{bmatrix} \alpha\gamma + (a,d) & \alpha c + \delta a - b \times d \\ \gamma b + \beta d + a \times c & (b,c) + \beta\delta \end{bmatrix}.$$

Letting $a_1 = (1,0,0)$, $a_2 = (0,1,0)$ and $a_3 = (0,0,1)$, the elements

$$(5.8) \qquad \begin{aligned} e_1 &= \begin{bmatrix} 1 & 0 \\ 0 & 0 \end{bmatrix}, \quad e_2 = \begin{bmatrix} 0 & 0 \\ 0 & 1 \end{bmatrix}, \\ u_i &= \begin{bmatrix} 0 & -a_i \\ 0 & 0 \end{bmatrix}, \quad v_i = \begin{bmatrix} 0 & 0 \\ a_i & 0 \end{bmatrix}, \quad i = 1, 2, 3 \end{aligned}$$

form a basis of C and give the multiplication table

$$(5.9) \qquad \begin{aligned} & e_i^2 = e_i (i = 1,2), \ e_1 u_i = u_i = u_i e_2, \ e_2 v_i = v_i = v_i e_1, \\ & u_i u_j = \varepsilon_{ijk} v_k, \ v_i v_j = \varepsilon_{ijk} u_k, \ u_i v_j = -\delta_{ij} e_1, \ v_i u_j = -\delta_{ij} e_2 \end{aligned}$$

for $i, j, k = 1, 2, 3$, and all other products are zero, where ε_{ijk} is a totally skew symmetric tensor in 1, 2, 3 .

Let U and V denote the subspaces spanned by the u_i's and v_i's , respectively. It follows from (5.9) that the Peirce decomposition of C relative to idempotents e_1, e_2 is given by

$$C = Fe_1 + U + V + Fe_2 , \tag{5.10}$$

with $U = e_1Ce_2$ and $V = e_2Ce_1$. Denote

$$D_0 = \{\sigma \in \text{Der } C \mid \sigma(e_1) = 0\} . \tag{5.11}$$

Since $\sigma(1) = 0$ for $\sigma \in$ Der C and $e_1 + e_2 = 1$, any derivation in D_0 vanishes on e_2 also. Thus, both U and V are stable under D_0 , and we denote the restriction of $\sigma \in D_0$ to U and V by

$$\sigma_U = \sigma|_U , \sigma_V = \sigma|_V . \tag{5.12}$$

Lemma 5.1. Let C be a split octonion algebra over F of characteristic ≠ 2,3 . Let U, V, D_0 and σ_U be the same as in (5.10)-(5.12). Then, the Lie algebra D_0 is isomorphic to $\mathfrak{sl}(3)$, and the mappings $\sigma \to \sigma_U$ and $\sigma \to \sigma_V$ are faithful irreducible representions of D_0 in U and V , respectively. Furthermore, U and V are contragredient to each other as $\mathfrak{sl}(3)$-modules.

Proof. The mappings $\sigma \to \sigma_U$ and $\sigma \to \sigma_V$ are clearly Lie homomorphisms of D_0 into $(\text{Hom}_F U)^-$ and $(\text{Hom}_F V)^-$. For each $\sigma \in D_0$, we identify σ_U and σ_V with their matrices relative to the bases $\{u_1,u_2,u_3\}$ of U and $\{v_1,v_2,v_3\}$ of V , given by (5.8). Thus, let

$$\sigma_U(u_j) = \sum_{i=1}^{3} \alpha_{ij}u_i , \sigma_V(v_j) = \sum_{i=1}^{3} \beta_{ij}v_i \tag{5.13}$$

for $\sigma \in D_0$, where $\alpha_{ij},\beta_{ij} \in F$. Applying σ to $u_iv_j = -\delta_{ij}e_1$ in

(5.9), we have $\sigma_U(u_i)v_j + u_i\sigma_V(v_j) = 0$ and by (5.13)

$$\sum_k \alpha_{ki}u_kv_j + \sum_k \beta_{kj}u_iv_k = 0 ,$$

which can combine with (5.9) to give $\sum_k \alpha_{ki}\delta_{kj}e_1 + \sum_k \beta_{kj}\delta_{ik}e_1 = 0$. Thus, $\alpha_{ij} = -\beta_{ji}$ for all $i, j = 1, 2, 3$ and

$$\sigma_V = -\sigma_U^t \tag{5.14}$$

where "t" denotes the transpose. Applying σ to $u_1u_2 = v_3$ in (5.9), by (5.13) and (5.14), we have $\sum_j \alpha_{j1}u_ju_2 + \sum_j \alpha_{j2}u_1u_j = -\sum_j \alpha_{3j}v_j$, which by (5.9) implies $\alpha_{11} + \alpha_{22} + \alpha_{33} = 0$. Thus, $\operatorname{tr}\sigma_U = \operatorname{tr}\sigma_V = 0$. This shows that $\sigma \to \sigma_U$ and $\sigma \to \sigma_V$ are injective linear mappings of D_0 into $\mathfrak{sl}(3)$.

In order to prove that the mappings are surjective, let $T = (\alpha_{ij})$ be any matrix in $\mathfrak{sl}(3)$. Define a linear transformation σ on C by

$$\sigma(e_i) = 0\ (i = 1,2),\ \sigma(u_j) = \sum_i \alpha_{ij}u_i,\ \sigma(v_j) = -\sum_i \alpha_{ji}v_i \tag{5.15}$$

for $i = 1, 2, 3$. One can easily check that σ acts as a derivation on the product uv for basis elements u, v . In fact, if one of u, v is e_i , then by (5.9) σ acts as a derivation on uv . Using the equations $u_iu_j + u_ju_i = 0 = v_iv_j + v_jv_i$, we see that σ acts on u_i^2 and v_i^2 as a derivation. Similarly, σ acts as a derivation on the products u_iu_j, v_iv_j, u_iv_j . Therefore, $\sigma \in D_0$ such that $\sigma_U = T$ and $\sigma_V = -T^t$. We have shown that both $\sigma \to \sigma_U$ and $\sigma \to \sigma_V$ are isomorphisms of D_0 to $\mathfrak{sl}(3)$ and are faithful, irreducible representations of D_0 in U and V . When $\mathfrak{sl}(3)$ acts on U and V as in (5.13) or (5.15), (5.14) shows that U and V are contragredient to each other. □

Let e_{ij} for $i, j = 1, 2, 3$ be the 3×3 matrix units, so that

$e_{11} - e_{22}$, $e_{22} - e_{33}$ and the e_{ij}'s for $i \neq j$ form a basis of $\mathfrak{sl}(3)$. Then, the action of e_{ij} on u_k and v_k in (5.15) is expressed as

(5.16) $$e_{ij} \cdot u_k = \delta_{jk} u_i, \ e_{ij} \cdot v_k = - \delta_{ik} v_j .$$

Let $s = (\alpha_{ij})$ be any matrix in $\mathfrak{sl}(3)$. When the u_i's and v_i's are identified with the usual 3×1 matrix units and 1×3 matrix units, from (5.16) we deduce

(5.17) $$s \cdot u = su, \ s \cdot v = - vs$$

for any vector $u \in U$, $v \in V$, where the right sides denote the usual matrix product.

Recall from (3.1) that for $x,y \in C$,

$$d(x,y) = ad_{[x,y]} + [ad_x, ad_y]$$

is a derivation of C^-. Since each ad_x is a derivation of C^+ by flexibility of C, $d(x,y)$ is a derivation of C and satisfies the property

(5.18) $$[\sigma, d(x,y)] = d(\sigma x, y) + d(x, \sigma y)$$

for all $x,y \in C$ and $\sigma \in \mathrm{Der}\ C$ (Sagle [1,p.453]). We give a proof of the well known theorem of Cartan and Jacobson.

Theorem 5.2. Let C be a split octonion algebra over a field F of characteristic $\neq 2, 3$, and let D_0 denote the set of derivations σ of C such that $\sigma(e_1) = 0$. Let U and V be the same as in (5.10). Then, $\mathrm{Der}\ C$ has the decomposition

(5.19) $$\mathrm{Der}\ C = D_0 \oplus d(e_1, U) \oplus d(e_2, V)$$

as the direct sum of irreducible D_0-submodules under the adjoint action, where $d(e_1,U) \cong U$ and $d(e_2,V) \cong V$ as $\mathfrak{sl}(3)$-modules. Furthermore, Der C is a 14-dimensional central simple Lie algebra of type G .

<u>Proof</u>. We use the basis of C given by (5.9). Denote D = Der C , and let σ be any derivation in D . Since $\sigma(1) = 0$, $\sigma(e_1) = -\sigma(e_2)$. From $e_i^2 = e_i$ $(i = 1,2)$, we have

$$e_1\sigma(e_1) + \sigma(e_1)e_1 = \sigma(e_1),\ e_2\sigma(e_2) + \sigma(e_2)e_2 = \sigma(e_2) .$$

Using $\sigma(e_2) = -\sigma(e_1)$, the last equation becomes $e_2\sigma(e_1) + \sigma(e_1)e_2 = \sigma(e_1)$, which together with the first equation implies that $\sigma(e_1) \in U + V$. Thus,

$$\sigma(e_1) = u - v,\ \sigma(e_2) = -u + v$$

for some $u \in U$, $v \in V$. For $u \in U$, $v \in V$, we compute

$$d(e_1,u)e_1 = -2u,\ d(e_2,v)e_2 = -2v,\ d(e_2,v)e_1 = 2v , \tag{5.20}$$

using (5.9) and $\sigma(e_2) = -\sigma(e_1)$. Hence, if $\sigma \in D$ and $\sigma(e_1) = u - v$, then $(\sigma + \frac{1}{2}d(e_1,u) + \frac{1}{2}d(e_2,v))(e_1) = 0$ and so

$$D = D_0 + d(e_1,U) + d(e_2,V) .$$

To show that the sum is direct, assume $\sigma_0 + d(e_1,u) + d(e_2,v) = 0$ for $\sigma_0 \in D_0$. When the left side of this is applied to e_1 , by (5.20) we obtain $-2u + 2v = 0$, and hence $u = v = 0$ and $\sigma_0 = 0$. Thus, the sum is direct

Using the representation of D_0 in U and V given by (5.12) and using (5.18), we get $[\sigma,d(e_1,u)] = d(e_1,\sigma_U(u))$ and $[\sigma,d(e_2,v)] = d(e_2,\sigma_V(v))$ for $\sigma \in D_0$. Since the mappings $u \to d(e_1,u)$ and $v \to d(e_2,v)$ are injective by (5.20), it follows from Lemma 5.1 that they are $\mathfrak{sl}(3)$-module isomorphisms of U and V onto $d(e_1,U)$ and $d(e_2,V)$. In particular, we have $\dim D = 8 + 3 + 3 = 14$.

In the remainder of the proof, we may assume that F is algebraically closed and show that D is simple, and so central simple. Denote $D_1 = d(e_1,U)$ and $D_2 = d(e_2,V)$, and let $D' = D_1 + D_2$, so that $D = D_0 \oplus D'$. Let J be a nonzero ideal of D. Since D_0 is simple by Lemma 5.1, either $D_0 \cap J = D_0$ or $D_0 \cap J = 0$. Since $D' \cap J$ is a D_0-submodule of D' and $D' = D_1 + D_2$ is an irreducible module direct sum, $D' \cap J = D'$, $D' \cap J = D_1$, $D' \cap J = D_2$, or $D' \cap J = 0$. Let π_0 and π' be the projections onto D_0 and D', respectively. Since π_0 and π' are D_0-module homomorphisms, we have $\pi_0(J) = D_0$ or 0, and $\pi'(J) = D'$, or D_1, or D_2, or 0. Assume first that $D_0 \cap J = 0$ and $D' \cap J = 0$. But then, both π_0 and π' are injective on J, so that $\dim J = \dim \pi_0(J) = \dim \pi'(J)$. Since the possibilities for $\dim \pi_0(J)$ are 8 and 0, and those for $\dim \pi'(J)$ are 6, 3 and 0, this case cannot occur. If $D_0 \cap J \neq 0$, then $J \supseteq D_0$ and $J \supseteq [D_0,D_i] = D_i$ by (5.9) and (5.18), and hence $J = D$.

Assume finally that $D' \cap J \neq 0$. Then, it suffices to suppose that $J \supseteq D_1$. Since F is algebraically closed, letting $e_i = e_{ii} (i = 1,2)$, $e_{12} = \sqrt{-1}\, u_1$ and $e_{21} = \sqrt{-1}\, v_1$, we see from (5.9) that C contains $F_2 = M(2,F)$, the 2×2 matrix algebra, as a subalgebra, and the e_{ij}'s are the matrix units of F_2. We note that $\operatorname{Der} F_2$ is inner, simple and isomorphic to $\mathfrak{sl}(2)$. Since $d(F_2,F_2) \subseteq \operatorname{Der} F_2$ and by (5.20) $d(e_{11},e_{12})$, $d(e_{12},e_{21})$, $d(e_{22},e_{21})$ are linearly independent, we have $d(F_2,F_2) = \operatorname{Der} F_2$. Thus, any derivation of F_2 can be extended to a derivation of C. Since $d(e_{11},e_{12}) \neq 0$ is in $J \cap \operatorname{Der} F_2$, the simplicity of $\operatorname{Der} F_2$ implies $d(F_2,F_2) \subseteq J$. But then, F_2 has a nonzero derivation σ such that $\sigma(e_{11}) = 0$, e.g., $\mathrm{ad}_{e_{11}}$, and σ can be extended to a derivation of C which must be conatined in J. Therefore, $D_0 \cap J \neq 0$ and $J = D$. Thus, D is simple. □

In the remainder of this chapter, we assume that F *denotes an algebraically closed field of characteristic* 0. We use the same notations as in Chapter 4. Thus, if A is a flexible Malcev-admissible algebra over F such that A^- is not solvable, then A has the decomposition

$$A = \sum_{i=1}^{n} S_i + \sum_{\Gamma\neq\phi} R_\Gamma + R_0 , \tag{5.21}$$

where $S = \Sigma S_i$ is a Levi factor of A^- and $R = \Sigma_{\Gamma\neq\phi} R_\Gamma + R_0$ is the radical of A^- (see (4.22)). As in (4.23), denote $R_\star = \Sigma_{\Gamma\neq\phi} R_\Gamma$.

5.2. Dimension 5

We determine all flexible Malcev-admissible algebras A of dimension 5 over F for which A^- is not solvable. Thus, A^- has the Levi decomposition $A^- = s\ell(2) + R$, where R^- is a 2-dimensional solvable Lie algebra. Hence, R is either abelian, or solvable with basis $\{u,v\}$ given by

$$[u,v] = u . \tag{5.22}$$

Assume first that R is not abelian. Since $[R,R] = Fu$ is an $s\ell(2)$-submodule, R is a sum of two trivial submodules and hence $[s\ell(2),R] = 0$. Thus, we are in the situation of Theorem 4.3 with $n = 1$, and the multiplication in A is described by (4.15). Since the center of

R^- is 0 and since the only flexible Malcev-admissible product on $s\ell(2)$ under our assumption is a nonzero multiple of [,] , by (4.15) we have $xy = \frac{1}{2}[x,y]$ for $x,y \in s\ell(2)$. Recall from Theorem 4.3 that the linear form $\tau = \tau_1$ on R vanishes on $[R,R] = Fu$. Hence, letting $\tau(v) = \alpha \in F$, (4.15) gives $xu = ux = 0$, $vx = xv = \alpha x$ for $x \in s\ell(2)$. Since R is a subalgebra of A by Theorem 4.3, it remains to determine the multiplication in R . Since R is flexible Lie-admissible, we use $ad_R \subseteq Der R$. The equation $[uv,v] = [u,v]v = uv$ gives $uv = \beta u$ and $vu = (\beta - 1)u$ for some $\beta \in F$ by (5.22). From this, we get $0 = [uv,u] = u[v,u] = - u^2$, or $u^2 = 0$. The equation $[v^2,u] = -(uv + vu)$ implies $v^2 = \lambda u + (2\beta - 1)v$ for some $\lambda \in F$, but then $v^2 v = vv^2$ forces $\lambda = 0$. Hence, R is given by

(5.23) $$uv = \beta u,\ vu = (2\beta - 1)u,\ v^2 = (2\beta - 1)v,\ u^2 = 0 .$$

Conversely, it is easy to see that any algebra B given by (5.23) is flexible Lie-admissible with $B^- \cong R^-$, and that the algebra A determined here is flexible Lie-admissible. Therefore, using the standard basis {e,h,f} of $s\ell(2)$ given by (5.1), we have the multiplication in A given by Table 5.1.

TABLE 5.1

	e	h	f	u	v
e	0	$-e$	$\frac{1}{2}h$	0	αe
h	e	0	$-f$	0	αh
f	$-\frac{1}{2}h$	f	0	0	αf
u	0	0	0	0	βu
v	αe	αh	αf	$(\beta-1)u$	$(2\beta-1)v$

Suppose next that R is abelian. There are two possibilities $[s\ell(2),R] = 0$ and $[s\ell(2),R] \neq 0$. Assume first that $[s\ell(2),R] = 0$. We are again in the case of Theorem 4.3 with $n = 1$. If $a_1 = 0$ in (4.15), then, letting $\tau(u) = \alpha$ and $\tau(v) = \beta$, we have

$$xy = \frac{1}{2}[x,y], \quad xu = ux = \alpha x, \quad xv = vx = \beta x$$

for $x,y \in s\ell(2)$, where $\{u,v\}$ is a basis of R and the product in R is any commutative product. If $a_1 \neq 0$ in (4.15), then we let $a = a_1 = \frac{1}{4}\gamma u$ for some $\gamma \neq 0$ in F and form a basis $\{u,v\}$ of R for which we set $\tau(u) = \alpha$ and $\tau(v) = \beta$. Recall from (4.19) the Killing form of $s\ell(2)$ relative to $\{e,h,f\}$. Thus, in these two cases, by (4.15) A is determined by Tables 5.2 and 5.3.

TABLE 5.2

	e	h	f	u	v
e	0	$-e$	$\frac{1}{2}h$	αe	βe
h	e	0	$-f$	αh	βh
f	$-\frac{1}{2}h$	f	0	αf	βf
u	αe	αh	αf	any commutative	
v	βe	βh	βf	product in R	

TABLE 5.3

	e	h	f	u	v
e	0	$-e$	$\frac{1}{2}h + \gamma u$	αe	βe
h	e	$2\gamma u$	$-f$	αh	βh
f	$-\frac{1}{2}h + \gamma u$	f	0	αf	βf
u	αe	αh	αf	any commutative	
v	βe	βh	βf	product in R	

We note that algebras in Tables 5.2 and 5.3 are flexible Lie-admissible by Theorem 4.3 and those of Table 5.3 are simple by Corollary 4.7(3) if $\alpha \neq 0$, $\beta \neq 0$, and R is simple.

We now consider the case where $[s\ell(2),R] \neq 0$ and R^- is abelian. Thus, R is an irreducible $s\ell(2)$-module under the adjoint action. Assume that R is a Lie module for $s\ell(2)$. Then, by (5.2) there is a basis $\{u,v\}$ of R such that

$$(5.24)\qquad \begin{aligned} &[h,u] = u,\ [h,v] = -v,\ [f,u] = v\ , \\ &[f,v] = 0,\ [e,u] = 0,\ [e,v] = u\ , \end{aligned}$$

since $R = V(1)$. By Clebsch-Gordan formula, $s\ell(2) \otimes s\ell(2) = V(2) \otimes V(2) = V(4) + V(2) + V(0)$ and $R \otimes s\ell(2) \cong s\ell(2) \otimes R = V(3) + V(1)$, which by Lemma 3.1 imply that the products between $s\ell(2)$ and $s\ell(2)$, and between $s\ell(2)$ and R are multiples of $[\ ,\ ]$. Since the attached minus algebra of A is the same as in A^-, we have $xy = \frac{1}{2}[x,y]$ and $xa = \frac{1}{2}[x,a]$ for $x,y \in s\ell(2)$ and $a \in R$. But, by Lemma 4.19 $RR = R_*R_* = 0$, and hence by (5.24) A is the unique algebra given by Table 5.4.

TABLE 5.4

	e	h	f	u	v
e	0	-e	$\frac{1}{2}h$	0	$\frac{1}{2}u$
h	e	0	-f	$\frac{1}{2}u$	$-\frac{1}{2}v$
f	$-\frac{1}{2}h$	f	0	$\frac{1}{2}v$	0
u	0	$-\frac{1}{2}u$	$-\frac{1}{2}v$	0	0
v	$-\frac{1}{2}u$	$\frac{1}{2}v$	0	0	0

We note in this case that A is a Lie algebra isomorphic to A^-.

Assume finally that R is an irreducible non-Lie Malcev module for $s\ell(2)$. Then, by (3.25) R has a basis $\{u,v\}$ such that

$$\begin{aligned} &[h,u] = 2u,\ [h,v] = -2v,\ [e,u] = 2v\ , \\ &[e,v] = 0,\ [f,v] = 2u,\ [f,u] = 0\ . \end{aligned} \tag{5.25}$$

Since R is isomorphic to V(1) as a $d(s\ell(2), s\ell(2))$-module (see Section 3.3), as above we have $xy = \frac{1}{2}[x,y]$ and $ax = \frac{1}{2}[a,x]$ for $x,y \in s\ell(2)$, $a \in R$. It follows from this and (5.25) that A is a non-Lie, Malcev algebra isomorphic to A^- and is given by Table 5.5.

TABLE 5.5

	e	h	f	u	v
e	0	$-e$	$\frac{1}{2}h$	v	0
h	e	0	$-f$	u	$-v$
f	$-\frac{1}{2}h$	f	0	0	u
u	$-v$	$-u$	0	0	0
v	0	v	$-u$	0	0

The algebra of Table 5.5 is the unique non-Lie, non-solvable Malcev algebra of dimension 5 which has been determined by Kuzmin [2] . We summarize our classification in

<u>Theorem 5.3</u>. Let A be a flexible Malcev-admissible algebra of dimension 5 over an algebraically closed field of characteristic 0 such that A^- is not solvable. Then, A is one of the algebras described by Tables 5.1-5.5, and all but the algebra of Table 5.5 are flexible Lie-admissible. □

5.3. DIMENSION 6

We determine all flexible Malcev-admissible algebras A of dimension 6 over F such that A^- is not solvable and such that the solvable radical R of A^- is nilpotent when R^- is not a direct summand of A^-. Thus, A^- has a Levi decomposition

$$A^- = \mathfrak{sl}(2) \oplus \mathfrak{sl}(2) \quad \text{or} \quad A^- = \mathfrak{sl}(2) + R ,$$

where R^- is a 3-dimensional Malcev algebra which must be Lie. Let $\tilde{M}_2$ denote the 2-dimensional non-Lie, $\mathfrak{sl}(2)$-module given by (5.25). We proceed by case. If $A^- = \mathfrak{sl}(2) + \mathfrak{sl}(2)$, then A is Lie and is isomorphic to A^- by Corollary 3.23. Hence, we begin with

<u>Case 1</u>. R^- is abelian and $[\mathfrak{sl}(2),R] \neq 0$. Thus, R decomposes as

$$R = V(1) + V(0),\ R = \tilde{M}_2 + V(0),\ \text{or}\ R = V(2) ,$$

as an $\mathfrak{sl}(2)$-module. In the first two cases, let $\{u,v\}$ be a basis of $V(1)$ and $\tilde{M}_2$ satisfying relations (5.24) and (5.25), respectively. Then, $[V(0),A^-] = 0$, and let $V(0) = Fw$. By Lemma 4.19, $0 = R_*R_* = V(1)V(1) = \tilde{M}_2\tilde{M}_2$, and by Lemma 4.12 $V(1)\mathfrak{sl}(2)$ and $\tilde{M}_2\mathfrak{sl}(2)$ are contained in R. Hence, it follows from (5.4) that $ax = - xa = \frac{1}{2}[a,x]$ for $x \in \mathfrak{sl}(2)$, $a \in V(1)$ or $a \in \tilde{M}_2$. Since $V(0)$ is a subalgebra of A, we can let $w^2 = \beta w$ for some $\beta \in F$. By Lemma 4.13(3), we have $xw = wx = \gamma x$ for $x \in \mathfrak{sl}(2)$ and $aw = wa = \delta a$ for $a \in V(1)$ or $a \in \tilde{M}_2$. Since $\mathfrak{sl}(2) \otimes \mathfrak{sl}(2) = V(4) + V(2) + V(0)$,

$$xy = \frac{1}{2}[x,y] + \alpha K(x,y)w,\ \alpha \in F$$

for $x,y \in s\ell(2)$, where the Killing form $K(\ ,\ )$ is given by (4.19). When $R = V(1) + V(0)$, we obtain $\gamma = \delta$ from (5.24) and flexibility $(v,e,w) + (w,e,v) = 0$. The same holds for the case $R = \tilde{M}_2 + V(0)$, using (5.25) and flexibility $(w,e,u) + (u,e,w) = 0$. Therefore, replacing w by $\frac{1}{4}w$ and using (4.19), (5.24) and (5.25), we see that A in these cases is given by Tables 5.6 and 5.7.

TABLE 5.6

	e	h	f	u	v	w
e	0	$-e$	$\frac{1}{2}h+\alpha w$	0	$\frac{1}{2}u$	γe
h	e	$2\alpha w$	$-f$	$\frac{1}{2}u$	$-\frac{1}{2}v$	γh
f	$-\frac{1}{2}h+\alpha w$	f	0	$\frac{1}{2}v$	0	γf
u	0	$-\frac{1}{2}u$	$-\frac{1}{2}v$	0	0	γu
v	$-\frac{1}{2}u$	$\frac{1}{2}v$	0	0	0	γv
w	γe	γh	γf	γu	γv	βw

TABLE 5.7

	e	h	f	u	v	w
e	0	$-e$	$\frac{1}{2}h+\alpha w$	v	0	γe
h	e	$2\alpha w$	$-f$	u	$-v$	γh
f	$-\frac{1}{2}h+\alpha w$	f	0	0	u	γf
u	$-v$	$-u$	0	0	0	γu
v	0	v	$-u$	0	0	γv
w	γe	γh	γf	γu	γv	βw

We note that the algebras of Table 5.6 are Lie-admissible and those of Table 5.7 are non-Lie, Malcev-admissible.

Consider next the case $R = V(2)$. Thus, R is isomorphic to $s\ell(2)$

as a module, and A is flexible Lie-admissible. This is the simplest case of Theorem 4.22, and $A = s\ell(2) \otimes P$. Since the product $x \# y$ defined by (3.19) is zero and $R_0 = 0$, we can assume that P is a 2-dimensional commutative associative algebra with basis $\{q,r\}$ and with product "$*$" given by

$$r * r = 0, \; q * r = \frac{1}{2}r, \; q * q = \frac{1}{2}q \; . \tag{5.26}$$

Therefore, by Theorem 4.22, $A = s\ell(2) \otimes P$ has the multiplication

$$(x \otimes a)(y \otimes b) = [x,y] \otimes a * b \tag{5.27}$$

for $x,y \in s\ell(2)$ and $a,b \in P$. Letting $u = e \otimes r$, $v = h \otimes r$, $w = f \otimes r$, and $e = e \otimes q$, $h = h \otimes q$, $f = f \otimes q$, we see from (5.26) and (5.27) that A is described by Table 5.8.

TABLE 5.8

	e	h	f	u	v	w
e	0	$-e$	$\frac{1}{2}h$	0	$-u$	$\frac{1}{2}v$
h	e	0	$-f$	u	0	$-w$
f	$-\frac{1}{2}h$	f	0	$-\frac{1}{2}v$	w	0
u	0	$-u$	$\frac{1}{2}v$	0	0	0
v	u	0	$-w$	0	0	0
w	$-\frac{1}{2}v$	w	0	0	0	0

In fact, the algebra of Table 5.8 is a Lie algebra isomorphic to A^-.

<u>Case 2</u>. R^- is abelian and $[s\ell(2),R] = 0$. In this case, A is determined by Theorem 4.3 with $n = 1$, while R is an arbitrary commutative algebra over F. If $a_1 = 0$ in (4.15), then we can choose a basis

$\{u,v,w\}$ of R such that $\tau(u) = \alpha \in F$ and $\tau(v) = \tau(w) = 0$. Thus, when $a_1 = 0$, A is given by

$$xy = -yx = \frac{1}{2}[x,y] ,$$

$$xu = ux = \alpha x,\ xv = vx = xw = wx = 0 , \tag{5.28}$$

$$R \text{ is any commutative algebra over } F ,$$

where $x,y \in \mathfrak{sl}(2)$ and $\alpha \in F$. Assume $a_1 \neq 0$. Then, we choose a basis $\{u = a_1, v, w\}$ of R. Using (4.15) and the Killing form of (4.19), we have A described by

$$ee = ff = 0,\ eh = -he = -e,\ hf = -fh = -f ,$$

$$ef = \frac{1}{2}h + u,\ fe = -\frac{1}{2}h + u,\ h^2 = 2u , \tag{5.29}$$

$$xu = ux = \alpha x,\ xv = vx = \beta x,\ xw = wx = \gamma x ,$$

$$R \text{ is any commutative algebra over } F ,$$

where $x \in \mathfrak{sl}(2)$ and $\alpha, \beta, \gamma \in F$. In algebras of (5.28), $\mathfrak{sl}(2)$ is an ideal, but algebras of (5.29) are simple by Corollary 4.7(3) if R is simple and one of α, β, γ is nonzero. In both cases, A is flexible Lie-admissible.

<u>Case 3</u>. R^- is nonabelian and $[\mathfrak{sl}(2), R] = 0$. Then, it is known (Jacobson [2,p.12]) that there is a basis $\{u,v,w\}$ of R such that R^- is one of the solvable Lie algebras given by

$$[v,w] = u,\ [u,v] = [u,w] = 0 , \tag{5.30}$$

$$[u,v] = u,\ [u,w] = [v,w] = 0 , \tag{5.31}$$

(5.32) $$[u,w] = u,\ [v,w] = \alpha v,\ [u,v] = 0,\ \alpha \neq 0 \text{ in } F ,$$

(5.33) $$[u,w] = u + \beta v,\ [v,w] = v,\ [u,v] = 0,\ 0 \neq \beta \in F .$$

All algebras but (5.30) are nilpotent. If R^- is given by (5.30), then $[R,R] = Fu = Z(A^-)$, the center of A^-. If R^- is given by (5.31), then $[R,R] = Fu$ but $Z(A^-) = Fw$. In cases of (5.32) and (5.33), $Z(A^-) = 0$ and $[R,R] = Fu + Fv$.

Consider cases of (5.30) and (5.31), so that $\tau(u) = 0$ in (4.15), and either $a_1 = 0$ or $a_1 = \alpha u$ for $\alpha \neq 0$ in F. Let $\tau(v) = \beta$ and $\tau(w) = \gamma$. If $a_1 = 0$, then in both cases A is described by

(5.34) $$\begin{aligned} &xy = -yx = \tfrac{1}{2}[x,y] ,\\ &xu = ux = 0,\ xv = vx = \beta x,\ xw = wx = \gamma x \end{aligned}$$

for $x,y \in \mathfrak{sl}(2)$. If $a_1 \neq 0$, then by (4.19) A is given by

(5.35) $$\begin{aligned} &ee = ff = 0,\ eh = -he = -e,\ hf = -fh = -f ,\\ &ef = \tfrac{1}{2}h + \alpha u,\ fe = -\tfrac{1}{2}h + \alpha u,\ h^2 = 2\alpha u ,\\ &xu = ux = 0,\ xv = vx = \beta x,\ xw = wx = \gamma x , \end{aligned}$$

where $x \in \mathfrak{sl}(2)$ and $0 \neq \alpha \in F$. Assume that R^- is given by (5.32) or (5.33). Since $Z(R^-) = 0$, $a_1 = 0$ in (4.15) and $\tau(u) = \tau(v) = 0$. Thus, for $x,y \in \mathfrak{sl}(2)$, A is described by

(5.36) $$\begin{aligned} &xy = -yx = \tfrac{1}{2}[x,y] ,\\ &xu = ux = vx = xv = 0,\ xw = wx = \alpha x,\ \alpha \in F . \end{aligned}$$

In the case where R^- is nonabelian and $[\mathfrak{sl}(2),R] = 0$, we have determined the algebra A up to the radical of A^-. Since R is a subalgebra of A, R is arbitrary subject only to flexible Lie-admissibility and conditions (5.30)-(5.33). Algebras of (5.34) are not simple while those of (5.35) are simple if R is simple and $\beta \neq 0$ or $\gamma \neq 0$ (Corollary 4.7). Each of Tables (5.34) and (5.35) gives rise to two non-isomorphic classes of algebras, according as case (5.30) or (5.31) holds.

Case 4. R^- is nonabelian nilpotent and $[\mathfrak{sl}(2),R] \neq 0$. Thus, R^- is given by (5.30) and $[R,R] = Fu$. Since $[R,R]$ is an $\mathfrak{sl}(2)$-submodule of R, $R_0 = Fu$ and hence R has the decomposition $R = V(1) + V(0)$ or $R = \tilde{M}_2 + V(0)$, where $V(0) = Fu$. We can choose a basis $\{v,w\}$ of $V(1)$ or $\tilde{M}_2$ such that $\{u,v,w\}$ is a basis of R satisfying (5.30) and (5.24) or (5.25). Consider first the case $R = V(1) + V(0)$. From the decomposition of $\mathfrak{sl}(2) \otimes V(1)$, we have $ax = -xa = \frac{1}{2}[a,x]$ for $x \in \mathfrak{sl}(2)$, $a \in V(1)$, so that $[a,x]$ is determined by (5.24). Since $0 \neq V(1)V(1) \subseteq [R,R] + R_0 = R_0 = Fu$ by Corollary 4.13(4) and both the mappings $a \otimes b \to ab$ and $a \otimes b \to [a,b]$ are elements of $\mathrm{Hom}_{\mathfrak{sl}(2)}(V(1) \otimes V(1), V(0))$, we obtain $vw = -wv = \frac{1}{2}[v,w] = \frac{1}{2}u$ and $v^2 = w^2 = 0$ from (5.30). From the decompositions of $\mathfrak{sl}(2) \otimes \mathfrak{sl}(2)$, $\mathfrak{sl}(2) \otimes V(0)$ and $V(1) \otimes V(0)$,

$$xy = \tfrac{1}{2}[x,y] + \alpha K(x,y)u, \quad xu = ux = \beta x, \quad au = ua = \gamma a \tag{5.37}$$

for $x,y \in \mathfrak{sl}(2)$, $a \in V(1)$ and $\alpha,\beta,\gamma \in F$. Since $V(0)$ is a subalgebra of A, $u^2 = \delta u$ for $\delta \in F$. But then, flexibility $(w,e,u) + (u,e,w) = 0$ implies $\beta = \gamma$, which together with $(vw)v = v(wv)$ gives $\beta = \gamma = 0$. Similarly, we have $\delta = 0$ from $(v,w,u) + (u,w,v) = 0$, Therefore,

$$\begin{aligned} &xy = \tfrac{1}{2}[x,y] + \alpha K(x,y)u, \quad xa = -ax = \tfrac{1}{2}[x,a], \\ &xu = ux = au = ua = u^2 = 0 \end{aligned} \tag{5.38}$$

for $x, y \in s\ell(2)$, $a \in V(1)$ and $\alpha \in F$. The same arguments apply to the case $R = \tilde{M}_2 + V(0)$, and (5.38) holds for this case also, except the fact that the products $xa = -ax = \frac{1}{2}[x,a]$ are determined by (5.25). Therefore, using the Killing form in (4.19) and (5.24), (5.25), we have A described by Tables 5.9 and 5.10.

TABLE 5.9

	e	h	f	u	v	w
e	0	$-e$	$\frac{1}{2}h+\alpha u$	0	0	$\frac{1}{2}v$
h	e	$2\alpha u$	$-f$	0	$\frac{1}{2}v$	$-\frac{1}{2}w$
f	$-\frac{1}{2}h+\alpha u$	f	0	0	$\frac{1}{2}w$	0
u	0	0	0	0	0	0
v	0	$-\frac{1}{2}v$	$-\frac{1}{2}w$	0	0	$\frac{1}{2}u$
w	$-\frac{1}{2}v$	$\frac{1}{2}w$	0	0	$-\frac{1}{2}u$	0

TABLE 5.10

	e	h	f	u	v	w
e	0	$-e$	$\frac{1}{2}h+\alpha u$	0	w	0
h	e	$2\alpha u$	$-f$	0	v	$-w$
f	$-\frac{1}{2}h+\alpha u$	f	0	0	0	v
u	0	0	0	0	0	0
v	$-w$	$-v$	0	0	0	$\frac{1}{2}u$
w	0	w	$-v$	0	$-\frac{1}{2}u$	0

Algebras in Table 5.9 are Lie-admissible but those in Table 5.10 are non-Lie, Malcev-admissible. We summarize the algebras determined in

Theorem 5.4. Let A be a flexible Malcev-admissible algebra over F of dimension 6 such that A^- is not solvable and such that the radical R^- of A is nilpotent if R^- is not a direct summand of A^-. Then, A is either a Lie algebra isomorphic to $\mathfrak{sl}(2) \oplus \mathfrak{sl}(2)$, or isomorphic to one of the algebras described by Tables 5.6-5.8, (5.28), (5.29), (5.34), (5.35), (5.36), and Tables 5.9 and 5.10. No algebras from different lists are isomorphic. Each of (5.34) and (5.35) gives two non-isomorphic classes of algebras, according as R^- is given by (5.30) or (5.31), and the algebras in (5.36) form two non-isomorphic classes of algebras, according as R^- is given by (5.32) or (5.33). All algebras except the ones given by Tables 5.7 and 5.10 are flexible Lie-admissible. □

5.4. Dimension 7

We determine all flexible Malcev-admissible algebras of dimension 7 over F such that A^- is not solvable and such that the radical R^- of A^- is abelian when a Levi factor of A^- is $\mathfrak{sl}(2)$. If A^- is semisimple, then since there is no semisimple Lie algebra of dimension 7 over F, it must be that $A^- \cong C_0^-$ and hence A is a Malcev algebra isomorphic to C_0^-. In the remaining cases, A^- has a Levi decomposition $A^- = \mathfrak{sl}(2) + \mathfrak{sl}(2) + R^-$ or $A^- = \mathfrak{sl}(2) + R^-$.

We consider first the case $A^- = \mathfrak{sl}(2) + \mathfrak{sl}(2) + R$. Then, $R = Fu$ is a trivial module for $\mathfrak{sl}(2) + \mathfrak{sl}(2)$, and hence is a subalgebra of A. Let $S_1 = \mathfrak{sl}(2)$ and $S_2 = \mathfrak{sl}(2)$, and let $\{e_i,h_i,f_i\}$ denote the standard basis of S_i. Then, the products in A are determined by (4.15) of Theorem 4.3. Thus, from (4.19) we obtain

$$(5.39)\quad \begin{aligned} &e_i^2 = f_i^2 = 0,\ e_i h_i = -h_i e_i = -e_i,\ h_i f_i = -f_i h_i = -f_i\ , \\ &e_i f_i = \tfrac{1}{2}h_i + \alpha_i u,\ f_i e_i = -\tfrac{1}{2}h + \alpha_i u,\ h_i^2 = 2\alpha_i u,\ i = 1,\ 2\ , \\ &u^2 = \beta u,\ xy = yx = 0 \quad \text{for} \quad x \in S_1,\ y \in S_2\ , \\ &x_i u = u x_i = \beta_i x_i,\ x_i \in S_i,\ i = 1,\ 2\ , \end{aligned}$$

where α_i, β, $\beta_i \in F$.

Consider now the case $A^- = s\ell(2) + R$ and R^- is abelian. Assume $[s\ell(2),R] = 0$. As above, we are in the situation of Theorem 4.3 with $n = 1$. Let $a_1 = \alpha v_0$ for some $0 \neq v_0 \in R$ and $\alpha \in F$ in (4.15). There is a basis $\{v_0,v_1,v_2,v_3\}$ of R such that $\tau(v_0) = \beta_0$, $\tau(v_1) = \beta_1$ but $\tau(v_2) = \tau(v_3) = 0$. Thus, A is described by

$$(5.40)\quad \begin{aligned} &ee = ff = 0,\ eh = -he = -e,\ hf = -fh = -f\ , \\ &ef = \tfrac{1}{2}h + \alpha v_0,\ fe = -\tfrac{1}{2}h + \alpha v_0,\ h^2 = 2\alpha v_0\ , \\ &xv_i = v_i x = \beta_i x\ (i = 0,1),\ xv_i = v_i x = 0\ (i = 2,3),\ x \in s\ell(2)\ , \\ &R \text{ is any commutative algebra over } F\ . \end{aligned}$$

Assume then that $[s\ell(2),R] \neq 0$. The irreducible $s\ell(2)$-module decomposition for R has the following possibilities.

$$(5.41)\qquad R = V(3)\ ,$$

$$(5.42)\qquad R = V(2) + V(0)\ ,$$

$$(5.43)\qquad R = V(1) + V(1)\ ,$$

$$(5.44)\qquad R = \tilde{M}_2 + \tilde{M}_2\ ,$$

(5.45) $$R = V(1) + \tilde{M}_2 ,$$

(5.46) $$R = V(1) + V(0) + V(0) ,$$

(5.47) $$R = \tilde{M}_2 + V(0) + V(0) .$$

The case (5.41) is rather routine. By Lemma 4.19 $RR = 0$, and from Clebsch-Gordan formula, it follows that A is a Lie algebra isomorphic to A^-. Thus,

(5.48) $$xy = -yx = \frac{1}{2}[x,y], \quad x,y \in A .$$

We note that the products between $s\ell(3)$ and $V(3)$ are determined by (5.2).

Assume (5.42) is the case. Since $s\ell(2) \cong V(2)$, A is described by the construction in Theorem 4.22. Thus, we can let $A = s\ell(2) \otimes P + Fu$, where $R_0 = Fu$ and $P = Fq + Fr$ is a 2-dimensional algebra over F with product "$*$" given by (5.26). The products in $s\ell(2) \otimes P$ are given by (4.48), where $x \# y = 0$ for all $x,y \in s\ell(2)$ and $\phi : P \times P \to R_0$ is a symmetric bilinear mapping satisfying the property $\phi(a * b,c) = \phi(a,b * c)$ for all $a,b,c \in P$. The product $(x \otimes a)u$ is described by (4.68), where T is a linear mapping of R_0 into P. Using (4.55) and (5.26), we have $\frac{1}{2}\phi(r,r) = \phi(q * r,r) = \phi(q,r * r) = 0$ and $\phi(q,r) = \frac{1}{2}\phi(q * q,r) = \frac{1}{2}\phi(q,q * r) = \frac{1}{4}\phi(q,r)$, so that $\phi(q,r) = 0$. Let $\phi(q,q) = \alpha u$, $T(u) = \beta q + \gamma r$ for some $\alpha,\beta \in F$, and identify $e \otimes q$, $h \otimes q$, $f \otimes q$, $e \otimes r$, $h \otimes r$, $f \otimes r$ with e, h, f, $\tilde{e}$, $\tilde{h}$, $\tilde{f}$, respectively. Noting that Fu is a subalgebra of A, so that $u^2 = \delta u$, the products in (4.48) and (4.68) translate into Table 5.11.

TABLE 5.11

	e	h	f	$\tilde{e}$	$\tilde{h}$	$\tilde{f}$	u
e	0	$-e$	$\frac{1}{2}h+\alpha u$	0	$-\tilde{e}$	$\frac{1}{2}\tilde{h}$	$\frac{1}{2}(\beta e+\gamma\tilde{e})$
h	e	$2\alpha u$	$-f$	$\tilde{e}$	0	$-\tilde{f}$	$\frac{1}{2}(\beta h+\gamma\tilde{h})$
f	$-\frac{1}{2}h+\alpha u$	f	0	$-\frac{1}{2}\tilde{h}$	$\tilde{f}$	0	$\frac{1}{2}(\beta f+\gamma\tilde{f})$
$\tilde{e}$	0	$-\tilde{e}$	$\frac{1}{2}\tilde{h}$	0	0	0	$\frac{1}{2}\beta\tilde{e}$
$\tilde{h}$	$\tilde{e}$	0	$-\tilde{f}$	0	0	0	$\frac{1}{2}\beta\tilde{h}$
$\tilde{f}$	$-\frac{1}{2}\tilde{h}$	$\tilde{f}$	0	0	0	0	$\frac{1}{2}\beta\tilde{f}$
u	$\frac{1}{2}(\beta e+\gamma\tilde{e})$	$\frac{1}{2}(\beta h+\gamma\tilde{h})$	$\frac{1}{2}(\beta f+\gamma\tilde{f})$	$\frac{1}{2}\beta\tilde{e}$	$\frac{1}{2}\beta\tilde{h}$	$\frac{1}{2}\beta\tilde{f}$	δu

We note that the algebra given by Table 5.11 is flexible Lie-admissible by Theorem 4.22.

Consider the cases (5.43)-(5.45), and let U, V denote the summands of R. We note first that $RR = 0$ by Lemma 4.19 and $xy = -yx = \frac{1}{2}[x,y]$ for $x,y \in s\ell(2)$. Since both U and V are not isomorphic to $s\ell(2)$, from Corollary 4.18

$$U \circ s\ell(2) \subseteq [s\ell(2) \circ s\ell(2), U] \subseteq U, \quad V \circ s\ell(2) \subseteq [s\ell(2) \circ s\ell(2), V] \subseteq V$$

in these cases. Hence, $s\ell(2)U \subseteq U$ and $s\ell(2)V \subseteq V$, which by Clebsch-Gordan formula implies $ax = -xa = \frac{1}{2}[a,x]$ for $x \in s\ell(2)$ and $a \in U$ or $a \in V$. Therefore, in cases (5.43)-(5.45), the products in A are given by

$$xy = -xy = \frac{1}{2}[x,y], \quad x,y \in A, \tag{5.49}$$

and hence A is isomorphic to A^-. The products between $s\ell(2)$ and R are described by (5.24) and (5.25). In case (5.43), A is a Lie algebra, and in cases (5.44) and (5.45) A is a non-Lie, Malcev algebra.

For the remaining cases (5.46) and (5.47), let $\{v_0,v_1,v_2,v_3\}$ be a basis for R such that $\{v_2,v_3\}$ is a basis for $R_0 = V(0) + V(0)$ and $\{v_0,v_1\}$ is the basis for $V(1)$ or $\tilde{M}_2$ satisfying (5.24) or (5.25), respectively. From $s\ell(2) \otimes s\ell(2) = V(4) + V(2) + V(0)$, we have $xy = \frac{1}{2}[x,y] + \alpha K(x,y)v_2 + \beta K(x,y)v_3$ for $x,y \in s\ell(2)$. Similarly, $V(1) \otimes s\ell(2) = V(3) + V(1)$ implies that $xa = - ax = \frac{1}{2}[x,a]$ for $x \in s\ell(2)$ and $a \in V(1)$ or $a \in \tilde{M}_2$. Now, $xv_2 = v_2x = \gamma x$, $av_2 = v_2a = \delta a$ for $x \in s\ell(2)$, $a \in V(1)$ or $a \in \tilde{M}_2$, $\gamma,\delta \in F$, but then the flexible law $(x,v_2,a) + (a,v_2,x) = 0$ gives $\gamma = \delta$. The same argument shows that $xv_3 = v_3x = \lambda x$ and $av_3 = v_3a = \lambda a$. Since $V(1)V(1) = \tilde{M}_2\tilde{M}_2 = 0$, using (4.19) the products in A are given by

$$e^2 = f^2 = 0,\ he = - eh = e,\ hf = - fh = - f,\ ef = \frac{1}{2}h + \alpha v_2 + \beta v_3\ ,$$

$$h^2 = 2(\alpha v_2 + \beta v_3),\ fe = - \frac{1}{2}h + \alpha v_2 + \beta v_3\ ,$$

$$\text{(5.50)} \qquad xa = - ax = \frac{1}{2}[x,a],\ x \in s\ell(2),\ a \in V(1) \text{ or } a \in \tilde{M}_2\ ,$$

$$v_0v_1 = v_1v_0 = v_0^2 = v_1^2 = 0\ ,$$

$$xv_2 = v_2x = \gamma x,\ xv_3 = v_3x = \lambda x,\ x \in s\ell(2) + V(1) \text{ or } x \in s\ell(2) + \tilde{M}_2\ ,$$

$$V(0) + V(0) \text{ is any commutative algebra over } F\ .$$

The products xa for $x \in s\ell(2)$, $a \in V(1)$ or $a \in \tilde{M}_2$ are determined by (5.24) or (5.25). We note that A in case (5.46) is Lie-admissible, but is non-Lie, Malcev-admissible in case (5.47).

Therefore, we have proved

<u>Theorem 5.5.</u> Let A be a flexible Malcev-admissible algebra over F of dimension 7 such that A^- is not solvable and such that the radical

of A^- is abelian when a Levi factor of A^- is $\mathfrak{sl}(2)$. Then, A is either isomorphic to C_0^-, or to one of the algebras described by (5.39), (5.40), (5.48), Table 5.11, (5.49), and (5.50). In cases (5.39), (5.40), (5.48) and Table 5.11, A is flexible Lie-admissible, and is a Lie algebra isomorphic to A^- for case (5.48). Relation (5.49) lists three non-isomorphic algebras A all isomorphic to A^-, according as case (5.43), (5.44), or (5.45), and the first case gives a Lie algebra and the remaining two cases are non-Lie, Malcev algebras. Relation (5.50) provides two non-isomorphic classes of algebras, and the first class is flexible Lie-admissible for case (5.46) while the second class is non-Lie, Malcev-admissible for case (5.47). □

The results in Theorems 5.3-5.5 have been proved independently by Myung [5] and Malek and Micali [1].

We conclude with the construction of a class of 7-dimensional non-associative algebras A over F which are acted on by $\mathfrak{sl}(3)$ as derivations, and which have the direct sum decomposition

$$A = U + V + Fa \,, \tag{5.51}$$

as $\mathfrak{sl}(3)$-modules where U and V are the contragredient 3-dimensional irreducible $\mathfrak{sl}(3)$-modules, and Fa is a trivial module. Denote $L = \mathfrak{sl}(3)$, and note by Lemma 5.1 that L is isomorphic to the Lie algebra D_0 of derivations of C which vanish on an idempotent $e_1 \neq 1$. We note also that the relations between the standard basis $e_0, e_{\pm i}$ $(i = 1,2,3)$ of C_0^- given by (3.24) and the basis of C given by (5.9) are $e_i = -u_i$, $e_{-j} = v_j$, and $e_0 = e_1 - e_2$, $i, j = 1, 2, 3$. Since $s \cdot e_0 = 0$ for all $s \in L$, we see that $A \cong C_0^-$ as an L-module. Thus, letting $a = e_1 - e_2$, the multiplication of C_0^- in (3.24) is described by

$$[a,u_i] = 2u_i,\ [a,v_j] = -2v_j,\ [u_i,v_j] = -\delta_{ij}a\ ,$$

(5.52)

$$[u_i,u_j] = 2\varepsilon_{ijk}v_k,\ [v_i,v_j] = 2\varepsilon_{ijk}u_k$$

for $i, j, k = 1, 2, 3$, where the u_i's and v_j's are bases for U and V. In this way, we identify A with C_0^- as an L-module.

As in Chapters 3 and 4, the condition that $L \subseteq \mathrm{Der}\,A$ imposes fairly restrictive conditions on A. For example, by (3.11) and (4.14), $Ua \subseteq U$ and there exists a constant $\nu \in F$ such that $u_i a = \nu u_i$. Similarly, there exist constants $\nu', \xi, \xi', \phi \in F$ such that $au_i = \nu' u_i$, $v_j a = \xi v_j$ and $av_j = \xi' v_j$, and $a^2 = \phi a$. Since we can let $U = V(\lambda_1)$ and $V = V(\lambda_2)$, from (5.5)

$$\mathrm{Hom}_L(U \otimes U,U) = \mathrm{Hom}_L(V \otimes V,V) = \mathrm{Hom}_L(U \otimes V,U) = \mathrm{Hom}_L(U \otimes V,V)$$

$$= \mathrm{Hom}_L(U \otimes U,Fa) = \mathrm{Hom}_L(V \otimes V,Fa) = 0\ ,$$

(5.53)

$$\dim \mathrm{Hom}_L(U \otimes V,Fa) = \dim \mathrm{Hom}_L(V \otimes U,Fa)$$

$$= \dim \mathrm{Hom}_L(U \otimes U,V) = \dim \mathrm{Hom}_L(V \otimes V,U) = 1\ .$$

It follows from (5.9) that the mappings $u_i \otimes u_j \to \varepsilon_{ijk}v_k$, $v_i \otimes v_j \to \varepsilon_{ijk}u_k$, $u_i \otimes v_j \to \delta_{ij}a$, and $v_i \otimes u_j \to \delta_{ij}a$ induce nonzero L-homomorphisms of $U \otimes U$, $V \otimes V$, $U \otimes V$, and $V \otimes U$ into V, U, Fa, and Fa, respectively. Hence, under the condition that $L \subseteq \mathrm{Der}\,A$, the multiplication in A is given by Table 5.12, where $\nu, \nu', \xi, \xi', \phi, \lambda, \rho, \sigma, \tau \in F$. Thus, A^- has multiplication given by Table 5.13.

TABLE 5.12

	u_j	v_j	a
u_i	$\lambda\varepsilon_{ijk}v_k$	$\sigma\delta_{ij}a$	νu_i
v_i	$\tau\delta_{ij}a$	$\rho\varepsilon_{ijk}u_k$	ξv_i
a	$\nu' u_j$	$\xi' v_j$	ϕa

TABLE 5.13

	u_j	v_j	a
u_i	$2\lambda\varepsilon_{ijk}v_k$	$(\sigma-\tau)\delta_{ij}a$	$(\nu-\nu')u_i$
v_i	$-(\sigma-\tau)\delta_{ij}a$	$2\rho\varepsilon_{ijk}u_k$	$(\xi-\xi')v_i$
a	$-(\nu-\nu')u_j$	$-(\xi-\xi')v_j$	0

The special case of the algebra of Domokos and Kövesi-Domokos [1] is obtained from Table 5.12 when $\nu = \nu' = \xi = \xi' = \phi = \lambda = \rho = \sigma = \tau = 1$. They claim erroneously that the algebra is Malcev-admissible. In fact, $[[u_1,u_2],[u_1,u_3]] = u_1$ but $[u_1,u_1,u_2,u_3] + [u_1,u_2,u_3,u_1] + [u_2,u_3,u_1,u_1] + [u_3,u_1,u_1,u_2] = 0 + 0 + 0 + 0 = 0$. Thus the Malcev identity (3.20) does not hold for A^- . The construction of Domokos and Kövesi-Domokos is based on the "triality rule" that the products of dynamical variables corresponding to quarks and antiquarks are singlets under SU(3) . Table 5.12 shows that the triality rule is regarded as a special case of the decomposition of $V \otimes U$ and our general results developed in Chapters 3 and 4.

Let $m(x,z,y,t) = 0$ denote the Malcev identity $[[x,z],[y,t]] - [x,y,z,t] - [y,z,t,x] - [z,t,x,y] - [t,x,y,z] = 0$, where $[x,y,z,t] = [[[x,y],z],t]$. We investigate which algebras of our class satisfy the

Malcev identity. This can generally be determined by substituting all possible combinations of basis elements into $m(x,z,y,t)$ and using Table 5.12 or 5.13. This results in a set of conditions on the contants. The computations are easier for A than for an arbitrary algebra of dimension 7 because of the symmetries possessed by A and because of known information about Malcev algebras.

Theorem 5.6. Let A be a 7-dimensional $\mathfrak{sl}(3)$-module over F decomposed as in (5.51), and let $L = \mathfrak{sl}(3)$. Then, A is a nonassociative algebra over F which is acted on by L as derivations if and only if the multiplication in A is given by Table 5.12. If A is Malcev-admissible, then A^- is either isomorphic to C_0^- or is solvable. Furthermore,

(1) If A^- is a nonsolvable Malcev algebra, then $\lambda\rho \neq 0$, $\sigma - \tau \neq 0$, $\xi - \xi' = -(\nu - \nu') \neq 0$, and the basis given by

$$\bar{u}_i = \sqrt[3]{\frac{2}{\lambda(\nu-\nu')(\sigma-\tau)}}\, u_i, \quad \bar{v}_j = \sqrt[3]{\frac{2}{\rho(\nu-\nu')(\sigma-\tau)}}\, v_j, \quad \bar{a} = \frac{-2}{\nu-\nu'}\, a$$

for $i, j = 1, 2, 3$ forms a standard basis satisfying (5.52).

(2) If A^- is a solvable Malcev algebra, then $\lambda\rho = (\sigma - \tau)(\xi - \xi')\cdot(\nu - \nu') = 0$, and $(A^-)^{(3)} = 0$, where $(A^-)^{(0)} = A^-$ and $(A^-)^{(n)} = [(A^-)^{(n-1)}, (A^-)^{(n-1)}]$.

(3) If A is flexible Malcev-admissible such that A^- is not solvable, then A is a Malcev algebra isomorphic to C_0^-.

Proof. The first part is immediate from the foregoing results. Let R be the solvable radical of A^-. It is well known that R is invariant under Der A^-, hence under L. Then L acts as derivations on

$S_1 = A^-/R$ and S_1 is semisimple or 0. If S_1 is semisimple, then the simple summands of S_1 are also L-modules, and L cannot annihilate any of these summands, since L annihilates only Fa. Thus, L acts faithfully on each simple summand. Since the derivations of a simple algebra are inner, the dimension of each simple summand is at least 7. Since dim A = 7 and C_0^- is the only simple Malcev algebra of dimension 7 over F, it must be that $A^- \cong S_1 \cong C_0^-$. We conclude from this that if A is Malcev-admissible, then either $A^- \cong C_0^-$ or A^- is solvable.

When A is Malcev-admissible, using Table 5.13 we obtain the following constant conditions corresponding to some specializations of the basis elements in the Malcev identity.

$$(5.54) \qquad m(u_1,u_2,u_1,u_3) = 0 \; ; \quad 2\lambda^2\rho = \lambda(\sigma - \tau)(\nu - \nu') \; ,$$

$$(5.55) \qquad m(v_1,v_2,v_1,v_3) = 0 \; ; \quad 2\lambda\rho^2 = -\rho(\sigma - \tau)(\xi - \xi') \; ,$$

$$(5.56) \qquad m(u_1,v_1,u_1,u_2) = 0 \; ; \quad -\lambda(\sigma - \tau)(\xi - \xi') = \lambda(\sigma - \tau)(\nu - \nu') \; .$$

(1) Assume that A^- is a nonsolvable Malcev algebra, so that $A^- \cong C_0^-$. If $\lambda = 0$, then Table 5.13 shows that U + Fa is an ideal of A^-. Similarly, if $\rho = 0$, then V + Fa is an ideal of A^-. Thus, $\lambda\rho \neq 0$, and this with (5.54) and (5.55) implies $\sigma - \tau \neq 0$, $\xi - \xi' = -(\nu - \nu') \neq 0$, and

$$(5.57) \qquad 2\lambda\rho = (\sigma - \tau)(\nu - \nu') \; .$$

We let

$$(5.58) \qquad \alpha = \sqrt[3]{\frac{2}{\lambda(\nu-\nu')(\sigma-\tau)}} \; , \quad \beta = \sqrt[3]{\frac{2}{\rho(\nu-\nu')(\sigma-\tau)}} \; , \quad \gamma = \frac{-2}{\nu-\nu'} \; .$$

Then, α, β, γ are scalars in F, since F is algebraically closed. From (5.57) and (5.58) we obtain

(5.59) $$\alpha\beta = \frac{2}{(\nu-\nu')(\sigma-\tau)} = \frac{-\gamma}{\sigma-\tau} , \quad \lambda\alpha^2 = \beta, \ \rho\beta^2 = \alpha .$$

Letting $\bar{u}_i = \alpha u_i$, $\bar{v}_j = \beta v_j$, $\bar{a} = \gamma a$, we compute from Table 5.13

$$[\bar{u}_i,\bar{u}_j] = \alpha^2[u_i,u_j] = \alpha^2\lambda\varepsilon_{ijk}v_k = \varepsilon_{ijk}\beta v_k = \varepsilon_{ijk}\bar{v}_k , \text{ by (5.59)} ,$$

$$[\bar{u}_i,\bar{v}_j] = \alpha\beta[u_i,v_j] = \alpha\beta(\sigma - \tau)\delta_{ij}a = - \gamma\delta_{ij}a = - \delta_{ij}\bar{a} , \text{ by (5.59)} ,$$

$$[\bar{a},\bar{u}_i] = \gamma\alpha[a,u_i] = - \gamma\alpha(\nu - \nu')u_i = 2\bar{u}_i , \text{ by (5.58)} .$$

Similarly, the remaining products in (5.52) hold.

(2) Assume that A^- is a solvable Malcev algebra. The computations above show that $\lambda\rho \neq 0$ if and only if $A^- \cong C_0^-$. Hence, under the assumption, we must have $\lambda\rho = 0$. Consider first the case $\sigma - \tau \neq 0$. Then, if $\gamma = 0$ and $\rho \neq 0$, then by (5.55) $\xi - \xi' = 0$. Since $m(v_1,u_1,v_1,v_2) = 0$ gives $-\rho(\sigma - \tau)(\nu - \nu') = \rho(\sigma - \tau)(\xi - \xi')$, we have $\nu - \nu' = 0$ also. Thus, $[A,A] \subseteq U + Fa$ and $(A^-)^{(2)} = 0$. If $\lambda \neq 0$ and $\rho = 0$, then we similarly obtain $\xi - \xi' = \nu - \nu' = 0$, and hence $(A^-)^{(2)} = 0$. If $\lambda = \rho = 0$, then it follows from Table 5.13 that $\nu - \nu' = 0$ or $\xi - \xi' = 0$. In the former case, $[A,A] \subseteq U + Fa$, $(A^-)^{(2)} \subseteq U$, and $(A^-)^{(3)} = 0$. In the latter case, $[A,A] \subseteq V + Fa$ and $(A^-)^{(3)} = 0$. Consider the remaining case of $\sigma - \tau = 0$. If $\lambda = 0$ and $\rho \neq 0$, then $[A,A] \subseteq U + V$, $(A^-)^{(2)} \subseteq U$, and $(A^-)^{(3)} = 0$, while if $\lambda \neq 0$ and $\rho = 0$, then $(A^-)^{(2)} \subseteq V$ and $(A^-)^{(3)} = 0$. The case of $\lambda = \rho = 0$ gives $(A^-)^{(2)} = 0$. In particular, we have shown that $\lambda\rho = (\sigma - \tau)(\nu - \nu')(\xi - \xi') = 0$ and $(A^-)^{(3)} = 0$.

(3) Since $A^- \cong C_0^-$, flexibility implies by Corollary 3.22 that A is a Malcev algebra isomorphic to A^-. □

5.5. Dimension 8

We determine flexible Malcev-admissible algebras of dimension 8 over F, when the radical R of A^- is small. Thus, we treat only the cases:

$$A^- = \mathfrak{sl}(2) + \mathfrak{sl}(2) + R, \quad A^- = C_0^- + R .$$

Since A in the latter case has been classified in Theorem 4.11(2), it suffices to determine A when $A^- = \mathfrak{sl}(2) \oplus \mathfrak{sl}(2) + R$. Thus, R^- is a 2-dimensional abelian or solvable Lie algebra with a basis $\{u,v\}$ given by (5.22).

Assume first that R^- is solvable with $[u,v] = u$, and let S_1, S_2 be the two copies of $\mathfrak{sl}(2)$. Then, $[S_1,R] = [S_2,R] = 0$, and hence A is determined by (4.15) in Theorem 4.3, where $\tau_i(u) = 0$ and $\tau_i(v) = \alpha_i \in F$, $i = 1, 2$. Hence, in this case, A is given by

$$\begin{aligned}
&xy = - yx = \tfrac{1}{2}[x,y], \quad x,y \in S_i, \quad i = 1, 2 ,\\
&xy = yx = 0, \quad x \in S_1, \quad y \in S_2 ,\\
&xu = ux = 0, \quad xv = vx = \alpha_i x, \quad x \in S_i, \quad i = 1, 2 ,\\
&uv = \beta u, \quad vu = (\beta - 1)u, \quad v^2 = (2\beta - 1)v, \quad u^2 = 0 .
\end{aligned} \tag{5.60}$$

The last multiplication in R is obtained from (5.23).

Assume next that R^- is abelian and $[S_1 + S_2,R] = 0$. Let $\{e_i,h_i,f_i\}$ be the standard basis of $S_i (i = 1,2)$. The multiplication in A is again determined by (4.15). We choose a basis $\{u_1,u_2\}$ of R and let $\tau_i(u_i) = \beta_i$, $\tau_i(u_2) = \gamma_i$ for $i = 1, 2$. Letting $a_i = \alpha_i u_i$ for $\alpha_i \in F$ in (4.15) and using the Killing form of $\mathfrak{sl}(2)$ in (4.19), we have A described by

$$e_i^2 = f_i^2 = 0,\ e_ih_i = -h_ie_i = -e_i,\ h_if_i = -f_ih_i = -f_i\ ,$$

$$e_if_i = \frac{1}{2}h_i + \alpha_iu_i,\ f_ie_i = -\frac{1}{2}h_i + \alpha_iu_i,\ h_i^2 = 2\alpha_iu_i\ ,$$

$$(5.61) \qquad x_iu_1 = u_1x_i = \beta_ix_i,\ x_iu_2 = u_2x_i = \gamma_ix_i,\ x_i \in S_i\ ,$$

$$xy = yx = 0,\ x \in S_1,\ y \in S_2\ ,$$

R is any commutative algebra over F .

In view of Corollary 4.7(3), we note that the algebra defined by (5.61) is simple if R is a simple commutative algebra, $\alpha_i \neq 0$, and $\beta_i \neq 0$ or $\gamma_i \neq 0$ for $i = 1, 2$.

Suppose now that R^- is abelian and $[S_1 + S_2, R] \neq 0$. We note first that $RR = 0$ by Lemma 4.19 and that $xy = -yx = \frac{1}{2}[x,y]$ for $x,y \in S_i (i = 1,2)$, by Lemma 4.2(1). Since R is an irreducible module for $S_1 + S_2$, R must be annihilated by one of S_1 and S_2 , say by S_2 , because R is the tensor product of two irreducible modules for $s\ell(2)$. Hence, Lemma 4.2(2) implies $S_1S_2 = S_2S_1 = 0$. Since $RS_1 \subseteq R$ and $S_1R \subseteq R$ by Lemma 4.12, $ax = -xa = \frac{1}{2}[a,x]$ for $a \in R,\ x \in S_1 + S_2$. Thus,

$$(5.62) \qquad xy = -yx = \frac{1}{2}[x,y],\ x,y \in A\ ,$$

and A is a Lie or Malcev algebra isomorphic to A^- , according as $R \cong V(1)$ or $R \cong M_2$ as an $s\ell(2)$-module. Therefore, we have

<u>Theorem 5.7</u>. Let A be a flexible Malcev-admissible algebra of dimension 8 over F such that the dimension of the radical R of A^- is less than or equal to 2. Then, A is isomorphic to one of the algebras given by (4.20) in Theorem 4.11, (5.60), (5.61), and (5.62). The algebras

described by (4.20) and the algebra of (5.62) with R corresponding to $\tilde{M}_2$ are non-Lie, Malcev-admissible, and the remaining algebras are Lie-admissible. □

If A is an algebra given by (4.20), then the Lie algebra G_2 clearly acts on A as derivations. Conversely, we can show that any nonassociative algebra A of dimension 8 over F which is nontrivially acted on by G_2 as derivations and for which A^- is a nonsolvable Malcev algebra is isomorphic to the algebra described in (4.20). Let A denote a nontrivial G_2-module of dimension 8 over F . Then, A has the decomposition $A = V(\lambda_1) + Fe$ as a G_2-module, where Fe is a trivial module. Since $V(\lambda_1) \cong C_0^-$ as a G_2-module, we use the decomposition $A = C_0^- + Fe$ with the multiplication in C_0^- denoted by [,] . Then, it follows from (5.6) that the condition $G_2 \subseteq \text{Der } A$ imposes the following restrictions on the products in A .

$$xy = \alpha[x,y] + \beta K(x,y)e \ ,$$

(5.63)

$$xe = \gamma x, \ ey = \gamma' y, \ e^2 = \delta e$$

for $x,y \in C_0^-$, α, β, γ, γ', $\delta \in F$, where K(,) is the Killing form on C_0^- .

<u>Theorem 5.8.</u> Let A be an 8-dimensional nontrivial G_2-module over F . Then, a product defined on A satisfies the condition $G_2 \subseteq \text{Der } A$ if and only if the product is given by (5.63). Furthermore,

(1) If A^- is a nonsolvable Malcev algebra, then $\alpha \neq 0$, $\gamma = \gamma'$, A is flexible, and is isomorphic to the algebra given by (4.20). If, in addition, $\beta \neq 0$ and $\gamma \neq 0$, then A is simple.

(2) If A^- is a solvable Malcev algebra, then $\alpha = 0$ and $(A^-)^{(2)} = 0$.

Proof. The first part is immediate from the foregoing remarks. We adopt the basis of C_0^- used in (5.52). The specialization of basis elements u_1, v_1, u_1, e in the Malcev identity in A^- gives $\alpha^2(\gamma - \gamma') = 0$ by (5.63). Assume that A^- is a nonsolvable Malcev algebra, and let S be a Levi factor of A^- . Then, S is a G_2-module. Hence, as in the proof of Theorem 5.6, by dimension count of simple summands of S, we have $S = C_0^-$ and $R = Fe$, the radical of A^- . If $\alpha = 0$, then $[A,A] \subseteq C_0^-$ and $[C_0^-,C_0^-] = 0$, so A^- is solvable. Thus, $\alpha \neq 0$ so $\gamma - \gamma' = 0$, and if we replace e_i with $e_i' = \alpha^{-1}e_i$, the products in (5.63) can be normalized to the one given by (4.20). Thus, letting $x + \lambda e$ denote an element of A for $x \in C_0^-$, $\lambda \in F$, we have the product in A expressed by

$$(x + \lambda e)(y + \mu e) = \frac{1}{2}[x,y] + \lambda y + \mu x + [\lambda\mu\delta + \beta K(x,y)]e ,$$

which is a special case of the product defined by (4.84) with $t(e) = \gamma^{-1}$, $(\, , \,) = \beta K(\, , \,)$, and $M = C_0^-$. Hence, it follows from Lemma 4.26(4) that A is flexible and from Theorem 4.27 that if $\beta \neq 0$ and $\gamma \neq 0$, A is simple, since $\beta K(\, , \,)$ is nondegenerate. If A^- is a solvable Malcev algebra, then since $\alpha^2(\gamma - \gamma') = 0$, $\alpha = 0$ by (5.63) and $(A^-)^{(2)} = 0$. □

5.6. DIMENSION 15 ; $s\ell(3) \subseteq$ DER A

In this section we construct a class of algebras A of dimension 15 over F which are acted on by $s\ell(3)$ as derivations and which have the decomposition

$$A = S + U + V + Fa \tag{5.64}$$

as an $s\ell(3)$-module, where S is isomorphic to $s\ell(3)$, U, V two non-isomorphic irreducible $s\ell(3)$-modules of dimension 3, and Fa a trivial module. Let $L = s\ell(3)$ and identify L with the Lie algebra of 3×3 trace zero matrices over F, so that the matrix units e_{ij}'s for $i \neq j$ and $e_{11} - e_{22}$, $e_{22} - e_{33}$ form a basis of L. The corresponding basis for S will be denoted by s_{ij} for $i \neq j$ and $s_{11} - s_{22}$, $s_{22} - s_{33}$. We can also identify U and V with the spaces spanned by the u_i and v_i in (5.8), respectively. Accordingly, the module actions of L on U and V are the same as in (5.16) or in (5.17). In the latter case, each element of U and V can be regarded as a 3×1 column matrix and a 1×3 row matrix over F.

In view of Lemma 5.1, L acts as derivations on the products in (5.9). The products in the submodule U + V + Fa have already been determined in Table 5.12, except for those in UV corresponding to component S of $U \otimes V$. But then, it is easy to see from (5.17) that the mapping $u_i \otimes v_j \to s_{ij} - \frac{1}{3}\delta_{ij}I$ is an L-homomorphism of $U \otimes V$ into S. (Although $I = s_{11} + s_{22} + s_{33}$ does not exist in S, $s_{ij} - \frac{1}{3}\delta_{ij}I$ has trace zero and so is an element of S.) Since $\dim \mathrm{Hom}_L(U \otimes V, S) = 1$ by (5.5), the mapping forms a basis of $\mathrm{Hom}_L(U \otimes V, S)$. Since Fa is a trivial module, for each of the modules S, U, V, Fa, both left and right multiplication by a are multiples of the identity mapping. The fact that

$$\mathrm{Hom}_L(S \otimes U,S) = \mathrm{Hom}_L(S \otimes U,V) = \mathrm{Hom}_L(S \otimes U,Fa) = 0 ,$$

$$\dim \mathrm{Hom}_L(S \otimes U,U) = \dim \mathrm{Hom}_L(S \otimes V,V) = 1$$

by (5.5) and the fact that the module actions of L and V induce L-homomorphisms $S \otimes U \to U$ and $S \otimes V \to V$ imply that $su = \zeta s \cdot u$ and $sv = \eta s \cdot v$ for some $\zeta, \eta \in F$ and all $s \in S, u \in U, v \in V$, where $s \cdot u$ denotes the module action. The same results hold for the products US and VS. It remains to determine products in SS. It follows from Lemma 3.7 that $\dim \mathrm{Hom}_L(S \otimes S,S) = 2$ and that the mappings $s \otimes t \to [s,t]$ and $s \otimes t \to s \# t$ defined by (3.19) form a basis of $\mathrm{Hom}_L(S \otimes S,S)$. Since $\mathrm{Hom}_L(S \otimes S,U) = \mathrm{Hom}_L(S \otimes S,V) = 0$ by (5.5) and since $\mathrm{Hom}_L(S \otimes S,Fa)$ is spanned by the Killing form $K(s,t)$ carried over to S under the isomorphism between L and S, we have

$$st = \alpha[s,t] + \beta s \# t + \gamma K(s,t)a$$

for some $\alpha, \beta, \gamma \in F$ and for all $s,t \in S$. Therefore, extending the products in $U + V + Fa$ of Table 5.12 to those in A, we obtain the multiplication in A given by Table 5.14, where s, t are elements of S, and the Greek lower case letters denote fixed scalars in F.

In the special case of $\mu = \mu' = 0$, the algebra $U + V + Fa$ constructed in Table 5.12 is a subalgebra of A. Hence, the algebra of Domokos and Kövesi-Domokos [1] is included in Table 5.14. The algebra used in Günaydin and Gürsey [1] is the Lie algebra G_2. To obtain this algebra from Table 5.14, we must suppress the element a by taking the subspace $S + U + V$, and by taking $\gamma = \sigma = \tau = 0$. Then, choosing

$$\alpha = \zeta = -\zeta' = \eta = -\eta' = \frac{1}{2},\ \beta = 0,\ \mu = -\mu' = -\frac{3}{2},\ \lambda = \rho = 1$$

makes $S + U + V$ into a copy of G_2 .

TABLE 5.14

	t	u_j	v_j	a
s	$\alpha[s,t] + \beta s \# t + \gamma K(s,t)a$	$\zeta s \cdot u_j$	$\eta s \cdot v_j$	θs
u_i	$\zeta' t \cdot u_i$	$\lambda \varepsilon_{ijk} v_k$	$\mu(s_{ij} - \frac{1}{3}\delta_{ij} I) + \sigma \delta_{ij} a$	νu_i
v_i	$\eta' t \cdot v_i$	$\mu'(s_{ji} - \frac{1}{3}\delta_{ij} I) + \tau \delta_{ij} a$	$\rho \varepsilon_{ijk} u_k$	ξv_i
a	$\theta' t$	$\nu' u_j$	$\xi' v_j$	ϕa

In the remainder of this section we investigate the resulting Lie structure of A^- when A is assumed to be Lie-admissible, and determine the resulting sets of conditions on the constants dictated by the structure of A^- . We also discuss the same problems under the assumption that A is flexible Lie-admissible. The computations are more complicated than for the Malcev-admissible algebra $U + V + Fa$ in Section 5.3, but easier for A than for an arbitrary algebra of dimension 15. We begin by assuming that A is Lie-admissible and prove

<u>Theorem 5.9</u>. Let A be a 15-dimensional $\mathfrak{sl}(3)$-module over F decomposed as in (5.64). Then, A is a nonassociative algebra over F which is acted on by $\mathfrak{sl}(3)$ as derivations if and only if the multiplication in A is given by Table 5.14. Furthermore, if A is Lie-admissible, then one of the following four cases holds:

(1) A^- is a simple Lie algebra of type A_3 .

(2) $S + U + V$ is a simple Lie subalgebra of A^- of type G_2, and Fa is the center of A^-.

(3) S is a simple Lie subalgebra of A^- of type A_2 and $U + V + Fa$ is the solvable radical of A^-.

(4) A^- is a solvable Lie algebra.

Proof. The first part follows from Table 5.14. Assume that A is Lie-admissible, and let S_1 be a nonzero Levi factor of A^-. By the same argument as in the proof of Theorem 5.6 and by a dimension count, we must have that S_1 is simple and $S \subseteq S_1$. Let R be the solvable radical of A^-. Since R is stable under $\operatorname{Der} A^-$ and S_1 is an L-module, S_1 and R are composed of some subsets of the modules $\{S,U,V,Fa\}$, while S_1 is complementary to R. We note from Table 5.14 that S_1 cannot equal $S + U$, $S + U + Fa$, $S + V$, or $S + V + Fa$, since in any of these cases either U or V is an ideal of S_1. Thus, whenever $U \subseteq S_1$, then $V \subseteq S_1$ and conversely. Since S is a Lie ideal of $S + Fa$, S_1 cannot be $S + Fa$. This leaves the three possibilities: (i) $S_1 = A^- = S + U + V + Fa$; (ii) $S_1 = S + U + V$; (iii) $S_1 = S$.

In case (i), A^- is a simple 15-dimensional Lie algebra, and by dimension count of simple Lie algebras, A^- must be of type A_3. In case (ii), S_1 is a simple 14-dimensional Lie algebra, which implies that S_1 is of type G_2. Since Fa is the radical of A^-, it is a 1-dimensional module for S_1 under the adjoint action. Hence $[S_1,Fa] = 0$ and Fa is the center of A^-. From this, we have $Fa \nsubseteq [A^-,A^-]$, so that Table 5.14 implies $\sigma - \tau = 0$, and $S + U + V$ is a Lie subalgebra of A^-, giving part (2). If case (iii) holds, then $U + V + Fa$ is the radical of A^- and part (3) holds in this case. Finally, $S_1 = 0$ implies part (4): A^- is a solvable Lie algebra. □

We note that, under the assumption of Lie-admissibility, the submodule $U + V + Fa$ cannot be a nonsolvable Lie subalgebra, since if $U + V + Fa$ were nonsolvable, then by Theorem 5.6(1) $(U + V + Fa)^-$ would be isomorphic to C_0^-, being non-Lie. To obtain more stringent information about four cases in Theorem 5.9, we make certain specializations of the standard basis elements of A in the Jacobi identity $J(x,y,z) = [[x,y],z] + [[y,z],x] + [[z,x],y] = 0$ in A^-. Each specialization results in a constant condition. However, it is worth noting beforehand that the symmetric products in Table 5.14 vanish in A^-, so there will be no restrictions imposed on β, γ, or ϕ. To facilitate computation, it is useful to note relation (5.16), Tables 5.13 and 5.14 along with the equations

(5.65)
$$[s_{ij}, s_{k\ell}] = 2\alpha(\delta_{jk}s_{i\ell} - \delta_{i\ell}s_{kj}) ,$$
$$[s_{ij}, u_k] = (\zeta - \zeta')\delta_{jk}u_i, \ [s_{ij}, v_k] = -(\eta - \eta')\delta_{ik}v_j .$$

We list those Jacobi specializations and corresponding conditions on scalars in Table 5.15 where we let an ordered triple x, y, z denote the Jacobi specialization $J(x,y,z) = 0$.

We note in Table 5.14 that if $\alpha \neq 0$, then we can replace each s in S with $s' = \frac{1}{2\alpha}s$, so that by (5.65) $[s'_{ij}, s'_{k\ell}] = \delta_{jk}s'_{i\ell} - \delta_{i\ell}s'_{kj}$ and the mapping $e_{ij} \to s'_{ij}$ gives a Lie algebra isomorphism of L onto S. Thus, whenever $\alpha \neq 0$, we assume that the s_{ij} have been normalized to the first equation of (5.65) with $2\alpha = 1$. It is also useful to note from Table 5.14 that $U + V + Fa$ is an ideal of A^- if $\mu - \mu' = 0$.

TABLE 5.15

	Jacobi Specialization	Constant Condition
(1)	s_{12}, s_{21}, u_1	$2\alpha(\zeta - \zeta') - (\zeta - \zeta')^2 = 0$
(2)	s_{12}, s_{21}, v_1	$2\alpha(\eta - \eta') - (\eta - \eta')^2 = 0$
(3)	s_{12}, s_{21}, a	$\alpha(\theta - \theta') = 0$
(4)	s_{12}, u_2, u_3	$[(\zeta - \zeta') - (\eta - \eta')]\lambda = 0$
(5)	s_{12}, v_1, v_3	$[(\zeta - \zeta') - (\eta - \eta')]\rho = 0$
(6)	s_{12}, u_2, v_3	$(\mu - \mu')[2\alpha - (\zeta - \zeta')] = 0$
(7)	s_{12}, v_1, u_3	$(\zeta - \zeta')[2\alpha - (\eta - \eta')] = 0$
(8)	s_{12}, u_2, v_1	$[(\zeta - \zeta') - (\eta - \eta')](\sigma - \tau) = 0$
(9)	s_{12}, u_2, a	$(\zeta - \zeta')(\theta - \theta') = 0$
(10)	s_{12}, v_1, a	$(\eta - \eta')(\theta - \theta') = 0$
(11)	u_1, u_2, u_3	$\lambda(\sigma - \tau) = 0$
(12)	v_1, v_2, v_3	$\rho(\sigma - \tau) = 0$
(13)	u_1, u_2, a	$\lambda[(\xi' - \xi) + 2(\nu - \nu')] = 0$
(14)	v_1, v_2, a	$\rho[2(\xi' - \xi) + (\nu - \nu')] = 0$
(15)	u_1, v_1, v_2	$\frac{4}{3}(\mu - \mu')(\eta - \eta') + (\sigma - \tau)(\xi' - \xi) + 4\rho\lambda = 0$
(16)	u_1, u_2, v_1	$\frac{4}{3}(\mu - \mu')(\zeta - \zeta') + (\sigma - \tau)(\nu - \nu') + 4\rho\lambda = 0$

In case (2) of Theorem 5.9, to obtain the constant conditions that normalize to a multiplication in G_2, we use the information about G_2 described in Lemma 5.1 and Theorem 5.2. Let U, V, D_0, $d(e_1,U)$, and $d(e_2,V)$ be the same as in Theorem 5.2, where D_0 is identified with S as an L-module, and $G_2 = D_0 + d(e_1,U) + d(e_2,V)$. For $s \in D_0$, $u \in U$, $v \in V$, by (5.18) we have $[s,d(e_1,u)] = d(e_1,s \cdot u)$ and $[s,d(e_2,v)] = d(e_2,s \cdot v)$. To determine the products $[d(e_1,U),d(e_1,U)]$,

$[d(e_2,V),d(e_2,V)]$ and $[d(e_1,U),d(e_2,V)]$, it is useful to note (5.9), (5.16), (5.18), (5.20), (5.52), and the identity

$$d(x,y)z = [x,y,z] + [z,y,x] + [x,z,y] \,, \tag{5.66}$$

where $[x,y,z] = [[x,y],z]$. In any alternative algebra, the identity

$$d(xy,z) + d(yz,x) + d(zx,y) = 0 \tag{5.67}$$

holds (Schafer [3,p.78]).

Let $u, u' \in U$ and $v, v' \in V$. By (5.66) and (5.9) $d(e_1,u)u' = [u',u] = -2uu'$ and by (5.20) $d(e_1,u)e_1 = -2u$, and hence by (5.18)

$$\begin{aligned} [d(e_1,u),d(e_1,u')] &= d(d(e_1,u)e_1,u') + d(e_1,d(e_1,u)u') \\ &= -2d(u,u') - 2d(e_1,uu') \\ &= -2d(u,u') + 2d(e_2,uu') \,, \end{aligned}$$

since $1 = e_1 + e_2$. But then, (5.67) gives $d(e_2,uu') = -d(uu',e_2) = d(u'e_2,u) + d(e_2u,u') = d(u',u) = -d(u,u')$. This and similar computations imply

$$[d(e_1,u),d(e_1,u')] = 4d(e_2,uu') = 2d(e_2,[u,u']) \,,$$

$$[d(e_2,v),d(e_2,v')] = 4d(e_1,vv') = 2d(e_1,[v,v']) \,.$$

For the remaining products, we note first that $d(e_1,u)e_2 = 2u$ by (5.66), and $d(e_1,u)v = 2[u,v]$, which implies $d(e_2,d(e_1,u)v) = 0$, since $[u,v] \in Fe_1 + Fe_2$ and $d(e_2,e_1) = 0$. Thus,

$$[d(e_1,u),d(e_2,v)] = 2d(u,v) \,. \tag{5.68}$$

Hence, it suffices to determine $d(u_i,v_j)$ as an element of $D_0 = S$. From (5.52) and (5.66) we obtain

$$d(u_i,v_i)u_i = -4u_i,\ d(u_i,v_i)u_k = 2u_k,\ i \neq k,$$

$$d(u_i,v_i)v_i = 4v_i,\ d(u_i,v_i)v_k = -2v_k,\ i \neq k,$$

which by (5.16) imply $d(u_i,v_i)u_k = -6(s_{ii} - \frac{1}{3}I) \cdot u_k$ and $d(u_i,v_i)v_k = -6(s_{ii} - \frac{1}{3}I) \cdot v_k$. Similarly, we compute for $i \neq j$

$$d(u_i,v_j)u_k = -6\delta_{jk}u_i = -6s_{ij} \cdot u_k,$$

$$d(u_i,v_j)v_k = 6\delta_{ik}v_j = -6s_{ij} \cdot v_k.$$

Hence, it follows from this and (5.68) that

$$[d(e_1,u_i),d(e_2,v_j)] = -12(s_{ij} - \tfrac{1}{3}\delta_{ij}I)$$

for all i, j = 1, 2, 3. We summarize these results in Table 5.16.

TABLE 5.16

	t	$d(e_1,u_j)$	$d(e_2,v_j)$
s	$[s,t]$	$d(e_1,s \cdot u_j)$	$d(e_2,s \cdot v_j)$
$d(e_1,u_i)$	$-d(e_1,t \cdot u_i)$	$4\varepsilon_{ijk}d(e_2,v_k)$	$-12(s_{ij} - \frac{1}{3}\delta_{ij}I)$
$d(e_2,v_i)$	$-d(e_2,t \cdot v_i)$	$12(s_{ji} - \frac{1}{3}\delta_{ij}I)$	$4\varepsilon_{ijk}d(e_1,u_k)$

Theorem 5.10. If $A = S + U + V + Fa$ is Lie-admissible, then after suitable normalizations, one of the following cases holds.

(1) A^- is a simple Lie algebra of type A_3, and there hold the constant conditions:

$$2\alpha = 1 = \zeta - \zeta' = \eta - \eta' = \mu - \mu' ,$$

$$\sigma - \tau = \frac{1}{3}, \ \xi - \xi' = \nu' - \nu = 4, \ \lambda = \rho = 0 .$$

(2) $S + U + V$ is a simple Lie subalgebra of A^- of type G_2 and Fa is the center of A^- . Moreover

$$2\alpha = \zeta - \zeta' = \eta - \eta' = 1, \ \lambda = \rho = 2, \ \mu - \mu' = -12 ,$$

$$\theta - \theta' = \xi - \xi' = \nu - \nu' = \sigma - \tau = 0 .$$

(3) S is a simple Lie subalgebra of A^- of type A_2 , and $U + V + Fa$ is the solvable radical of A^- . Moreover,

$$2\alpha = \zeta - \zeta' = \eta - \eta' = 1, \ \mu - \mu' = \theta - \theta' = 0 ,$$

and either

(i) $\sigma - \tau \neq 0$ and $\lambda = \rho = \xi - \xi' = \nu - \nu' = 0$, or

(ii) $\lambda\rho = \sigma - \tau = 0$.

(4) A^- is a solvable Lie algebra with the conditions

$$\alpha = \zeta - \zeta' = \eta - \eta' = \lambda\rho = (\sigma - \tau)(\xi - \xi') = (\sigma - \tau)(\nu - \nu') = 0 .$$

Proof. (1) This is case (1) of Theorem 5.9. As noted above, since A^- is a simple Lie algebra of type A_3 , $\mu - \mu' \neq 0$, and since A_3 has no abelian 8-dimensional subalgebras, it must be that $\alpha \neq 0$. Hence, by (6), (7) in Table 5.15 and by our normalization process; $2\alpha = 1$, we obtain $\zeta - \zeta' = \eta - \eta' = 1$. In order that $Fa \subseteq [A^-,A^-]$, it must be that $\sigma - \tau \neq 0$. Hence, by (11) and (12), $\lambda = \rho = 0$. We now make substitution $v_i' = \frac{1}{\mu-\mu'} v_i$, and regard S as those 4×4 trace zero matrices having zeros in the fourth row and column. We also think of u_1, u_2, u_3, v_1', v_2', v_3' as being s_{14}, s_{24}, s_{34}, s_{41}, s_{42}, s_{43}. Under these substitutions, we have $s_{11} - s_{44} = [s_{14}, s_{41}] = [u_1, v_1'] = \frac{1}{\mu-\mu'}[u_1, v_1] = s_{11} - \frac{1}{3}I + (\sigma - \tau)(\mu - \mu')^{-1}a$, so that the substitution $a' = 3(\sigma - \tau)(\mu - \mu')^{-1}a$ gives $a' = s_{11} + s_{22} + s_{33} - 3s_{44}$. Hence, our normalization must be that $\mu - \mu' = 1$, and so $\sigma - \tau = \frac{1}{3}$. After these normalizations and identification, we find that A^- has the same multiplication as A_3.

(2) Since $S + U + V$ is a subalgebra of type G_2 and Fa is the center of A^-, $\mu - \mu' \neq 0$ and $\theta - \theta' = \xi - \xi' = \nu - \nu' = 0$. Since there are no abelian subalgebra of G_2 of dimension 8, $\alpha \neq 0$ and so we again have the normalization $2\alpha = 1$, $\zeta - \zeta' = \eta - \eta' = 1$. From (15) of Table 5.15 and the fact that $\sigma - \tau = 0$, we deduce $\lambda\rho \neq 0$. To normalize the multiplication in $S + U + V$ to the one given by Table 5.16, we replace u_i and v_i with $2/(\lambda^2\rho)^{-\frac{1}{3}} u_i$ and $2/(\lambda\rho^2)^{-\frac{1}{3}} v_i$. This gives the new values $\lambda = \rho = 2$, which by (15) of Table 5.15 imply $\mu - \mu' = -12$. Thus, the algebra in this case is just the one described by Table 5.16.

(3) Since S is a subalgebra of type A_2, $2\alpha = 1$, so $\zeta - \zeta' = \eta - \eta' = 1$ as before. From the fact that $R = U + V + Fa$ is the radical of A^-, we deduce $\mu - \mu' = 0$. It follows from (3) of Table 5.15 that $\theta - \theta' = 0$. At this point, case (3) splits into two subcases depending on whether $\sigma - \tau \neq 0$ or $\sigma - \tau = 0$. Consider first $\sigma - \tau \neq 0$. Then, by (11) and (12) of Table 5.15, $\lambda = \rho = 0$, and by (15), (16), $\xi - \xi' = \nu - \nu' = 0$. If $\sigma - \tau = 0$, then (15) gives $\lambda\rho = 0$. Thus, we obtain the desired conditions on scalars.

(4) Since A^- is solvable, $\alpha = 0$ and so by (1), (2) of Table 5.15, $\zeta - \zeta' = \eta - \eta' = 0$. Solvability of A^- also imples $\lambda\rho = 0$,

and thus by (15) and (16), $(\sigma - \tau)(\xi - \xi') = (\sigma - \tau)(\nu - \nu') = 0$, giving the desired conditions. □

Remark 5.1. (i) In case (2) of Theorem 5.10, as noted by Benkart and Osborn [2], using $1/(\lambda^2\rho)^{-\frac{1}{3}} u_i$ and $1/(\lambda\rho^2)^{-\frac{1}{3}} v_i$ in place of u_i and v_i results in the new values $\lambda = \rho = 1$, so $\mu - \mu' = -3$. The algebra in this case is the one used by Günaydin and Gürsey [1,Eq.(7.2)].

(ii) In case (3) of Theorem 5.10, if $\sigma - \tau \neq 0$, then by (11) and (12) of Table 5.15 $\lambda = \rho = 0$ and hence by (15) and (16) $\xi - \xi' = \nu - \nu' = 0$. This means that $[R,R] = Fa$, where Fa is the center of A^-. Assume $\sigma - \tau = 0$, so that $(R^-)^{(1)} \subseteq U + V$. Relation (15) implies $\lambda\rho = 0$, and hence $(R^-)^{(2)} \subseteq U$ or V. If λ or ρ is nonzero, then (13) or (14) gives a relationship between $\xi - \xi'$ and $\nu - \nu'$.

(iii) In case (4) of Theorem 5.10, the situations $\sigma - \tau \neq 0$ and $\sigma - \tau = 0$ are quite different. In the first case, $\xi - \xi' = \nu - \nu' = 0$ and $\lambda = \rho = 0$. Thus, $[R,R] \subseteq S + Fa$ and $(R^-)^{(2)} \subseteq S$, so $(R^-)^{(3)} = 0$. If $\sigma - \tau = 0$, then by (15) $\lambda\rho = 0$ and hence $[R,R] \subseteq S + U + V$, $(R^-)^{(2)} \subseteq U$ or V, giving $(R^-)^{(3)} = 0$. □

We finally consider those constant conditions which result from the assumption that A is flexible Lie-admissible. Here we establish

Theorem 5.11. If $A = S + U + V + Fa$ is flexible Lie-admissible over F with A^- nonsolvable, then A is described by one of the following four cases.

(a) A is an algebra with multiplication given by

$$xy = \frac{1}{2}[x,y] + \beta x \# y ,$$

defined on $s\ell(4)$, the 4×4 matrices of trace zero over F , where $\beta \in F$, and $x \# y = xy + yx - \frac{1}{2}(\mathrm{tr}\, xy)I$.

(b) $S_1 = S + U + V$ is a subalgebra of A^- which is a simple Lie algebra of type G_2 . Moreover, $S_1 \circ S_1 \subseteq Fa$, and

$$\theta = \theta' = \nu = \nu' = \xi = \xi' .$$

(c) S is a subalgebra of A^- of type A_2 . Moreover,

$$\alpha = \zeta = -\zeta' = \eta = \eta' = \sigma = -\tau = \frac{1}{2} ,$$

$$\beta = \theta = \theta' = \lambda = \mu = \mu' = \nu = \nu' = \xi = \xi' = \rho = \phi = 0 .$$

(d) S is a subalgebra of A^- of type A_2 , and the following constant conditions hold.

$$2\alpha = \zeta - \zeta' = \eta - \eta' = 1,\ \mu - \mu' = \sigma = \tau = \lambda\rho = 0 ,$$

$$\gamma(\xi - \xi') = \frac{4}{3}\beta = \gamma(\nu' - \nu),\ -\eta' = \frac{1}{2} + \beta = \zeta ,$$

$$\lambda(3\nu - \nu' - 2\xi) = 0,\ \lambda(3\nu' - \nu - 2\xi') = 0 ,$$

$$\rho(3\xi - \xi' - 2\nu) = 0,\ \rho(3\xi' - \xi - 2\nu') = 0,\ \nu + \zeta(\nu' - \nu) = \theta ,$$

$$\phi(\nu' - \nu) = (\nu + \nu')(\nu' - \nu),\ \phi(\xi' - \xi) = (\xi' - \xi)(\xi' + \xi) .$$

Proof. Since A^- is not solvable, either case (1), (2), or (3) of Theorem 5.10 holds. In the first two cases A satisfies the hypothesis of Theorem 4.3. If $A^- = A_3$, then A is clearly as described in case (a). If $A^- = S_1 + Fa$ where S_1 is of type G_2 , then since the "#" product on G_2 is zero, $S_1 \circ S_1 \subseteq Fa$ by Theorem 4.3. It also follows from Theorem 4.3 that $xa = ax = \delta x$ for $x \in S_1$ and a fixed $\delta \in F$. Hence, $\delta = \theta = \theta' = \nu = \nu' = \xi = \xi'$, giving part (b).

To obtain cases (c), (d), we need to recall Lemma 1.5 that A is flexible Lie-admissible if and only if

$$[xy,z] = [x,z]y + x[y,z] , \tag{5.69}$$

and that A is flexible if and only if

$$[x \circ y,z] = [x,z] \circ y + x \circ [y,z] , \tag{5.70}$$

where $x \circ y = \frac{1}{2}(xy + yx)$. Since A^- is assumed to be nonsolvable, only case (3) of Theorem 5.10 remains to be investigated. Thus, we proceed by two subcases: $\sigma - \tau \neq 0$, $\sigma - \tau = 0$.

We first consider the case where $\sigma - \tau \neq 0$ and make specializations of identity (5.70). As before, we let an ordered triple x,y,z denote the x, y, z - specialization into (5.70). In obtaining the resulting conditions on the constants, since we are in the situation of subcase (i) in (3) of Theorem 5.10, we use the constant conditions: $1 = 2\alpha = \zeta - \zeta' = \eta - \eta'$ and $0 = \theta - \theta' = \lambda = \rho = \xi - \xi' = \nu - \nu'$, toegther with the fact that R is a subalgebra of A , which follows from Corollary 4.14(1). The latter implies that $\mu = \mu' = 0$ by Table 5.14. Since $\sigma - \tau \neq 0$, we may assume that a has been normalized, so that $\sigma - \tau = 1$. We compute some x, y, z - specializations of identity (5.70) as listed in Table 5.17.

It follows from Table 5.17 that every constant except γ is specified, and the values are given by

$$\begin{gathered} \frac{1}{2} = \alpha = \zeta = -\zeta' = \eta = -\eta' = \sigma = -\tau , \\ 0 = \beta = \theta = \theta' = \lambda = \mu = \mu' = \nu = \nu' = \xi = \xi' = \rho = \phi . \end{gathered} \tag{5.71}$$

TABLE 5.17

x,y,z - Specialization of (5.70)	Constant Condition
u_1, v_1, u_1	$\nu = \nu' = 0$
v_1, u_1, v_1	$\xi = \xi' = 0$
v_1, a, u_1	$\phi = 0$
v_1, s_{23}, u_1	$\theta = \theta' = 0$
s_{12}, s_{21}, u_1	$\frac{1}{3}\beta = \zeta + \zeta'$
s_{32}, s_{21}, u_1	$\beta = \zeta + \zeta'$
s_{21}, s_{12}, v_1	$\frac{1}{3}\beta = \eta + \eta'$
$s_{11} - s_{22}, u_1, v_1$	$(\zeta + \zeta')(\sigma - \tau) = \sigma + \tau$

To have the constant γ appearing in the expression $[x \circ y,z]$, it must be that $x,y \in S$. But since the product of a with anything is 0 by (5.71), γ disappears. To obtain γ appearing in $[x,z] \circ y$ or $x \circ [y,z]$, we must have $x,y,z \in S$. However, by Theorem 4.3 $S + Fa$ is flexible Lie-admissible, regardless of the choice of γ. Hence, γ is arbitrary and case (c) is verified.

Assume now that $\sigma - \tau = 0$. Hence, by the constant conditions in subcase (ii) of (3) in Theorem 5.10, we have $2\alpha = \zeta - \zeta' = \eta - \eta' = 1$ and $\theta - \theta' = \lambda\rho = 0$. Since R is a subalgebra of A, again $\mu = \mu' = 0$. To deduce the desired conditions on scalars, we compute certain x,y,z - specializations of (5.69) as described in Table 5.18.

TABLE 5.18

x,y,z - Specialization of (5.69)	Constant Condition
s_{12}, u_2, v_1	$\tau = 0$
$s_{11} - s_{22}, s_{11} - s_{22}, u_3$	$\frac{4}{3}\beta = \gamma(\nu' - \nu)$
$s_{11} - s_{22}, s_{11} - s_{22}, v_3$	$\frac{4}{3}\beta = \gamma(\xi - \xi')$
s_{12}, s_{23}, u_3	$\frac{1}{2} + \beta = \zeta$
s_{12}, s_{23}, v_1	$\frac{1}{2} + \beta = - \eta'$
u_1, a, u_2	$\lambda(3\nu - \nu' - 2\xi) = 0$
a, u_1, u_2	$\lambda(3\nu' - \nu - 2\xi') = 0$
v_1, a, v_2	$\rho(3\xi - \xi' - 2\nu) = 0$
a, v_1, v_2	$\rho(3\xi' - \xi - 2\nu') = 0$
s_{12}, a, u_2	$\theta = \nu + \zeta(\nu' - \nu)$
s_{12}, a, v_1	$\theta = \xi + \eta(\xi' - \xi)$
a, a, u_1	$\phi(\nu' - \nu) = (\nu' - \nu)(\nu + \nu')$
a, a, v_1	$\phi(\xi' - \xi) = (\xi' - \xi)(\xi + \xi')$

Table 5.18 lists all the desired conditions on scalars, and hence case (d) is verified. □

It is interesting to note that the algebras described by case (c) of Theorem 5.11 do not arise from any of the constructions given in Chapter 4 (Theorems 4.3 and 4.22). In fact, $\zeta - \zeta' = 1$, so that $R \neq R_0$, and $[u_i, v_i] = a$, so R^- is not abelian.

Theorems 5.10 and 5.11 are contained in the work of Benkart and Osborn [2].

5.7. Dimension 15; $G_2 \subseteq$ Der A

We consider classes of 15-dimensional algebras A over F such that $G_2 \subseteq \text{Der } A$ and such that $A = V(\lambda_1) + V(\lambda_1) + Fw$ or $A = V(\lambda_2) + Fw$ as a G_2-module where Fw is a trivial module. We can identify $V(\lambda_1)$ with C_0^- and $V(\lambda_2)$ with G_2. In the latter case, the module action is the adjoint action, and in case $V(\lambda_1) = C_0^-$, G_2 acts on C_0^- as inner derivations. We consider first the case $A = C_0^- + C_0^- + Fw$. As in Section 4.4, the two copies of C_0^- can be identified with the tensor products $C_0^- \otimes q$ and $C_0^- \otimes r$, so that

$$A = C_0^- \otimes P + Fw , \tag{5.72}$$

where P is a 2-dimensional trivial G_2-module spanned by q, r. Using similar arguments as in Section 4.4 and the decomposition (5.6), we find that there exist a product "$*$" on P and constants $\beta, \gamma, \delta, \delta', \mu_0, \mu, \mu_0', \mu', \nu_0, \nu, \nu_0', \nu', \phi \in F$ such that the product in A is given by

$$(5.73)\quad \begin{aligned}
(x \otimes q)(y \otimes q) &= [x,y] \otimes q * q + \beta K(x,y)w , \\
(x \otimes r)(y \otimes r) &= [x,y] \otimes r * r + \gamma K(x,y)w , \\
(x \otimes q)(y \otimes r) &= [x,y] \otimes q * r + \delta K(x,y)w , \\
(y \otimes r)(x \otimes q) &= [y,x] \otimes r * q + \delta' K(x,y)w ,
\end{aligned}$$

$$(x \otimes q)w = \mu_0 x \otimes q + \mu x \otimes r,\ w(x \otimes q) = \mu_0' x \otimes q + \mu' x \otimes r ,$$

$$(x \otimes r)w = \nu_0 x \otimes q + \nu x \otimes r,\ w(x \otimes r) = \nu_0' x \otimes q + \nu' x \otimes r,\ w^2 = \phi w .$$

Conversely, G_2 acts as derivations on any algebra defined by (5.73).

Theorem 5.12. Let A be a 15-dimensional G_2-module with the decomposition given by (5.72) as a G_2-module. Then, A is a nonassociative algebra over F which is acted on by G_2 as derivations if and only if the multiplication in A is given by (5.73). Moreover, if A is Malcev-admissible, then one of the following holds:

(1) $C_0^- \otimes q$ is a simple, non-Lie, Malcev algebra isomorphic to C_0^- and $C_0^- \otimes r + Fw$ is the solvable radical of A^-.

(2) $C_0^- \otimes P$ is a semisimple subalgebra of A^- which is isomorphic to the direct sum of two copies of C_0^-, and Fw is the center of A^-.

(3) A^- is a solvable Malcev algebra.

Proof. Assume that A^- is a nonsolvable Malcev algebra, and let S_1 denote a Levi factor of A^-. Then, $S_1 \cong A^-/R$, where the solvable radical R is a G_2-module, so that S_1 is regarded as a G_2-module and is acted on by G_2 as derivations. Thus, each simple summand of S_1 is a faithful G_2-module, since Fw is the only trivial submodule of A^-. Hence, each simple summand of S_1 must contain a copy of C_0^-. By dimension count, this implies that S_1 has at most two simple summands. If S_1 is simple, then it must be that $S_1 \cong C_0^-$, since there is no 14-dimensional simple, non-Lie, Malcev algebra over F. This gives case (1). If S_1 consists of two simple summands, then $S_1 = C_0^- \otimes P$, and Fw is the center of A^-, giving case (2). □

We make further analysis only for cases (1) and (2) in terms of constant conditions. Recall first the Killing form $K(x,y)$ of C_0^- from (4.21):

(5.74) $$K(a,a) = 24,\ K(u_i,v_i) = K(v_i,u_i) = -12,\ i = 1, 2, 3,$$

and all other entries are zero, where $a, u_i, v_i (i = 1,2,3)$ are the basis elements of C_0^- given by (5.52).

Theorem 5.13. Let $A = C_0^- \otimes P + Fw$ be Malcev-admissible such that A^- is not solvable. Then, one of the following cases holds.

(1) $C_0^- \otimes q$ is a simple subalgebra of A^- and $R = C_0^- \otimes r + Fw$ is the radical of A^- such that $(R^-)^{(3)} = 0$. Moreover, for some $\sigma_0, \sigma, \sigma' \in F$

$$q * q = \frac{1}{2}q,\ r * r = 0,\ q * r = \sigma_0 q + \sigma r,\ r * q = -\sigma_0 q + \sigma' r\ ,$$

$$\mu_0 - \mu_0' = \mu - \mu' = \nu_0 - \nu_0' = 0\ ,$$

and either

(i) $\sigma + \sigma' = 1$ and $\delta - \delta' = \nu - \nu' = 0$, or

(ii) $\sigma + \sigma' = 0$ and $(\delta - \delta')(\nu - \nu') = 0$.

(2) $C_0^- \otimes P$ is a semisimple subalgebra of A^- and Fw is the center of A^-. Moreover,

$$q * q = \frac{1}{2}q,\ r * r = \frac{1}{2}r,\ q * r = -r * q = \sigma_0 q + \sigma r\ ,$$

$$\delta - \delta' = \mu_0 - \mu_0' = \mu - \mu' = \nu_0 - \nu_0' = \nu - \nu' = \sigma + \sigma' = 0\ .$$

Proof. (1) This is case (1) of Theorem 5.12. Write $x = x \otimes q$ and $x' = x \otimes r$ for $x \in C_0^-$. Since $[C_0^- \otimes q, C_0^- \otimes q] = C_0^- \otimes q$, $q \otimes q = \alpha q$ for some $\alpha \neq 0$ in F. Replacing q with $(2\alpha)^{-1}q$ gives $q * q = \frac{1}{2}q$. Since R is a solvable ideal of A^-, by (5.73) $r * r = 0$. Similarly, from $[C_0^- \otimes q, w] \subseteq C_0^- \otimes r$ and $[C_0^- \otimes r, w] \subseteq C_0^- \otimes r$, we obtain $\mu_0 - \mu_0' = \nu_0 - \nu_0' = 0$. These with (5.73) give the multiplication in A^- as in Table 5.19.

TABLE 5.19

	y	y'	w
x	$[x,y]$	$(\sigma + \sigma')[x,y]' + (\delta - \delta')K(x,y)w$	$(\mu - \mu')x'$
x'	$(\sigma + \sigma')[x,y]' - (\delta - \delta')K(x,y)w$	0	$(\nu - \nu')x'$
w	$-(\mu - \mu')y'$	$-(\nu - \nu')y'$	0

Using Table 5.19, (5.52) and (5.74), the specializations of u_1, w, u_2, u_3 and a, w, a, u_1 into the Malcev identity $m(x,z,y,t) = 0$ imply $(\mu - \mu')[2(\sigma + \sigma') + 1] = 0$ and $(\mu - \mu')[2(\sigma + \sigma') - 1] \times [\sigma + \sigma' + 1] = 0$, which force $\mu - \mu' = 0$. We compute the specialization of a, a', u_1', w to obtain $(\delta - \delta')(\nu - \nu') = 0$, and this can combine with the specialization u_1', w, a, v_1 to give $(\sigma + \sigma')(\nu - \nu')(\sigma + \sigma' - 1) = 0$. On the other hand, substituting a', w, v_2, v_3 into the Malcev identity implies $(\sigma + \sigma')(\nu - \nu')(2(\sigma + \sigma') + 1) = 0$. Hence, it must be that $(\sigma + \sigma')(\nu - \nu') = 0$. The specialization of u_1, v_1', v_1, u_1 now gives

$$(\sigma + \sigma')^3 - (\sigma + \sigma')^2 = 0 = (\delta - \delta')(\sigma + \sigma') .$$

Thus, if $\sigma + \sigma' \neq 0$, then $\sigma + \sigma' = 1$, and this in turn implies $\delta - \delta' = \nu - \nu' = 0$. If $\sigma + \sigma' = 0$, then the constant condition is $(\delta - \delta')(\nu - \nu') = 0$. Finally, from $[C_0^- \otimes q, C_0^- \otimes r] \subseteq C_0^- \otimes r$, it follows that $q * r + r * q \in Fr$. Hence, case (1) is verified.

(2) Assume that $C_0^- \otimes P$ is a semisimple subalgebra of A^- and Fw is the center of A^-. In this case, the simple summands of $C_0^- \otimes P$

are $C_0^- \otimes q$ and $C_0^- \otimes r$. Thus, we have the normalizations $q * q = \frac{1}{2}q$, $r * r = \frac{1}{2}r$, $q * r = - r * q = \sigma_0 q + \sigma r$, and the equation $\delta - \delta' = 0$. Since Fw is the center of A^-, the second set of conditions follows. □

Under the added assumption of flexibility, we determine the algebra A for part (i) of case (1), and case (2) of Theorem 5.13.

Theorem 5.14. Let $A = C_0^- \otimes P + Fw$ be flexible Malcev-admissible such that A^- is not solvable. Denote $x = x \otimes q$ and $x' = x \otimes r$ for $x \in C_0^-$.

(1) Assume that $C_0^- \otimes q$ is a simple subalgebra of A^- and $C_0^- \otimes r + Fw$ is the radical of A^-. If $\sigma + \sigma' = 1$, then A is the algebra described by

$$
\begin{aligned}
&a^2 = -2\alpha w, \quad w^2 = \lambda w, \\
&au_i = -u_i a = u_i, \quad av_i = -v_i a = -v_i', \\
&u_i v_i = -\frac{1}{2}a + \alpha w, \quad v_i u_i = \frac{1}{2}a + \alpha w, \\
&u_i u_j = -u_j u_i = \varepsilon_{ijk} v_k, \quad v_i v_j = \varepsilon_{ijk} u_k, \\
&au_i' = -u_i' a = u_i', \quad av_i' = -v_i' a = -v_i', \\
&u_i v_i' = -v_i' u_i = u_i' v_i = -v_i u_i' = -\frac{1}{2}a', \\
&u_i u_j' = -u_j' u_i = \varepsilon_{ijk} v_k', \quad v_i v_j' = -v_j' v_i = \varepsilon_{ijk} u_k', \\
&u_i w = wu_i = \mu_0 u_i + \mu u_i', \quad v_i w = wv_i = \mu_0 v_i + \mu v_i', \\
&u_i' w = wu_i' = \nu u_i', \quad v_i' w = wv_i' = \nu v_i', \\
&aw = wa = \mu_0 a + \mu a', \quad a'w = wa' = \nu a',
\end{aligned} \tag{5.75}
$$

for $i, j = 1, 2, 3$, and all other products are zero, where $\alpha, \lambda, \mu_0, \mu, \nu \in F$ and $a, u_i, v_i (i = 1,2,3)$ are the basis elements of C_0^- given by (5.52).

(2) If $C_0^- \otimes P$ is a semisimple subalgebra of A^- and Fw is the center of A^- , then A is the one described by (4.15) of Theorem 4.3 with $S_1 = C_0^- \otimes q$, $S_2 = C_0^- \otimes r$, $x * y = \frac{1}{2}[x,y]$, and $a_i = \alpha_i w (i = 1,2)$ for $\alpha_i \in F$.

Proof. (1) If $\sigma + \sigma' = 1$, then by case (1) of Theorem 5.13, $\delta - \delta' = \nu - \nu' = 0$, and hence R^- is abelian. Thus, we are in the situation of Theorem 4.22. By Lemmas 4.19 and 4.21, $(C_0^- \otimes r)(C_0^- \otimes r) = 0$ and $q * r = r * q = \frac{1}{2}r$, which imply that $\sigma_0 = 0$, $\sigma = \sigma' = \frac{1}{2}$ and $\gamma = 0$. From the same argument as in case (5.42) of Section 5.3, it also follows that $\delta = \delta' = 0$. Since Fw is a trivial module, $(C_0^- \otimes r)w \subseteq C_0^- \otimes r$ (Lemma 4.13(3)), and hence $\nu_0 = \nu_0' = 0$. Letting $\alpha = -12\beta$, from (5.52), (5.73) and (5.74) we obtain the multiplication given by (5.75).

(2) Since A satisfies the hypothesis of Theorem 4.3, A is described by (4.15) of that theorem. □

We finally consider the case where $A = V(\lambda_2) + Fw$. When $V(\lambda_2)$ is identified with G_2 , $V(\lambda_2)$ is the adjoint module for G_2 . Since $\dim \mathrm{Hom}_{G_2}(V(\lambda_2) \otimes V(\lambda_2), V(\lambda_2)) = \dim \mathrm{Hom}_{G_2}(V(\lambda_2) \otimes V(\lambda_2), Fw) = 1$, the multiplication in A has the form

$$xy = \alpha[x,y] + \beta K(x,y)w ,$$

(5.76)

$$xw = \gamma x, \quad wx = \gamma' x, \quad w^2 = \lambda w$$

for $x \in V(\lambda_2)$ and $\alpha, \beta, \gamma, \gamma', \lambda \in F$, where $K(,)$ is the Killing form of G_2 .

<u>Theorem 5.15</u>. The algebra $A = V(\lambda_2) + Fw$ is acted on by G_2 as derivations if and only if the multiplication in A is given by (5.76).

(1) Assume that A^- is a nonsolvable Malcev algebra. Then, $\alpha \neq 0$, $\gamma = \gamma'$, and A is a flexible Lie-admissible algebra described by (4.15) of Theorem 4.3, with $\alpha = \frac{1}{2}$, $n = 1$, and $a_1 = \beta w$. If, in addition, $\beta \neq 0$, then A is simple.

(2) If A^- is a solvable Malcev algebra, then $\alpha = 0$ and $(A^-)^{(2)} = 0$.

<u>Proof</u>. We use Table 5.16 for the multiplication in G_2. Specializing the elements s_{12}, w, s_{12}, w into the Malcev identity and using (5.76), we get $\alpha(\gamma - \gamma')^2 = 0$. The remaining proof is identical with that of Theorem 5.8. □

We remark that any algebra A described by (5.75) is flexible Malcev-admissible with abelian radical $C_0^- \otimes r + Fw$ by Theorem 4.22 and is not simple, since $C_0^- \otimes r$ is an ideal of A.

BIBLIOGRAPHY

Albert, A. A.

[1] On the power-associativity of rings, Brasil. Math. 2(1948), 1-13.

[2] Power-associative rings, Trans. Amer. Math. Soc. 64(1948), 552-593.

Albert, A. A. and Frank, M. S.

[1] Simple Lie algebras of characteristic p, Univ. e Politec. Torino Read. Sem. Mat. 14(1954-55), 117-139.

Barnes, D. W.

[1] On Cartan subalgebras of Lie algebras, Math. Z. 101(1967), 350-355.

Benkart, G. M.

[1] The construction of examples of Lie-admissible algebras, Hadronic J. 5(1982), 431-493.

[2] Power-associative Lie-admissible algebras, J. Algebra 90(1984), 37-58.

Benkart, G. M., Britten, D. J. and Osborn, J. M.

[1] Flexible Lie-admissible algebras with the solvable radical of A^- abelian and Lie algebras with nondegenerate forms, Hadronic J. 4 (1981), 274-326.

[2] Real flexible division algebras, Canad. J. Math. 34(1982), 550-588.

Benkart, G. M. and Osborn, J. M.

[1] Flexible Lie-admissible algebras, J. Algebra 71(1981), 11-31.

[2] Real division algebras and other algebras motivated by physics, Hadronic J. 4(1981), 392-443.

[3] An investigation of real division algebras using derivations, Pacific J. Math. 96(1981), 265-300.

[4] Power-associative products on matrices, Hadronic J. 5(1982), 1859-1892.

Block, R. E.

[1] New simple Lie algebras of prime characteristic, Trans. Amer. Math. Soc. 89(1958), 421-449.

[2] Determination of A^+ for the simple flexible algebras, Proc. Nat. Acad. Sci. U.S.A. 61(1968), 394-397.

Bourbaki, N.

[1] Groupes et Algèbres de Lie, Chapter 1, Hermann, Paris, 1960.

Carlsson, R.

[1] Malcev-Moduln, J. Reine Angew, Math. 281(1976), 199-210.

[2] The first Whitehead lemma for Malcev algebras, Proc. Amer. Math. Soc. 58(1976), 79-84.

Divinsky, N. J.

[1] Rings and Radicals, University of Toronto Press, Toronto, Canada, 1965.

Djoković, D. Ž.

[1] Classification of some 2-graded Lie algebras, J. Pure Appl. Algebra 7(1976), 217-320.

Domokos, G. and Kövesi-Domokos, S.

[1] The algebra of color, J. Math. Phys. 19(1978), 1477-1481.

Eder, G.

[1] Physical implications of a Lie-admissible spin algebra, Hadronic J. 4(1981), 2018-2032.

[2] Lie-admissible spin algebra for arbitrary spin and the interaction of neutrons with the electric field of atoms, Hadronic J. 5(1982), 750-770.

Faulkner, J. R.

[1] Identity classification in triple systems,(to appear).

Filippov, V. T.

[1] Central simple Mal'tsev algebras, Algebra i Logika 15(1976), 235-242 = Algebra and Logic 15(1977), 147-151.

Gainov, A. T.

[1] Binary Lie algebras of the lowest ranks, Algebra i Logika, Seminar 2(1963), 21-40.

Gell-Mann, M.

[1] Symmetries of baryons and mesons, Phys. Rev. 125(1962), 1067-1084.

Gerstenhaber, M.

[1] On nilalgebras and linear varieties of nilpotent matrices. II, Duke Math. J. 27(1960), 21-31.

Gerstenhaber, M. and Myung, H. C.

[1] On commutative power-associative nilalgebras of low dimension, Proc. Amer. Math. Soc. 48(1975), 29-32.

Goto, M. and Grosshans, F. D.

[1] Semisimple Lie Algebras, Marcel Dekker, New York, 1978.

Grishkov, A. N.

[1] Analogue of Levi's theorem for Malcev algebras, Algebra i Logika 16(1977), 389-396 = Algebra and Logic 16(1978), 260-265.

Günaydin, M. and Gürsey, F.

[1] Quark structure and octonions, J. Math. Phys. 14(1973), 1651-1667.

[2] Quark statistics and octonions, Phys. Rev. D9(1974), 3387-3391.

Hawkins, T.

[1] Wilhelm Killing and the structure of Lie algebras, Archive for History of Exact Sciences 26(1982), 127-192.

Herstein, I. N.

[1] Topics in Algebra, Zerox College Publishing, Waltham, Mass., 1964.

[2] Topics in Ring Theory, University of Chicago Press, Chicago, 1969.

Howe, R.

[1] Very basic Lie theory, Amer. Math. Monthly 90(1983), 600-623.

Humphreys, J. E.

[1] Introduction to Lie Algebras and Representation Theory, Springer-Verlag, New York, 1972.

Jacobson, N.

[1] Classes of restricted Lie algebras of characteristic p, II, Duke Math. J. 10(1943), 107-121.

[2] Lie Algebras, Interscience Tracts in Pure and Appl. Math., No. 10, Interscience, New York, 1962.

[3] Structure and Representations of Jordan Algebras, Amer. Math. Soc. Colloq. Publ., vol. 39, Amer. Math. Soc., Providence, R.I., 1968.

[4] Basic Algebra I, second edition, Freeman and Company, New York, 1985.

[5] Exceptional Lie Algebras, Marcel Dekker, New York, 1971.

Jordan, P., von Neumann, J., and Wigner, E.P.

[1] On an algebraic generalization of the quantum mechanical formalism, Ann. of Math. 36(1934), 29-64.

Kac, V. G.

[1] Infinite Dimensional Lie Algebras, Birkhäuser, Boston, Mass., 1983.

Kleinfeld, E.

[1] A note on Moufang-Lie rings, Proc. Amer. Math. Soc. 9(1958), 72-74.

[2] A characterization of the Cayley numbers, M. A. A. Studies in Math. Vol. 2 (1963), 126-143.

Ko, Y. and Myung, H. C.

[1] On a class of Malcev-admissible algebras, J. Korean Math. Soc. 19 (1982), 47-54.

Krämer, M.

[1] Eine Klassifikation Bestimmter Untergruppen Kompakter Zusammen-Hängender Lie Gruppen, Comm. in Algebra 3(1975), 691-737.

Kruse, R. L. and Price, D. T.

[1] Nilpotent Rings, Gordon and Breach, New York, 1969.

Kuzmin, E. N.

[1] Mal'cev algebras and their representations, Algebra i Logika 7(4) (1968), 48-69 = Algebra and Logic 7(1968), 233-244.

[2] Mal'cev algebras of dimension five over a field of zero characteristic, Algebra i Logika 9(1970), 691-700 = Algebra and Logic 9(1970), 416-421.

[3] Levi's theorem for Mal'cev algebras, Algebra i Logika 16(1977), 424-431 = Algebra and Logic 16(1978), 286-291.

Laufer, P. J. and Tomber, M. L.

[1] Some Lie admissible algebras, Canad. J. Math. 14(1962), 287-292.

Leadley, J. D. and Ritchie, R. W.

[1] Conditions for the power-associativity of algebras, Proc. Amer. Math. Soc. 11(1960), 399-405.

Loos, O.

[1] Über einer beziehung zwischen Malcev-Algebren and Lie-Tripelsystemen, Pacific J. Math. 18(1966), 553-562.

Malcev, A.

[1] Analytic loops, Mat. Sb. 78(1955), 569-578.

Malek, A. A.

[1] Sur les algèbres de Mal'cev-admissibles, Atti. Accad. Naz. Lincei Rend. Cl. Sci. Fis. Mat. Natur. 68(1980), 390-396.

[2] The conjugacy of Cartan subalgebras of a solvable Malcev algebra, J. Algebra 85(1983), 323-332.

Malek, A. A. and Micali, A.

[1] Sur les algebres Malcev-admissibles, Algebras Groups Geom. 2(1985).

McCrimmon, K.

[1] Noncommutative Jordan rings, Trans.Amer.Math.Soc. 158(1971), 1-33.

[2] Jordan algebras and their applications, Bull. Amer. Math. Soc. 84 (1978), 612-627.

[3] The Russian revolution in Jordan algebras, Algebras Groups Geom. 1(1984), 1-64.

Myung, H. C.

[1] A note on Lie-admissible nilalgebras, Proc. Amer. Math. Soc. 31(1972), 95-96.

[2] Some classes of flexible Lie-admissible algebras, Trans. Amer. Math. Soc. 167(1972), 79-88.

[3] A class of almost commutative nilalgebras, Canad. J. Math. 26(1974), 1192-1198.

[4] Flexible Lie-admissible algebras with nil-basis, Hadronic J. 2(1979), 360-368.

[5] Flexible Malcev-admissible algebras, Hadronic J. 4(1981), 2033-2136.

[6] Lie Algebras and Flexible Lie-Admissible Algebras, Hadronic Press, Nonatum, Mass., 1982.

[7] The exponentiation and deformations of Lie-admissible algebras, Hadronic J. 5(1982), 771-903.

[8] A Malcev-admissible mutation of an alternative algebra, Bull. Korean Math. Soc. 20(1983), 37-43.

[9] Power-associative products on octonions, Comm. in Algebra 13(1985), 1681-1697.

[10] (Editor), Mathematical Studies on Lie-Admissible Algebras, Vols. I-VII, Hadronic Press, Nonantum, Mass., 1985-1986, (to appear).

Myung, H. C., Okubo, S. and Santilli, R. M., Editors

[1] Applications of Lie-Admissible Algebras in Physics, Vols. I and II, Hadronic Press, Nonantum, Mass., 1979.

Oehmke, R. H.

[1] On flexible algebras, Ann. of Math. (2) 68(1958), 221-230.

[2] On commutative algebras of degree two, Trans. Amer. Math. Soc. 105 (1962), 292-313.

[3] Nodal noncommutative Jordan algebras, Trans. Amer. Math. Soc. 112 (1964), 416-431.

[4] Commutative power-associative algebras of degree one, J. Algebra 14(1970), 326-332.

Okubo, S.

[1] Casimir invariants and vector operators in simple and classical Lie algebras, J. Math. Phys. 18(1977), 2382-2394.

[2] Gauge groups without triangular anomaly, Phys. Rev. D16(1977), 3528-3534.

[3] Octonions as traceless 3 × 3 matrices via a flexible Lie-admissible algebra, Hadronic J. 1(1978), 1432-1465.

[4] Deformation of the Lie-admissible pseudo-octonion algebra into the octonion algebra, Hadronic J. 1(1978), 1383-1431.

[5] Pseudo-quaternion and pseudo-octonion algebras, Hadronic J. 1(1978), 1250-1278.

[6] Non-associative quantum mechanics via flexible Lie-admissible algebras, Proc. of the Third Workshop on Current Problems in High Energy Particle Theory, held at Florence, Italy, (1979), Edit. R. Casabuoni, G. Domokos and S. Kövesi-Domokos, John Hopkins Univ. Press, Baltimore, MD. (1979), 103-120.

[7] Octonion, pseudo-alternative, and Lie-admissible algebras, Hadronic J. 2(1979), 39-66.

[8] A generalization of Hurwitz theorem and flexible Lie-admissible algebra, Hadronic J. 3(1979), 1-52.

[9] Dimension and classification of general composition algebras, Hadronic J. 4(1981), 216-273.

[10] Non-associative quantum mechanics and strong correspondence principle, Hadronic J. 4(1981), 608-633.

[11] Classification of flexible composition algebras, I, Hadronic J. 5 (1982), 1564-1612.

[12] Classification of flexible composition algebras, II, Hadronic J. 5(1982), 1613-1626.

[13] Representations and classifications of compact gauge groups for unified weak and electromagnetic interactions, Phys. Rev. D18(1978), 3792-3808.

[14] Some classes of flexible Lie-Jordan-admissible algebras, Hadronic J. 4(1981), 354-391.

Okubo, S. and Myung, H. C.

[1] On the classification of simple flexible Lie-admissible algebras, Hadronic J. 2(1979), 504-567.

[2] Some new classes of division algebras, J. Algebra 67(1980), 479-490.

[3] Adjoint operators in Lie algebras and the classification of simple flexible Lie-admissible algebras, Trans. Amer. Math. Soc. 264(1981), 459-472.

Okubo, S. and Osborn, J. M.

[1] Algebras with nondegenerate associative symmetric bilinear forms permitting composition, Comm. in Algebra 9(1981), 1233-1261.

[2] Algebras with nondegenerate associative symmetric bilinear forms permitting composition (II), Comm. in Algebra 9(1981), 2015-2073.

O'Raifeartaigh, L.

[1] Broken symmetry, Group Theory and its Applications, (E. M. Loebl, ed.), Academic Press, New York, 1968.

Osborn, J. M.

[1] Quadratic division algebras, Trans. Amer. Math. Soc. 105(1962), 202-221.

[2] Varieties of algebras, Advances in Math. 8(1972), 163-369.

[3] What are nonassociative algebras?, An invited address at the AMS 817th Meeting, Chicago, March 1985.

Patera, J., Sharp, R. T., Winternitz, P., and Zassenhaus, H.

[1] Invariants of real low order Lie algebras, J. Math. Phys. 17(1976), 986-994.

Petersson, H. P.

[1] Quasi-composition algebras, Math. Semi. Univ. Hamburg 35(1971), 215-222.

Ravisankar, T. S.

[1] On Malcev algebras, Pacific J. Math. 42(1972), 227-234.

Ree, R.

[1] On generalized Witt algebras, Trans. Amer. Math. Soc. 83(1956), 510-546.

Sagle, A. A.

[1] Malcev algebras, Trans. Amer. Math. Soc. 101(1961), 426-458.

[2] Simple Malcev algebras over field of characteristic zero, Pacific J. Math. 12(1962), 1057-1078.

[3] On derivations of semi-simple Malcev algebras, Portugal. Math. 21 (1962), 107-109.

[4] Anti-commutative algebras with an invariant form, Canad. J. Math. 16(1964), 370-378.

[5] On simple extended Lie algebras over fields of characteristic zero, Pacific J. Math. 15(1965), 621-648.

[6] Simple algebras which generalize the Jordan algebra M_3^8, Canad. J. Math. 17(1966), 550-558.

Sagle, A. A. and Walde, R. E.

[1] Introduction to Lie Groups and Lie Algebras, Academic Press, New York, 1973.

Santilli, R. M.

[1] Imbedding of a Lie algebra in nonassociative structures, Nuovo Cimento A (10) 51(1967), 570-576.

[2] An introduction to Lie-admissible algebras, Nuovo Cimento, Suppl. (1) 6(1968), 1225-1249.

[3] Dissipativity and Lie-admissible algebras, Meccanica 1(1969), 3-11.

[4] Haag theorem and Lie-admissible algebras, Analytic Methods in Mathematical Physics, R. P. Gilbert and R. G. Newton, Editors, Gordon and Breach, New York, 1970.

[5] Lie-Admissible Approach to the Hadronic Structure, Vols. I and II, Hadronic Press, Nonantum, Mass., 1978 and 1982.

Schafer, R. D.

[1] Noncommutative Jordan algebras of characteristic 0, Proc. Amer. Math. Soc. 6(1955), 472-475.

[2] Nodal noncommutative Jordan algebras and simple Lie algebras of characteristic p, Trans. Amer. Math. Soc. 94(1960), 310-326.

[3] An Introduction to Nonassociative Algebras, Academic Press, New York, 1966.

Seligman, G. B.

[1] Modular Lie Algebras, Ergebnisse der Mathematik and ihrer Grenzgebiete, Band 40, Springer-Verlag, New York, 1967.

[2] Rational Methods in Lie Algebras, Marcel Dekker, New York, 1976.

Shapiro, D. B.

[1] A private communication.

Sherk, J.

[1] An introduction to the theory of dual models and strings, Rev. Mod. Phys. 47(1975), 123-164.

Shestakov, I. P.

[1] Finite-dimensional algebras with a nil-basis, Algebra i Logika 10 (1971), 87-99.

Suttles, D.

[1] A counterexample to a conjecture of Albert, Notices Amer. Math. Soc. 19(1972), A-566.

Tomber, M. L.

[1] A short history of nonassociative algebras, Hadronic J. 2(1979), 1252-1387.

[2] Jacobson-Witt algebras and Lie-admissible algebras, Hadronic J. 4(1981), 183-197.

Weiner, L. M.

[1] Algebras based on linear functions, Math. Mag. 28(1954), 9-12.

[2] Lie admissible algebras, Univ. Nac. Tucuman, Rev. Ser. A. 11(1957), 10-24.

Wigner, E. P.

[1] On representations of certain finite groups, Amer. J. Math. 63(1941), 57-63.

Winter, D. J.

[1] Abstract Lie Algebras, MIT Press, Cambridge, Mass., 1972.

INDEX OF SYMBOLS

INDEX OF TERMINOLOGY

Progress in Mathematics
Edited by J. Coates and S. Helgason

Progress in Physics
Edited by A. Jaffe and D. Ruelle

- A collection of research-oriented monographs, reports, notes arising from lectures or seminars,
- Quickly published concurrent with research,
- Easily accessible through international distribution facilities,
- Reasonably priced,
- Reporting research developments combining original results with an expository treatment of the particular subject area,
- A contribution to the international scientific community: for colleagues and for graduate students who are seeking current information and directions in their graduate and post-graduate work.

Manuscripts

Manuscripts should be no less than 100 and preferably no more than 500 pages in length.

They are reproduced by a photographic process and therefore must be typed with extreme care. Symbols not on the typewriter should be inserted by hand in indelible black ink. Corrections to the typescript should be made by pasting in the new text or painting out errors with white correction fluid.

The typescript is reduced slightly (75%) in size during reproduction; best results will not be obtained unless the text on any one page is kept within the overall limit of 6 × 9½ in (16 × 24 cm). On request, the publisher will supply special paper with the typing area outlined.

Manuscripts should be sent to the editors or directly to:
Birkhäuser Boston, Inc., P.O. Box 2007,
Cambridge, MA 02139 (USA)

Progress in Mathematics

1 GROSS. Quadratic Forms in Infinite-Dimensional Vector Spaces
2 PHAM. Singularités des Systèmes Différentiels de Gauss-Manin
3 OKONEK/SCHNEIDER/SPINDLER. Vector Bundles on Complex Projective Spaces
4 AUPETIT. Complex Approximation, Proceedings, Quebec, Canada, July 3-8, 1978
5 HELGASON. The Radon Transform
6 LION/VERGNE. The Weil Representation, Maslov Index and Theta Series
7 HIRSCHOWITZ. Vector Bundles and Differential Equations Proceedings, Nice, France, June 12-17, 1979
8 GUCKENHEIMER/MOSER/NEWHOUSE. Dynamical Systems, C.I.M.E. Lectures, Bressanone, Italy, June, 1978
9 SPRINGER. Linear Algebraic Groups
10 KATOK. Ergodic Theory and Dynamical Systems I
11 BALSLEV. 18th Scandinavian Congress of Mathematicians, Aarhus, Denmark, 1980
12 BERTIN. Séminaire de Théorie des Nombres, Paris 1979-80
13 HELGASON. Topics in Harmonic Analysis on Homogeneous Spaces
14 HANO/MARIMOTO/MURAKAMI/OKAMOTO/OZEKI. Manifolds and Lie Groups: Papers in Honor of Yozo Matsushima
15 VOGAN. Representations of Real Reductive Lie Groups
16 GRIFFITHS/MORGAN. Rational Homotopy Theory and Differential Forms
17 VOVSI. Triangular Products of Group Representations and Their Applications
18 FRESNEL/VAN DER PUT. Géometrie Analytique Rigide et Applications
19 ODA. Periods of Hilbert Modular Surfaces
20 STEVENS. Arithmetic on Modular Curves
21 KATOK. Ergodic Theory and Dynamical Systems II
22 BERTIN. Séminaire de Théorie des Nombres, Paris 1980-81
23 WEIL. Adeles and Algebraic Groups
24 LE BARZ/HERVIER. Enumerative Geometry and Classical Algebraic Geometry
25 GRIFFITHS. Exterior Differential Systems and the Calculus of Variations
26 KOBLITZ. Number Theory Related to Fermat's Last Theorem
27 BROCKETT/MILLMAN/SUSSMAN. Differential Geometric Control Theory
28 MUMFORD. Tata Lectures on Theta I
29 FRIEDMAN/MORRISON. Birational Geometry of Degenerations
30 YANO/KON. CR Submanifolds of Kaehlerian and Sasakian Manifolds
31 BERTRAND/WALDSCHMIDT. Approximations Diophantiennes et Nombres Transcendants
32 BOOKS/GRAY/REINHART. Differential Geometry
33 ZUILY. Uniqueness and Non-Uniqueness in the Cauchy Problem
34 KASHIWARA. Systems of Microdifferential Equations
35 ARTIN/TATE. Arithmetic and Geometry: Papers Dedicated to I.R. Shafarevich on the Occasion of His Sixtieth Birthday, Vol. I
36 ARTIN/TATE. Arithmetic and Geometry: Papers Dedicated to I.R. Shafarevich on the Occasion of His Sixtieth Birthday, Vol. II
37 DE MONVEL. Mathématique et Physique
38 BERTIN. Séminaire de Théorie des Nombres, Paris 1981-82
39 UENO. Classification of Algebraic and Analytic Manifolds
40 TROMBI. Representation Theory of Reductive Groups

41 STANLEY. Combinatorics and Commutative Algebra
42 JOUANOLOU. Théorèmes de Bertini et Applications
43 MUMFORD. Tata Lectures on Theta II
44 KAC. Infinite Dimensional Lie Algebras
45 BISMUT. Large Deviations and the Malliavin Calculus
46 SATAKE/MORITA. Automorphic Forms of Several Variables Taniguchi Symposium, Katata, 1983
47 TATE. Les Conjectures de Stark sur les Fonctions L d'Artin en s=0
48 FRÖHLICH. Classgroups and Hermitian Modules
49 SCHLICHTKRULL. Hyperfunctions and Harmonic Analysis on Symmetric Spaces
50 BOREL, ET AL. Intersection Cohomology
51 BERTIN/GOLDSTEIN. Séminaire de Théorie des Nombres, Paris 1982-83
52 GASQUI/GOLDSCHMIDT. Déformations Infinitesimales des Structures Conformes Plates
53 LAURENT. Théorie de la Deuxième Microlocalisation dans le Domaine Complèxe
54 VERDIER/LE POTIER. Module des Fibres Stables sur les Courbes Algébriques Notes de l'Ecole Normale Supérieure, Printemps, 1983
55 EICHLER/ZAGIER. The Theory of Jacobi Forms
56 SHIFFMAN/SOMMESE. Vanishing Theorems on Complex Manifolds
57 RIESEL. Prime Numbers and Computer Methods for Factorization
58 HELFFER/NOURRIGAT. Hypoellipticité Maximale pour des Opérateurs Polynomes de Champs de Vecteurs
59 GOLDSTEIN. Séminaire de Théorie des Nombres, Paris 1983-84
60 PROCESI. Geometry Today: Giornate Di Geometria, Roma, 1984
61 BALLMANN/GROMOV/SCHROEDER. Manifolds of Nonpositive Curvature
62 GUILLOU/MARIN. A la Recherche de la Topologie Perdue
63 GOLDSTEIN. Séminaire de Théorie des Nombres, Paris 1984-85
64 MYUNG. Malcev-Admissible Algebras
65 GRUBB. Functional Calculus of Pseudo-Differential Boundary Problems